Inverse problems in vibration

MECHANICS: DYNAMICAL SYSTEMS
Editors: L Meirovitch and G Æ Oravas

E.H. Dowell, Aeroelasticity of Plates and Shells. 1974. ISBN 90-286-0404-9.

D.G.B. Edelen, Lagrangian Mechanics of Nonconservative Nonholonomic Systems. 1977. ISBN 90-286-0077-9.

J.L. Junkins, An Introduction to Optical Estimation of Dynamical Systems. 1978. ISBN 90-286-0067-1.

E.H. Dowell et al., A Modern Course in Aeroelasticity. 1978. ISBN 90-286-0057-4.

L. Meirovitch, Computational Methods in Structural Dynamics. 1980. ISBN 90-286-0580-0.

B. Skalmierski and A. Tylikowski, Stochastic Processes in Dynamics. 1982. ISBN 90-247-2686-7.

P.C. Müller and W.O. Schiehlen, Linear Vibrations. 1985. ISBN 90-247-2983-1.

Gh. Buzdugan, E. Mihăilescu and M. Radeş, Vibration Measurement. 1986. ISBN 90-247-3111-9.

G.M.L. Gladwell, Inverse Problems in Vibration. 1986. ISBN 90-247-3408-8.

Inverse problems in vibration

by

G.M.L. Gladwell

University of Waterloo
Faculty of Engineering
Waterloo, Ontario, Canada

1986 **MARTINUS NIJHOFF PUBLISHERS**
a member of the KLUWER ACADEMIC PUBLISHERS GROUP
DORDRECHT / BOSTON / LANCASTER

Distributors

for the United States and Canada: Kluwer Academic Publishers, 101 Philip Drive, Assinippi Park, Norwell, MA 02061, USA
for the UK and Ireland: Kluwer Academic Publishers, MTP Press Limited, Falcon House, Queen Square, Lancaster LA1 1RN, UK
for all other countries: Kluwer Academic Publishers Group, Distribution Center, P.O. Box 322, 3300 AH Dordrecht, The Netherlands

Library of Congress Cataloging in Publication Data

Gladwell, G. M. L.
Inverse problems in vibration.

(Mechanics, dynamical systems ; 9)
Bibliography: p.
1. Vibration. 2. Inverse problems (Differential equations) I. Title. II. Series: Monographs and textbooks on mechanics of solids and fluids. Mechanics, dynamical systems ; 9.
QA865.G53 1987 531'.32 86-21783
ISBN 90-247-3408-8

ISBN 90-247-3408-8 (this volume)
ISBN 90-247-2736-7 (series)

Copyright

PRINTED IN THE NETHERLANDS

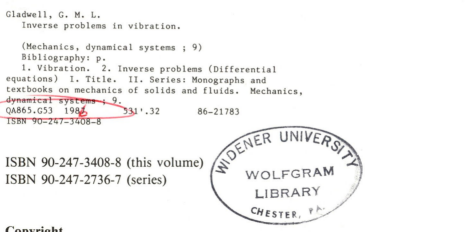

All appearance indicates neither a total exclusion nor a manifest presence of divinity, but the presence of a God who hides Himself. Everything bears this character.

Pascal's Pensées

PREFACE

The last thing one settles in writing a book is what one should put in first.
Pascal's Pensées

Classical vibration theory is concerned, in large part, with the infinitesimal (i.e., linear) undamped free vibration of various discrete or continuous bodies. One of the basic problems in this theory is the determination of the natural frequencies (eigenfrequencies or simply eigenvalues) and normal modes of the vibrating body. A body which is modelled as a *discrete* system of rigid masses, rigid rods, massless springs, etc., will be governed by an ordinary matrix differential equation in time t. It will have a finite number of eigenvalues, and the normal modes will be vectors, called eigenvectors. A body which is modelled as a continuous system will be governed by a partial differential equation in time and one or more spatial variables. It will have an infinite number of eigenvalues, and the normal modes will be functions (eigenfunctions) of the space variables.

In the context of this classical theory, inverse problems are concerned with the construction of a model of a given type; e.g., a mass-spring system, a string, etc., which has given eigenvalues and/or eigenvectors or eigenfunctions; i.e., given *spectral* data. In general, if some such spectral data is given, there can be no system, a unique system, or many systems, having these properties. In this book we shall be concerned exclusively with a stricter class of inverse problems, the so called reconstruction problems. Here the spectral data is such that there is one and only one vibrating system of the specified type which has the given spectral properties. For these problems, which are basically problems in applied mathematics, not engineering, there are always three questions to be asked, and answered:

1) What spectral data is necessary and sufficient to ensure that the system, if it exists at all, is unique?
2) What are the necessary and sufficient conditions which must be satisfied by this data to ensure that it does correspond to a realistic system; i.e., one with positive masses, lengths, cross-sectional areas etc?
3) How can the (unique) system be reconstructed?

My interest in inverse problems was sparked by acquiring a copy of the translation of Gantmakher and Krein's beautiful and difficult book *Oscillation Matrices and Kernels and Small Oscillations of Mechanical Systems.* During the first ten years that I owned the book I made a number of attempts to master it, without much success. One thing that I did understand and enjoy was their reconstruction of the positions and masses of a set of beads on a stretched string from a knowledge of the fixed-fixed and fixed-free spectra. In their reconstruction, the unknown quantities appear as the coefficients in a continued fraction representation of the ratio of two polynomials constructed from the given spectra. As a mathematician I was thrilled that a concept so esoteric and apparently useless as a continued fraction should appear (naturally) in the solution of a problem in mechanics.

Krein's continued fraction solution did not provide a stable numerical procedure for finding the positions and masses. That was not to come until the (equally beautiful) work of Golub (1973), using the Lanczos algorithm, and de Boor and Golub (1978). These papers are described in Section 4.2.

Krein's research on inverse problems for both discrete and continuous systems started in the 1930's. For continuous problems, the methods he used were mathematically deep, and did not win acceptance by later workers. Instead, the researchers found their inspiration in the work of Marchenko (1950) and particularly Gel'fand-Levitan (1951). These papers were concerned primarily with the inverse scattering problem, but the latter treated an inverse vibration problem (as defined above) as a special case. In this book we are concerned exclusively with inverse vibration problems and refer to results in the vast literature of inverse scattering only when it has relevance to our restricted interest.

In writing this book I have aimed to provide an introduction to the subject, and make no claim to completeness. On the one hand I have paid only scant attention to mathematical rigour; this is particularly clear in the chapters concerned with continuous systems. In many cases a more rigourous treatment is available in the original papers, by Barcilon, Burridge, Hochstadt, Levitan, McLaughlin, etc. On the other hand I have treated only the very simplest vibrating systems, discrete systems or one-dimensional continuous systems governed by second or fourth order differential equations. I have restricted myself in this way because there is at present no definitive treatment for inverse vibration problems in two or three dimensions. Sabatier (1978) for example, shows that complete answers have not yet been given to any of the three questions listed above even for seemingly simple higher-dimensional problems. I believe that progress in that area will come only when more is known about the qualitative properties of such systems; see Gladwell (1985).

This book is divided into two parts; Chapters 1-7 are concerned with discrete systems; Chapters 8-10 with continuous ones. In each part there is an alternation between theory and application. Thus Chapter 1 introduces matrices, which are then applied in Chapter 2. Chapter 3 studies Jacobian (i.e. tridiagonal) matrices, which are the class of matrix appearing in the inverse problems of Chapter 4. Chapter 5 is difficult; it draws largely on Gantmakher and Krein's book. On first reading, the reader is advised to note merely the principal results concerning oscillatory matrices; these are applied in Chapter 6, and particularly in Chapter 7.

Chapter 8 draws on Courant and Hilbert (1953) to give the basic properties of symmetric integral equations, and on Gantmakher and Krein to provide the extra properties of the eigenvalues and eigenfunctions of oscillatory kernels. The results of this chapter are basic to the study of the Sturm-Liouville systems of Chapter 9, and particularly for the Euler-Bernoulli problem of Chapter 10.

The manuscript of the book was read in part by A.H. England of Nottingham University, Dajun Wang of Beijing University, and three enthusiastic graduate students, Don Metzger, Tom Lemczyk, and Steve Dods from the University of Waterloo. They eliminated many errors and suggested many improvements. The book was produced in the office of the Solid Mechanics Division of the University of Waterloo. I am grateful to the Solid Mechanics Division Publication Officer, D.E. Grierson, to Assistant Pam McCuaig who typeset the final copy, and to Linda Strouth who typed the original manuscript.

I would be most grateful to be notified of any errors in the text and examples, and any omissions in the bibliography.

CONTENTS

CHAPTER 4 - INVERSION OF DISCRETE SECOND-ORDER SYSTEMS

CHAPTER 5 - FURTHER PROPERTIES OF MATRICES

CHAPTER 6 - SOME APPLICATIONS OF THE THEORY OF OSCILLATORY MATRICES

CHAPTER 7 - THE INVERSE PROBLEM FOR THE DISCRETE VIBRATING BEAM

CHAPTER 1

MATRIX ANALYSIS

It is a bad sign when, on seeing a person,
you remember his book.
Pascal's Pensées

1.1. Introduction

The analysis of Chapters 2 - 4 relies heavily on matrix analysis. In this section we shall present the basic definitions and properties of matrices and provide proofs of some of the particular theorems that will be used later. Since matrix analysis now has an established position in Engineering and Science, it will be assumed that the reader has had some exposure to it: the presentation in the early stages will therefore be brief. The reader may supplement the treatment here with standard texts, in particular Bishop, Gladwell and Michaelson (1965).

1.2. Basic definitions and notation

A matrix is a rectangular array of numbers together with a set of rules which specify how the numbers are to be manipulated.

The matrix **A** is said to have *order* $m \times n$ if it has m *rows* and n *columns*. Thus

$$\mathbf{A} = \begin{bmatrix} a_{11} & a_{12} & \cdots & a_{1n} \\ a_{21} & a_{22} & \cdots & a_{2n} \\ \cdot & \cdot & & \cdot \\ a_{m1} & a_{m2} & \cdots & a_{mn} \end{bmatrix} . \quad (1.2.1)$$

The element in row i and column j is a_{ij}, and $\mathbf{A}$ is often written simply as

$$\mathbf{A} = (a_{ij}) . \quad (1.2.2)$$

Two matrices **A**, **B** are said to be *equal* if they have the same order m×n, and if

$$a_{ij} = b_{ij} , \quad (i=1,2,...,m; \; j=1,2,...,n); \quad (1.2.3)$$

Then we write

$$\mathbf{A} = \mathbf{B} . \quad (1.2.4)$$

The *transpose* of the matrix **A** is the n×m matrix, denoted $\mathbf{A}^T$ (or sometimes $\mathbf{A}'$), whose rows are the columns of **A**. We note that the transpose of $\mathbf{A}^T$ is **A**; we say that **A** and $\mathbf{A}^T$ are *transposes* (of each other), and write this

$$(\mathbf{A}^T)^T = \mathbf{A} . \quad (1.2.5)$$

For example

$$\mathbf{A} = \begin{bmatrix} 1 & 2 & -4 \\ 2 & 6 & 7 \end{bmatrix} , \; \mathbf{A}^T = \begin{bmatrix} 1 & 2 \\ 2 & 6 \\ -4 & 7 \end{bmatrix} \quad (1.2.6)$$

are transposes.

If m=n then the m×n matrix **A** is said to be a *square* matrix of order n. We denote this by writing $\mathbf{A} = (a_{ij})_1^n$. A square matrix that is equal to its transpose is said to be *symmetric*; in this case

$$\mathbf{A} = \mathbf{A}^T , \quad (1.2.7)$$

or alternatively

$$a_{ij} = a_{ji} , \quad (i,j=1,2,...,n) . \quad (1.2.8)$$

The matrix

$$\mathbf{A} = \begin{bmatrix} 1 & 2 & 9 \\ 2 & 4 & 6 \\ 9 & 6 & 3 \end{bmatrix} \quad (1.2.9)$$

is symmetric. The square matrix **A** is said to be *diagonal* if it has non-zero elements only on the *principal diagonal* running from top left to bottom right. We write

$$\mathbf{A} = \begin{bmatrix} a_{11} & 0 & 0 & \cdots & 0 \\ 0 & a_{22} & 0 & \cdots & 0 \\ 0 & 0 & a_{33} & \cdots & 0 \\ 0 & 0 & 0 & \cdots & a_{nn} \end{bmatrix} = diag(a_{11},a_{22},...,a_{nn}). \tag{1.2.10}$$

The *unit* matrix of order n is

$$\mathbf{I}=\mathbf{I}_n=diag(1,1,...,1)\ . \tag{1.2.11}$$

The elements of this matrix are denoted by the *Kronecker delta*

$$\delta_{ij} = \begin{cases} 1 & i=j \\ 0 & i \neq j \end{cases} . \tag{1.2.12}$$

The *zero* matrix of order m×n is the matrix with all its m×n elements zero.

A matrix with 1 column and n rows is called a *column vector* of order n, and is written

$$\mathbf{x} = \begin{bmatrix} x_1 \\ x_2 \\ \\ x_n \end{bmatrix} = \{x_1,x_2,...,x_n\}\ . \tag{1.2.13}$$

The transpose of a column vector is a *row vector*, written

$$\mathbf{x}^T=[x_1,x_2,...,x_n]\ . \tag{1.2.14}$$

Two matrices **A,B** may be added or subtracted if and only if they have the same order m×n. Their sum and difference are matrices **C** and **D** respectively of the same order m×n, the elements of which are

$$c_{ij}=a_{ij}+b_{ij}\ \ ,\ \ d_{ij}=a_{ij}-b_{ij}\ . \tag{1.2.15}$$

We write,

$$\mathbf{C}=\mathbf{A}+\mathbf{B}\ \ ,\ \ \mathbf{D}=\mathbf{A}-\mathbf{B}\ . \tag{1.2.16}$$

The product of a matrix **A** by a *number* (or *scalar*) k is the matrix $k\mathbf{A}$ with elements ka_{ij}.

Two matrices **A** and **B** can be *multiplied* in the sense **AB** only if the number of columns of **A** is equal to the number of rows of **B**. Thus if **A** has order m×n, **B** has order n×p then

$$\mathbf{AB}=\mathbf{C}\ . \tag{1.2.17}$$

where **C** has order m×p. We write

$$\mathbf{A}(m \times n) \times \mathbf{B}(n \times p) = \mathbf{C}(m \times p)\ . \tag{1.2.18}$$

The element in row i and column j of C is c_{ij}, and is equal to the sum of the elements of row i of **A** multiplied by the corresponding elements of column j of **B**. Thus

$$c_{ij} = a_{i1}b_{1j} + a_{i2}b_{2j} + \ldots + a_{in}b_{nj} = \sum_{k=1}^{n} a_{ik}b_{kj}\ , \tag{1.2.19}$$

and for example

$$\begin{bmatrix} 2 & 3 & 1 \\ 1 & 6 & 7 \\ & & \end{bmatrix} \begin{bmatrix} 1 & 2 & 1 & 0 \\ 0 & 1 & -1 & 0 \\ -1 & 0 & 2 & 1 \end{bmatrix} = \begin{bmatrix} 1 & 7 & 1 & 1 \\ -6 & 8 & 9 & 7 \\ & & & \end{bmatrix}. \tag{1.2.20}$$

The most important consequence of this definition is that matrix multiplication is (in general) *non-commutative*, i.e.,

$$\mathbf{AB} \neq \mathbf{BA}\ . \tag{1.2.21}$$

Indeed if **A** is (m×n) and **B** is (n×p) then **BA** cannot be formed at all unless $m = p$. Even when $m = p$, the two matrices are not necessarily equal, as is shown by the example

$$\mathbf{A} = \begin{bmatrix} 1 & 1 \\ 0 & -1 \end{bmatrix},\ \mathbf{B} = \begin{bmatrix} 1 & 0 \\ -1 & 1 \end{bmatrix},$$

$$\mathbf{AB} = \begin{bmatrix} 0 & 1 \\ 1 & -1 \end{bmatrix},\ \mathbf{BA} = \begin{bmatrix} 1 & 1 \\ -1 & -2 \end{bmatrix}. \tag{1.2.22}$$

In addition, this definition implies that there are *divisors of zero*; i.e., there can be non-zero matrices **A**,**B** such that

$$\mathbf{AB} = \mathbf{0}\ . \tag{1.2.23}$$

An example is provided by

$$\begin{bmatrix} 1 & 1 \\ 2 & 2 \end{bmatrix} \begin{bmatrix} 1 & 1 \\ -1 & -1 \end{bmatrix} = \begin{bmatrix} 0 & 0 \\ 0 & 0 \end{bmatrix}. \tag{1.2.24}$$

The product of $\mathbf{A}(m \times n)$ and a column vector $\mathbf{x}(n \times 1)$ is a column vector $\mathbf{y}(m \times 1)$ with elements

$$y_i = a_{i1}x_1 + a_{i2}x_2 + \ldots + a_{in}x_n\ ,\ \ (i = 1,2,\ldots,m). \tag{1.2.25}$$

This means that the set of m equations

$$\begin{aligned} a_{11}x_1+a_{12}x_2+...+a_{1n}x_n&=y_1\ ,\\ a_{21}x_1+a_{22}x_2+...+a_{2n}x_n&=y_2\ ,\\ a_{m1}x_1+a_{m2}x_2+...+a_{mn}x_n&=y_m\ , \end{aligned} \tag{1.2.26}$$

may be written as the single matrix equation

$$\begin{bmatrix} a_{11} & a_{12} & \cdots & a_{1n} \\ a_{21} & a_{22} & \cdots & a_{2n} \\ \cdot & \cdot & \cdots & \cdot \\ a_{m1} & a_{m2} & \cdots & a_{mn} \end{bmatrix} \begin{bmatrix} x_1 \\ x_2 \\ \\ x_n \end{bmatrix} = \begin{bmatrix} y_1 \\ y_2 \\ \\ y_m \end{bmatrix}, \tag{1.2.27}$$

or

$$\mathbf{Ax}=\mathbf{y}\ . \tag{1.2.28}$$

The product of an (n×1) column vector $\mathbf{x}$ and its transpose $\mathbf{x}^T$ (1×n) is an n×n symmetric matrix

$$\mathbf{xx}^T = \begin{bmatrix} x_1^2 & x_1x_2 & \cdots & x_1x_n \\ x_2x_1 & x_2^2 & \cdots & x_2x_n \\ x_nx_1 & x_nx_2 & \cdots & x_n^2 \end{bmatrix}. \tag{1.2.29}$$

On the other hand, the product of $\mathbf{x}^T$ (1×n) and $\mathbf{x}$ (n×1) is a (1×1) matrix, i.e., a scalar

$$\mathbf{x}^T\mathbf{x}=x_1^2+x_2^2+...+x_n^2. \tag{1.2.30}$$

This quantity, which is positive if and only if the x_i (assumed to be real) are not all zero, is called the square of the L_2 *norm* of $\mathbf{x}$, i.e.,

$$||\mathbf{x}||_2^2=\mathbf{x}^T\mathbf{x}\ ,\quad ||\mathbf{x}||_2=(x_1^2+x_2^2+...+x_n^2)^{1/2}. \tag{1.2.31}$$

The scalar (or dot) product of $\mathbf{x}$ and $\mathbf{y}$ is defined to be

$$\mathbf{x}^T\mathbf{y}=\mathbf{y}^T\mathbf{x}=x_1y_1+x_2y_2+...+x_ny_n\,. \tag{1.2.32}$$

Two vectors are said to be *orthogonal* if

$$\mathbf{x}^T\mathbf{y}=0\ . \tag{1.2.33}$$

It has been noted that matrix multiplication is *non-commutative*. This hold even if the matrices are square (see (1.2.22)) or symmetric, as illustrated by

$$\begin{bmatrix} 1 & 2 \\ 2 & 2 \end{bmatrix} \begin{bmatrix} 1 & -1 \\ -1 & 1 \end{bmatrix} = \begin{bmatrix} -1 & 1 \\ 0 & 0 \end{bmatrix},\quad \begin{bmatrix} 1 & 1- \\ -1 & 1 \end{bmatrix} \begin{bmatrix} 1 & 2 \\ 2 & 2 \end{bmatrix} = \begin{bmatrix} -1 & 0 \\ 1 & 0 \end{bmatrix}. \tag{1.2.34}$$

This example, which shows that the product of two symmetric matrices is not (necessarily) symmetric, hints also that there might be a relation between the products **AB** and **BA**. This result is sufficiently important to be called:

Theorem 1.2.1

$$(\mathbf{AB})^T = \mathbf{B}^T \mathbf{A}^T \ , \tag{1.2.35}$$

so that when **A**, **B**, *are* symmetric (as in (1.2.34)), *then*

$$(\mathbf{AB})^T = \mathbf{BA} \ . \tag{1.2.36}$$

Proof

Consider the element in row i, column j on each side of (1.2.35). Suppose **A** is (m×n), **B** is (n×p), then **AB** is m×p and $(\mathbf{AB})^T$ is p×m. Then

$$((\mathbf{AB})^T)_{ij} = (\mathbf{AB})_{ji} = \sum_{k=1}^{n} a_{jk} b_{ki} \ ,$$

and

$$\begin{aligned}(\mathbf{B}^T\mathbf{A}^T)_{ij} &= (\text{ row i of } \mathbf{B}^T) \times (\text{ column j of } A^T) \\ &= (\text{ column i of } \mathbf{B}) \times (\text{ row j of } A) \\ &= \sum_{k=1}^{n} b_{ki} a_{jk} \quad \blacksquare\end{aligned} \tag{1.2.37}$$

Examples 1.2

1. If

$$\mathbf{A} = \begin{bmatrix} 1 & 2 & 3 \\ 2 & 3 & 5 \\ 3 & 5 & 8 \end{bmatrix}$$

   find a square matrix **B** such that **AB**=0. Show that if a_{33} is changed then the only possible matrix **B** would be the zero matrix.

2. Show that, whatever the matrix **A**, the two matrices $\mathbf{AA}^T$ and $\mathbf{A}^T\mathbf{A}$ are symmetric. Are these two matrices equal?

3. Show that if **A**, **B** are square and of order n and **A** is symmetric, then $\mathbf{BAB}^T$ and $\mathbf{B}^T\mathbf{AB}$ are symmetric.

4. Show that if **A**, **B**, **C** can be multiplied in the order **ABC**, then $(\mathbf{ABC})^T = \mathbf{C}^T\mathbf{B}^T\mathbf{A}^T$.

1.3. Matrix inversion and determinants

We shall now be concerned almost exclusively with *square* matrices. The *determinant* of a (square) matrix is defined to be

$$|\mathbf{A}| = \Sigma \pm a_{1i_1} a_{2i_2} \cdots a_{ni_n} ; \tag{1.3.1}$$

where the suffices i_1, i_2,...,i_n are a permutation of the numbers 1,2,3,...,n, the sign is + if the permutation is even and – if it is odd, and the summation is carried out over all $n!$ permutations of 1,2,3,...,n. We note that each product in the sum contains just one element from each row and just one element from each column of **A**. Thus for 2×2 and 3×3 matrices respectively

$$\begin{vmatrix} a_{11} & a_{12} \\ a_{21} & a_{22} \end{vmatrix} = a_{11}a_{22} - a_{12}a_{21} , \tag{1.3.2}$$

$$\begin{vmatrix} a_{11} & a_{12} & a_{13} \\ a_{21} & a_{22} & a_{23} \\ a_{31} & a_{32} & a_{33} \end{vmatrix} = \begin{matrix} a_{11}a_{22}a_{33} & + & a_{12}a_{23}a_{31} & + & a_{11}a_{21}a_{32} \\ -a_{11}a_{23}a_{32} & - & a_{12}a_{21}a_{33} & - & a_{13}a_{22}a_{31} \end{matrix} . \tag{1.3.3}$$

The permutation i_1,i_2,...,i_n is even or odd according to whether it may be obtained from 1,2,...,n by means of an even or an odd numer of interchange, respectively. Thus 1,3,2,4 and 2,3,1,4 are respectively odd and even permutations of 1,2,3,4 because

$$(1,2,3,4) \rightarrow (1,3,2,4) .$$

$$(1,2,3,4) \rightarrow (2,1,3,4) \rightarrow (2,3,1,4)$$

We now list some of the properties of determinants.

Lemma 1.3.1. *If two rows (or columns) of* **A** *are interchanged, the determinant retains its numerical value, but changes sign.* If the new matrix is called **B** then

$$b_{1i} = a_{2i} \ , \ b_{2i} = a_{1i} \ , \ b_{ji} = a_{ji} \ , \ (j = 3,4,...,n)$$

and

$$\begin{aligned} |\mathbf{B}| &= \Sigma \pm b_{1i_1} b_{2i_2} b_{3i_3} \cdots b_{ni_n} , \\ &= \Sigma \pm a_{2i_1} a_{1i_2} a_{3i_3} \cdots a_{ni_n} , \\ &= \Sigma \pm a_{1i_2} a_{2i_1} a_{3i_3} \cdots a_{ni_n} . \end{aligned} \tag{1.3.4}$$

But if i_1,i_2,i_3,...,i_n is even (odd) then i_2,i_1,i_3,...,i_n is odd (even), so that each term in $|\mathbf{B}|$ appears in $|\mathbf{A}|$ (and vice versa) with the opposite sign, so that $|\mathbf{B}| = -|\mathbf{A}|$.

Lemma 1.3.2 *If two rows (columns) of* **A** *are identical then* $|\mathbf{A}| = 0$. If the two rows (columns) are interchanged, then, on the one hand, $|\mathbf{A}|$ is unchanged, while on the other, Lemma 1.3.1, $|\mathbf{A}|$ changes sign. Thus $|\mathbf{A}| = -|\mathbf{A}|$ and hence $|\mathbf{A}|=0$.

Lemma 1.3.3. *If one row (column) of* **A** *is multiplied by k then the determinant is multiplied by k.* Each term in the expansion is multiplied by k.

Lemma 1.3.4. *If two rows (columns) of* **A** *are proportional, then* $|\mathbf{A}|=0$. This follows from Lemmas 1.3.1, 1.3.3.

Lemma 1.3.5. *If one row (column) of* **A** *is added to another row (column) then the determinant is unchanged.* If the matrix **B** is obtained, say, by adding row 2 to row 1 then

$$b_{1i}=a_{1i}+a_{2i}\ ,\ b_{ji}=a_{ji}\ ,\ j=2,3,\ldots,n\ .$$

Thus

$$\begin{aligned}|B| &= \Sigma \pm b_{1i_1}b_{2i_2}b_{3i_3}\cdots b_{ni_n} = \\ &= \Sigma \pm (a_{1i_1}+a_{2i_1})a_{2i_2}a_{3i_3}\cdots a_{ni_n}\ , \\ &= \Sigma \pm a_{1i_1}a_{2i_2}a_{3i_3}\cdots a_{ni_n} \pm \Sigma a_{2i_1}a_{2i_2}a_{3i_3}\cdots a_{ni_n}\ ,\end{aligned}$$

and the first sum is $|\mathbf{A}|$ while the second, having its first and second rows equal, is zero.

Lemma 1.3.6. *If a linear combination of rows (columns) of* **A** *is added to another row (column) then the determinant is unchanged.* This follows directly from Lemma 1.3.5. We may now prove:

Theorem 1.3.1. If the rows (columns) of **A** *are linearly dependent then* $|\mathbf{A}|=0$.

Proof

Denote the rows by $\mathbf{a}_1^T, \mathbf{a}_2^T, \ldots, \mathbf{a}_n^T$. Then, by hypotheses, there are scalars $c_1, c_2, \ldots, c_n$, *not all zero*, such that

$$c_1\mathbf{a}_1^T+c_2\mathbf{a}_2^T+\cdots+c_n\mathbf{a}_n^T=0\ . \tag{1.3.5}$$

There is a c_i not zero; let it be c_m. Then

$$-\mathbf{a}_m^T=\sum_{\substack{i=1\\ i\neq m}}^{n}(c_i/c_m)\mathbf{a}_i^T\ . \tag{1.3.6}$$

If the sum on the right is added to row m of **A**, the new matrix has a zero row so that its determinant, which by Lemma 1.3.6 is $|\mathbf{A}|$, is zero ■

Before proving the converse of this theorem we need some more notation. The coefficient of a_{ij} in the expansion of $|\mathbf{A}|$ given in (1.3.1) is called the *cofactor* of a_{ij} and is called A_{ij}. Thus in the expansion (1.3.3) we may write

$$
\begin{aligned}
|\mathbf{A}| &= a_{11}(a_{22}a_{33}-a_{23}a_{32})+a_{12}(a_{23}a_{31}-a_{21}a_{33})+ \\
&+ a_{13}(a_{21}a_{32}-a_{22}a_{31}) , \\
&= a_{11}\begin{vmatrix} a_{22} & a_{23} \\ a_{32} & a_{33} \end{vmatrix} - a_{12}\begin{vmatrix} a_{21} & a_{23} \\ a_{31} & a_{33} \end{vmatrix} + a_{13}\begin{vmatrix} a_{21} & a_{22} \\ a_{31} & a_{32} \end{vmatrix} ,
\end{aligned} \tag{1.3.7}
$$

so that

$$
A_{11}=a_{22}a_{33}-a_{23}a_{32} , \quad A_{12}=a_{23}a_{31}-a_{21}a_{33}, \text{ etc.} \tag{1.3.8}
$$

The general result is:

Lemma 1.3.7

$$
A_{ij}=(-)^{i+j}\hat{a}_{ij} , \tag{1.3.9}
$$

where $\hat{a}_{ij}$ *is the determinant of the (n-1) order matrix obtained by deleting row i and column j of* **A**. *We now prove*

Theorem 1.3.2.

$$
\sum_{k=1}^{n} a_{ik}A_{jk} = |\mathbf{A}|\,\delta_{ij} , \tag{1.3.10}
$$

$$
\sum_{k=1}^{n} A_{ki}a_{kj} = |\mathbf{A}|\,\delta_{ij} . \tag{1.3.11}
$$

Proof

When $i=j$, so that $\delta_{ij}=1$, this theorem merely states the definition of A_{ik}. When $i \neq j$, so that $\delta_{ij}=0$, then the theorem states that the determinant of the matrix with rows (columns) i and j equal, is zero ■

For an example consider $n=3$, $i=2$, $j=1$, then

$$
\sum_{k=1}^{3} a_{2k}A_{1k} = a_{21}\begin{vmatrix} a_{22} & a_{23} \\ a_{32} & a_{33} \end{vmatrix} - a_{22}\begin{vmatrix} a_{21} & a_{23} \\ a_{31} & a_{33} \end{vmatrix} + a_{22}\begin{vmatrix} a_{21} & a_{22} \\ a_{31} & \\ a_{32} & \end{vmatrix} \equiv 0 . \tag{1.3.12}
$$

We may now prove:

Theorem 1.3.3. *If* $|\mathbf{A}|=0$, *then the rows (columns) of* **A** *are linearly dependent.*

Proof

This theorem is somewhat difficult. We shall prove it by induction. Let $\mathbf{a}_1^T$, $\mathbf{a}_2^T,\ldots,\mathbf{a}_n^T$ be the rows of **A**, and suppose that $|\mathbf{A}|=0$. *Either* each set of $(n-1)$ vectors selected from $\mathbf{a}_1^T$, $\mathbf{a}_2^T,\ldots,\mathbf{a}_n^T$ is a linearly dependent set, in which case the complete set is linearly dependent as required, *or* there is a set of $(n-1)$ vectors, which without loss of generality we may take to be $\mathbf{a}_1^T$, $\mathbf{a}_2^T,\ldots$, $\mathbf{a}_{n-1}^T$, which is linearly independent. In this case the $(n-1)$ vectors $\mathbf{b}_1^T$, $\mathbf{b}_2^T,\ldots,$ $\mathbf{b}_{n-1}^T$ of order $(n-1)$, obtained by deleting the first elements of these vectors, will be linearly independent. Now suppose that the theorem holds for $(n-1)$. This means that the determinant of the $(n-1)$ vectors $\mathbf{b}_1^T$, $\mathbf{b}_2^T,\ldots,$ $\mathbf{b}_{n-1}^T$ will be non-zero. (For if the determinant were zero then, by the theorem, assumed to be true for $(n-1)$, the vectors $\mathbf{b}_i^T$ would be linearly dependent). But the determinant of the matrix made up of the $\mathbf{b}_i^T$ is

$$\hat{\mathbf{a}}_{1,n} = \begin{vmatrix} a_{21} & a_{22} & \cdots & a_{2,n-1} \\ a_{31} & a_{32} & \cdots & a_{3,n-1} \\ \cdot & \cdot & \cdots & \\ a_{n1} & a_{n2} & \cdots & a_{n,n-1} \end{vmatrix} = (-)^{n+1}A_{1n} \neq 0. \tag{1.3.13}$$

Now we use Theorem 1.3.2 with $j=1$ to deduce

$$\sum_{k=1}^{n} A_{1k}a_{jk} = 0\ , \quad (j=1,2,\ldots,n)\ . \tag{1.3.14}$$

This equation states that

$$A_{11}\mathbf{a}_1^T + A_{12}\mathbf{a}_2^T + \ldots + A_{1n}\mathbf{a}_n^T = \mathbf{0}. \tag{1.3.15}$$

Since $A_{1n} \neq 0$, this means that $\mathbf{a}_1^T$, $\mathbf{a}_2^T,\ldots,\mathbf{a}_n^T$ are linearly dependent. Thus if the theorem is true for $n-1$, then it is true for n. The final step is to note that the theorem holds (trivially) for $n=1$ ■

Theorem 1.3.4 *The matrix equations*

$$\mathbf{Ax} = \mathbf{0}\ , \quad \mathbf{y}^T\mathbf{A} = \mathbf{0} \tag{1.3.16}$$

have non-trivial solutions $\mathbf{x}$ *and* $\mathbf{y}^T$ *respectively if and only if* $|\mathbf{A}| = 0$.

Proof

Both results may be proved similarly; we consider the second. If the rows of $\mathbf{A}$ are $\mathbf{a}_1^T$, $\mathbf{a}_2^T,\ldots;$ $\mathbf{a}_n^T$, then

$$\mathbf{y}^T\mathbf{A} \equiv y_1\mathbf{a}_1^T + y_2\mathbf{a}_2^T + \ldots + y_n\mathbf{a}_n^T\ . \tag{1.3.17}$$

Thus there is a solution if and only if the $(\mathbf{a}_i^T)_1^n$ are linearly dependent. By Theorems 1.3.1 and 1.3.3, this occurs if and only if $|\mathbf{A}| = 0$. ■

Theorem 1.3.5. *If* $\mathbf{A}$, $\mathbf{B}$ *are of order* n, *then*

$$|\mathbf{AB}| = |\mathbf{A}| \cdot |\mathbf{B}|\ . \tag{1.3.18}$$

The proof of this result is left to Ex. 1.3.6.

The matrix $\mathbf{A}$ is said to be *singular* if $|\mathbf{A}|=0$, *non-singular* is $|\mathbf{A}|\neq 0$.

Theorem 1.3.6 *The matrix* $\mathbf{A}$ *has a (unique) inverse* $\mathbf{A}^{-1}$ *if and only if* $\mathbf{A}$ *is non-singular.*

Proof

Suppose $\mathbf{A}$ is non-singular, then $|\mathbf{A}|\neq 0$ and equation (1.3.10) states that

$$\sum_{k=1}^{n} a_{ik}(A_{jk}/|\mathbf{A}|)=\delta_{ij}=\sum_{k=1}^{n}(A_{ki}/|\mathbf{A}|)a_{kj}\ . \tag{1.3.19}$$

This means that if $\mathbf{R}$ is the matrix with elements

$$r_{kj}=A_{jk}/|\mathbf{A}|\ , \tag{1.3.20}$$

then

$$\mathbf{AR}=\mathbf{I}=\mathbf{RA}\ . \tag{1.3.21}$$

If there were another matrix $\mathbf{S}$ such that

$$\mathbf{AS}=\mathbf{I}\ , \tag{1.3.22}$$

then multiplication on the left would give

$$\mathbf{RAS}=\mathbf{R}\ ; \tag{1.3.23}$$

but $\mathbf{RA}=\mathbf{I}$ so that $\mathbf{RAS}=\mathbf{IS}=\mathbf{S}$ and $\mathbf{S}=\mathbf{R}$. The unique matrix $\mathbf{R}$ is called $\mathbf{A}^{-1}$. Thus

$$\mathbf{AA}^{-1}=\mathbf{I}=\mathbf{A}^{-1}\mathbf{A}\ . \tag{1.3.24}$$

If $\mathbf{A}$ is singular then it has *no* inverse. For suppose

$$\mathbf{AR}=\mathbf{I}\ , \tag{1.3.25}$$

then $|\mathbf{AR}|=|\mathbf{I}|=1=|\mathbf{A}|\cdot|\mathbf{R}|=0$, which is impossible. ■

Theorem 1.3.7. *The equation*

$$\mathbf{Ax}=\mathbf{b} \tag{1.3.26}$$

either *has a unique solution, if* $\mathbf{A}$ *is non-singular; or if* $\mathbf{A}$ *is singular, it has a solution only for certain* $\mathbf{y}$*; in the latter case it has an infinity of solutions.*

Proof

If $\mathbf{A}$ is non-singular then

$$\mathbf{A}^{-1}(\mathbf{Ax})=\mathbf{x}=\mathbf{A}^{-1}\mathbf{b}\ , \tag{1.3.27}$$

is the unique solution. If $\mathbf{A}$ is singular then there is (one or more) $\mathbf{y}$ such that

$$\mathbf{y}^T\mathbf{A}=0\ . \tag{1.3.28}$$

Then,

$$\mathbf{y}^T\mathbf{A}\mathbf{x}=0=\mathbf{y}^T\mathbf{b}\ , \tag{1.3.29}$$

so that (1.3.26) has a solution only if **b** is orthogonal to any solution **y** of (1.3.28). If **b** satisfies this condition then there is at least one solution $\mathbf{x}^o$. Since **A** is singular there are one or more linearly independent vectors $(\mathbf{x}^{(i)})_1^m$ satisfying $\mathbf{A}\mathbf{x}^{(i)}=\mathbf{0}$, and

$$\mathbf{x}=\mathbf{x}^o+\sum_{i=1}^{m}\alpha_i\mathbf{x}^{(i)}\ , \tag{1.3.30}$$

is the general solution of (1.3.26). ■

Examples 1.3

1. Show that if $\mathbf{A}=diag(a_{11},a_{22},...,a_{nn})$, then $|\mathbf{A}|=a_{11}a_{22}...a_{nn}$.
2. Show that if $(a_{1i})_2^n=0$, then

   $$|\mathbf{A}|=a_{11}\Delta_{11}\ .$$

3. One way of calculating the value of a determinant is to use Gaussian elimination. This is a systematic application of Lemmas 1.3.1 - 1.3.6 and the results of Ex. 1.3.1, 1.3.2. Thus for example, by subtracting 2 (row 1) and 3 (row 1) from row 2 and row 3 respectively, we find

   $$\begin{vmatrix}1&3&2\\2&5&6\\3&4&7\end{vmatrix}=\begin{vmatrix}1&3&2\\0&-1&2\\0&-5&1\end{vmatrix}=-\begin{vmatrix}1&3&2\\0&1&-2\\0&-5&1\end{vmatrix}=-\begin{vmatrix}1&3&2\\0&1&-2\\0&0&-9\end{vmatrix}=9\ .$$

4. Use Gaussian elminination to show that

   $$\begin{vmatrix}1&2&4&8\\0&1&3&2\\1&2&5&6\\-1&3&4&7\end{vmatrix}=\begin{vmatrix}1&2&4&8\\0&1&3&2\\0&0&1&-2\\0&0&0&-9\end{vmatrix}=9\ .$$

5. If

   $$\mathbf{A}=\begin{bmatrix}1&3&2\\2&5&6\\3&4&7\end{bmatrix}$$

   find $\mathbf{A}^{-1}$. Check that $\mathbf{A}\mathbf{A}^{-1}=\mathbf{A}^{-1}\mathbf{A}=\mathbf{I}$.
6. To prove $|\mathbf{AB}|=|\mathbf{A}|\cdot|\mathbf{B}|$ consider the $2n\times 2n$ matrix

$$\mathbf{C} = \begin{bmatrix} \mathbf{A} & \mathbf{0} \\ \mathbf{I} & \mathbf{B} \end{bmatrix} .$$

Show that $|\mathbf{C}| = |\mathbf{A}| \cdot |\mathbf{B}|$. Now subtract multiples of rows $(n+1)$ to 2n from rows 1 to n to delete all elements in the top n×n part of **C**. The elements in the top right will be those of $-\mathbf{AB}$.

1.4. Eigenvalues and eigenvectors

If **A**, **C** are square matrices of order n then the equation

$$\mathbf{Cx} = \lambda \mathbf{Ax} , \tag{1.4.1}$$

will have a non-trivial solution **x** (i.e., one for which $||\mathbf{x}|| \neq 0$) if and only if the scalar λ satisfies the determinantal equation

$$|\mathbf{C} - \lambda \mathbf{A}| = 0 . \tag{1.4.2}$$

The roots of this equation are called the *eigenvalues* of the matrix pair **C**, **A**; they may be complex and are labelled $(\lambda_i)_1^n$ i.e., $\lambda_1, \lambda_2, \ldots, \lambda_n$. The corresponding solutions of (1.4.1) are called *eigenvectors* and are labelled $\mathbf{u}^{(i)}$, thus

$$\mathbf{Cu}^{(i)} = \lambda_i \mathbf{Au}^{(i)} . \tag{1.4.3}$$

In many mathematical texts, attention is focussed almost exclusively on the case in which **A** is the unit matrix, so that (1.4.1) becomes

$$\mathbf{Cx} = \lambda \mathbf{x} , \quad \mathbf{x} \neq \mathbf{0} . \tag{1.4.4}$$

In this case λ is said to be an *eigenvalue* of **C**. In applications to mechanics, two matrices **C**, **A** occur and it will be convenient to develop the theory corresponding to two matrices. The general theory of eigenvalues for equation (1.4.1) or (1.4.4) is difficult; here we shall assume that **A** and **C** are *symmetric*; in this case we have:

Theorem 1.4.1. *The eigenvalues and eigenvectors for a real symmetric matrix pair are real.*

Proof

Suppose λ, **x**, possibly complex, are an eigen-pair of the matrix pair **A**, **C** so that (1.4.1) holds. Multiply both sides by $\bar{\mathbf{x}}^T$, the complex conjuate of $\mathbf{x}^T$ to obtain the scalar equation

$$\bar{\mathbf{x}}^T \mathbf{Cx} = \lambda \bar{\mathbf{x}}^T \mathbf{Ax} . \tag{1.4.5}$$

The quantities $\bar{\mathbf{x}}^T \mathbf{Ax}$ and $\bar{\mathbf{x}}^T \mathbf{Cx}$ are both *real*. This is so because $\bar{\mathbf{x}}^T \mathbf{Ax}$, for instance, is a scalar so that it is equal to its own transpose. Thus

$$a = \bar{\mathbf{x}}^T \mathbf{Ax} = (\bar{\mathbf{x}}^T \mathbf{Ax})^T = \mathbf{x}^T \mathbf{A}^T \bar{\mathbf{x}} = \mathbf{x}^T \mathbf{A} \bar{\mathbf{x}} = \bar{a} ; \tag{1.4.6}$$

but if $a = \bar{a}$, then a is real. Thus λ is a quotient of two real quantities and is therefore real. Since λ is real, $\mathbf{x}$, obtained by solving a set of simultaneous linear equations with real coefficients, is also real. Therefore $\bar{\mathbf{x}}^T = \mathbf{x}^T$ and

$$\lambda = \frac{\mathbf{x}^T \mathbf{C}\mathbf{x}}{\mathbf{x}^T \mathbf{A}\mathbf{x}}. \tag{1.4.7}$$

The quantity $\mathbf{x}^T\mathbf{A}\mathbf{x}$, when written out in full, is

$$\begin{aligned}\mathbf{x}^T\mathbf{A}\mathbf{x} &= a_{11}x_1^2 + 2a_{12}x_1x_2 + \ldots + 2a_{1n}x_1x_n + \\ &+ a_{22}x_2^2 + \ldots + 2a_{2n}x_2x_n + \ldots + a_{nn}x_n^2.\end{aligned} \tag{1.4.8}$$

It is called a *quadratic form*. In many physical applications it represents a quantity, such as kinetic energy, which is always positive – or zero when all the $(x_i)_1^n$ are zero. In this case we say that the matrix $\mathbf{A}$ is *positive-definite*. Sometimes the quadratic form is never negative, but can be zero without all the $(x_i)_1^n$ being zero; in this case $\mathbf{A}$ is said to be *positive semi-definite*. (See Ex. 1.4.1).

Theorem 1.4.2 *Let $(D_i)_1^n$ denote the principal minors of* $\mathbf{A}$, *so that*

$$D_1 = a_{11}, \quad D_2 = \begin{vmatrix} a_{11} & a_{12} \\ a_{21} & a_{22} \end{vmatrix}, \ldots, D_n = \begin{vmatrix} a_{11} & a_{12} & \cdots & a_{1n} \\ a_{21} & a_{22} & \cdots & a_{2n} \\ \cdot & \cdot & \cdots & \cdot \\ a_{n1} & a_{n2} & \cdots & a_{nn} \end{vmatrix}. \tag{1.4.9}$$

$\mathbf{A}$ *will be positive-definite if and only if $(D_i)_1^n > 0$;* $\mathbf{A}$ *will be positive semi-definite if and only if $(D_i)_1^{n-1} \geq 0,\ D_n = 0$.* This will not be proved until Chapter 5. Since $D_n = |\mathbf{A}|$, any positive-definite matrix must be non-singular, while any positive semi-definite matrix must be singular.

Theorem 1.4.3. *If* $\mathbf{A}$ *and* $\mathbf{C}$ *are positive-definite, then equation (1.4.1) will have n positive eigenvalues, although they need not be distinct. One at least of these will be zero if* $\mathbf{C}$ *is only positive semi-definite.*

Proof

Equation (1.4.2) may be expanded in terms of the coefficients $c_{ij} - \lambda a_{ij}$: the result is an nth degree equation for λ, namely

$$\Delta(\lambda) \equiv |\mathbf{C} - \lambda\mathbf{A}| \equiv \Delta_0 + \Delta_1\lambda + \Delta_2\lambda^2 + \ldots + \Delta_n\lambda^n = 0. \tag{1.4.10}$$

Most of the coefficients Δ_k are complicated functions of the a_{ij} and c_{ij}, but the first and last may be easily identified, namely

$$\Delta_0 = |\mathbf{C}|, \quad \Delta_n = (-)^n |\mathbf{A}|. \tag{1.4.11}$$

This shows that when $\mathbf{C}$ is positive semi-definite, so that $|\mathbf{C}| = 0$, then $\lambda = 0$ is an eigenvalue. In addition, since $\mathbf{A}$ is positive definite, $|\mathbf{A}| \neq 0$, and equation (1.4.10) is a proper equation with degree n and n roots $(\lambda_i)_1^n$, one of which will be zero when

$|\mathbf{C}|=0$. The positiveness of the eigenvalues follows directly from (1.4.7); both numerator and denominator will be positive. ■

The eigenvalues of (1.4.1) will henceforth be labelled in increasing order, so that

$$0 \le \lambda_1 \le \lambda_2 \cdots \le \lambda_n \;. \tag{1.4.12}$$

Theorem 1.4.4. *Eigenvectors* $\mathbf{u}^{(1)}$, $\mathbf{u}^{(2)}$ *corresponding to two different eigenvalues* λ_1, λ_2 $(\lambda_1 \neq \lambda_2)$ *of the symmetric matrix pair* $(\mathbf{C},\mathbf{A})$ *are orthogonal w.r.t. both to* $\mathbf{A}$ *and to* $\mathbf{C}$, *i.e.,*

$$\mathbf{u}^{(1)T}\mathbf{A}\mathbf{u}^{(2)} = 0 = \mathbf{u}^{(1)T}\mathbf{C}\mathbf{u}^{(2)} \;. \tag{1.4.13}$$

Proof

By definition

$$\mathbf{C}\mathbf{u}^{(1)} = \lambda_1 \mathbf{A}\mathbf{u}^{(1)} \;, \quad \mathbf{C}\mathbf{u}^{(2)} = \lambda_2 \mathbf{A}\mathbf{u}^{(2)} \;. \tag{1.4.14}$$

Transpose the first equation and multiply it on the right by $\mathbf{u}^{(2)}$; multiply the second equation on the left by $\mathbf{u}^{(1)T}$, to obtain

$$(\mathbf{u}^{(1)T}\mathbf{C})\mathbf{u}^{(2)} = \lambda_1(\mathbf{u}^{(1)T}\mathbf{A})\mathbf{u}^{(2)} \;,$$
$$\mathbf{u}^{(1)T}(\mathbf{C}\mathbf{u}^{(2)}) = \lambda_2 \mathbf{u}^{(1)T}(\mathbf{A}\mathbf{u}^{(2)}) \;. \tag{1.4.15}$$

Subtract these two scalar equations to yield

$$(\lambda_1 - \lambda_2)\mathbf{u}^{(1)T}\mathbf{A}\mathbf{u}^{(2)} = 0 \;. \tag{1.4.16}$$

But $\lambda_1 \neq \lambda_2$ so that $\mathbf{u}^{(1)T}\ \mathbf{A}\mathbf{u}^{(2)} = 0$. ■

An important corollary of this result is

Theorem 1.4.5. *If the symmetric matrix pair* $(\mathbf{C},\mathbf{A})$ *has distinct eigenvalues* $(\lambda_i)_1^n$ *and if* $\mathbf{A}$ *is positive definite, then the eigenvectors* $\mathbf{u}^{(i)}$ *are linearly independent, and therefore span the space of n-vectors.*

Proof

Suppose, if possible, that

$$\alpha_1\mathbf{u}^{(1)} + \alpha_2\mathbf{u}^{(2)} + \ldots + \alpha_n\mathbf{u}^{(n)} = \mathbf{0} \;. \tag{1.4.17}$$

Multiply on the right by $\mathbf{u}^{(m)T}\mathbf{A}$ to obtain

$$\sum_{i=1}^{n} \alpha_i \mathbf{u}^{(m)T}\mathbf{A}\mathbf{u}^{(i)} = 0 \;. \tag{1.4.18}$$

Of the n terms in the sum only the mth is non-zero, since the other products relate to eigenvectors for different eigenvalues $\lambda_i \neq \lambda_m$. Thus

$$\alpha_m \mathbf{u}^{(m)T}\mathbf{A}\mathbf{u}^{(m)}=0 \;, \tag{1.4.19}$$

but **A** is positive definite so that $\mathbf{u}^{(m)T}\mathbf{A}\mathbf{u}^{(m)}>0$ and hence $\alpha_m=0$. This holds for each $m=1,2,...,n$ so that the $\mathbf{u}^{(i)}$ are linearly independent. ■

If the symmetric matrix pair (**C**,**A**) has an r-fold multiple eigenvalue, so that r eigenvalues λ_1, λ_2,..., λ_r are equal, then it may be shown (Bishop, Gladwell and Michaelson (1965, Section 4.4)) that there is an r-dimensional space of eigenvectors corresponding to this multiple eigenvalue and that r eigenvectors $(\mathbf{u}^{(i)})_1^r$ may be found in this space which are orthogonal w.r.t. **A** and to **C**. In practice, it is found that the case of multiple eigenvalues does not appear in inverse problems. Considerations of multiple eigenvalues will therefore be postponed to Chapter 4; until then it will be assumed that

$$0\leq\lambda_1<\lambda_2<\cdots<\lambda_n \;. \tag{1.4.20}$$

The eigenvalue equation (1.4.3) determines only the *shape* of the vector $\mathbf{u}^{(i)}=\{u_{1i},u_{2i},...,u_{ni}\}$, i.e., the set of ratios $u_{1i}:u_{2i}:...u_{ni}$. In computation, the actual magnitudes of the components are chosen so that one of them, say the largest (in magnitude) is unity. For theoretical work, and this will be the choice here, the magnitudes are chosen so that

$$\mathbf{u}^{(i)T}\mathbf{A}\mathbf{u}^{(i)}=a_i \;, \tag{1.4.21}$$

where a_i is an arbitrary positive number. Equations (1.4.1) and (1.4.21) imply

$$\mathbf{u}^{(i)T}\mathbf{C}\mathbf{u}^{(i)}=\lambda_i a_i=c_i \;. \tag{1.4.22}$$

In this book we are not concerned with methods for computing eigenvalues and eigenvectors. Various iterative and direct methods were discussed at length in Bishop, Gladwell and Michaelson (1965). We emphasize here that we are concerned only with those properties of the solution of the equations which will be useful in solving the inverse problems.

Eigenvalue analysis will be taken up again in Section 2.5.

Examples 1.4

1. Show that

$$[x_1\;\; x_2\;\; x_3\;\; x_4]\begin{bmatrix} 1 & -1 & & \\ -1 & 2 & -1 & \\ & -1 & 2 & -1 \\ & & -1 & 1 \end{bmatrix}\begin{bmatrix} x_1 \\ x_2 \\ x_3 \\ x_4 \end{bmatrix} \equiv (x_1-x_2)^2+(x_2-x_3)^2+(x_3-x_4)^2$$

is a positive semi-definite quadratic form.

2. Verify the conditions (1.4.9) for $n=2$ by writing

$$\begin{aligned}\mathbf{x}^T\mathbf{A}\mathbf{x} &= a_{11}x_1^2+2a_{12}x_1x_2+a_{22}x_2^2\\ &= a_{11}\left\{\left[x_1+\frac{a_{12}}{a_{11}}x_2\right]^2+\frac{(a_{22}a_{11}-a_{12}^2)}{a_{11}^2}x_2^2\right\}.\end{aligned}$$

Extend this result to $n=3$.

3. Find the eigenvalues and eigenvectors of the pair

$$\mathbf{C} = \begin{bmatrix} 2 & -1 & 0 \\ -1 & 2 & -1 \\ 0 & -1 & 2 \end{bmatrix}, \mathbf{A} = \begin{bmatrix} 1 & & \\ & 1 & \\ & & 1 \end{bmatrix}.$$

3. [Hint: replace the eigen-equation by the equivalent recurrence relation $-x_{r-1}+(2-\lambda)x_r-x_{r+1}=0$, $x_o=0=x_4$ and seek a solution of the form $x_r=A\cos r\theta+B\sin r\theta$].

4. Show that if the symmetric matrix $\mathbf{A}$ has positive eigenvalues $(\lambda_i)_1^n$, then it is positive definite.

CHAPTER 2

VIBRATIONS OF DISCRETE SYSTEMS

Our nature consists in motion;
complete rest is death.
Pascal's Pensées

2.1 Introduction

The formulation and solution of the equations governing the motion of a discrete vibrating system, i.e., one which has a finite number of degrees of freedom, have been fully considered elsewhere. See for example, Bishop and Johnson (1960) and Bishop, Gladwell and Michaelson (1965). In this chapter we shall give a brief account of those parts of the theory that will be needed for the solution of inverse problems.

Throughout this book we shall be concerned with the *infinitesimal* vibration of a *conservative* system about some datum configuration, which will usually be an equilibrium position.

Before embarking on a general discussion we shall first formulate the equations of motion for some simple vibrating systems.

2.2 Vibration of some simple systems

Figure 2.2.1 shows a vibrating system consisting of N masses connected by linear springs of stiffnesses $(k_r)_1^N$. The whole lies in a straight line on a smooth horizontal table and is excited by forces $(F_r(t))_1^N$.

Newton's equations of motion for the system are

$$m_r \ddot{u}_r = F_r + \theta_{r+1} - \theta_r \ , \quad r = 1,2,...,N-1 \ , \qquad (2.2.1)$$

$$m_N \ddot{u}_N = F_N - \theta_N \ , \qquad (2.2.2)$$

where · denotes differentiation with respect to time. Hooke's law states that the spring forces are given by

$$\theta_r = k_r(u_r - u_{r-1}) \ , \quad r = 1,2,...,N \ . \qquad (2.2.3)$$

f the left hand end is pinned then

$$u_o = 0 \ . \qquad (2.2.4)$$

Forced vibration analysis concerns the solution of these equations for given forcing functions $F_r(t)$. *Free* vibration analysis consists in finding solutions to the equations which require no external excitation, i.e., $F_r(t) \equiv 0$, $r = 1,2,...,N$, and which satisfy the stated end conditions.

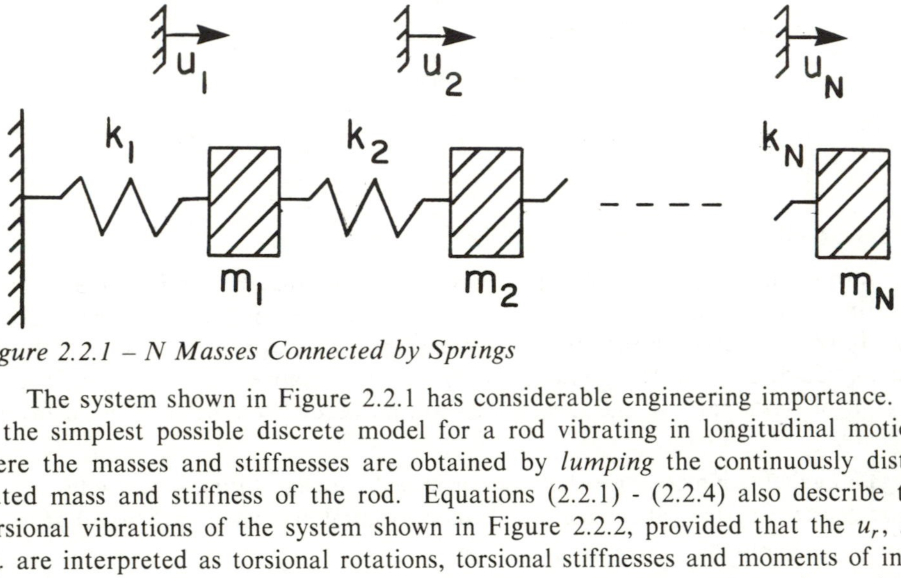

Figure 2.2.1 – N Masses Connected by Springs

The system shown in Figure 2.2.1 has considerable engineering importance. It is the simplest possible discrete model for a rod vibrating in longitudinal motion. Here the masses and stiffnesses are obtained by *lumping* the continuously distributed mass and stiffness of the rod. Equations (2.2.1) - (2.2.4) also describe the torsional vibrations of the system shown in Figure 2.2.2, provided that the u_r, k_r, m_r are interpreted as torsional rotations, torsional stiffnesses and moments of inertia respectively. Such a discrete system provides a simple model for the torsional vibrations of a rod with a continuous distribution of inertia and stiffness.

There is a third system which is mathematically equivalent to equations (2.2.1) - (2.2.4). This is the transverse motion of the string shown in Figure 2.2.3 which is pulled taut by a tension T and which is loaded by masses $(m_r)_1^N$. If in accordance with the assumption of infinitesimal vibration, the string departs very little from the straight line equilibrium position, then the equation governing the motion of mass m_r may be derived by considering Figure 2.2.4.

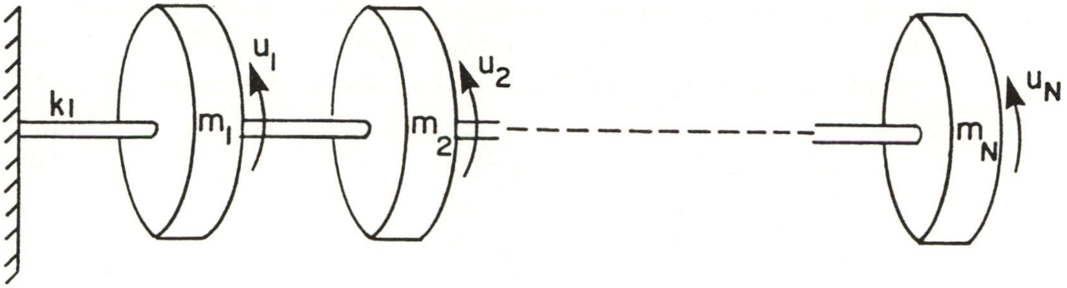

Figure 2.2.2 – A Torsionally Vibrating System

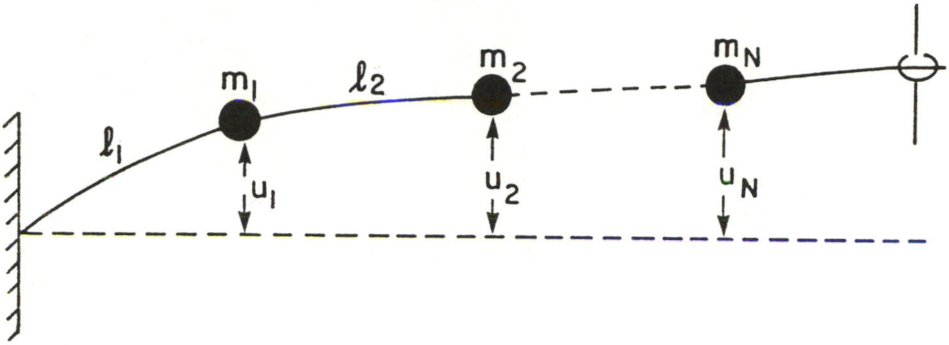

Figure 2.2.3 – N Masses on a Taut String

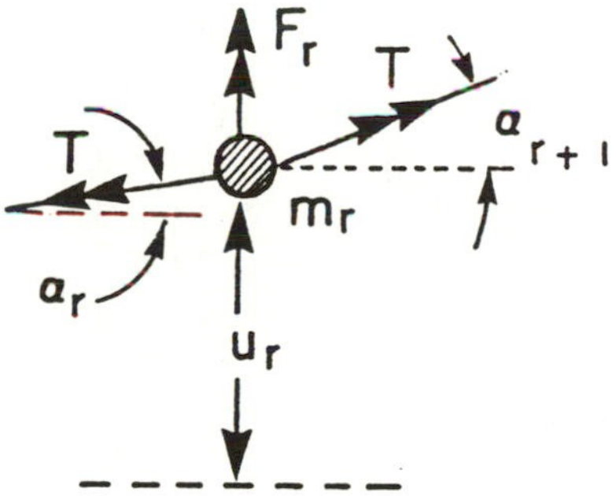

Figure 2.2.4 – The Forces Acting on the Mass m_r

Newton's equation of motion yields

$$m_r \ddot{u}_r = F_r + T\sin\alpha_{r+1} - T\sin\alpha_r \ , \tag{2.2.5}$$

$$= F_r + \theta_{r+1} - \theta_r \ , \tag{2.2.6}$$

where

$$\theta_r = T\alpha_r = k_r(u_r - u_{r-1}) \ , \quad k_r = T/\ell_r \ . \tag{2.2.7}$$

The analogue of the free end for the string is that the last segment of the string is attached to a massless ring which may slide on a smooth vertical rod.

In order to express equations (2.2.1) - (2.2.3) in matrix form we use (2.2.3) to obtain

$$m_r \ddot{u}_r = F_r + k_{r+1}u_{r+1} - (k_{r+1} + k_r)u_r + k_r u_{r-1} \ , \tag{2.2.8}$$

$$m_N \ddot{u}_N = F_N - k_N u_N + k_N u_{N-1} \ , \tag{2.2.9}$$

which yields

$$\begin{bmatrix} m_1 & & & \\ & m_2 & & \\ & & \cdot & \\ & & & m_N \end{bmatrix} \begin{bmatrix} \ddot{u}_1 \\ \ddot{u}_2 \\ \\ \ddot{u}_N \end{bmatrix} + \begin{bmatrix} k_1+k_2 & -k_2 & 0 & \cdots & 0 & 0 \\ -k_2 & k_2+k_3 & -k_3 & \cdots & 0 & 0 \\ \cdot & \cdot & \cdot & \cdots & \cdot & \cdot \\ 0 & 0 & 0 & \cdots & -k_N & k_N \end{bmatrix} \begin{bmatrix} u_1 \\ u_2 \\ \cdot \\ k_N \end{bmatrix} = \begin{bmatrix} F_1 \\ F_2 \\ \cdot \\ F_N \end{bmatrix} . \tag{2.2.10}$$

This equation may be written

$$\mathbf{A}\ddot{\mathbf{u}} + \mathbf{C}\mathbf{u} = \mathbf{F} \ , \tag{2.2.11}$$

where the matrices **A**, **C** are called respectively the *inertia* (or *mass*) and the *stiffness* matrices of the system. Note that both **A** and **C** are *symmetric*; this is a property shared by the matrices corresponding to any conservative system. We note also that both **A** and **C** are positive-definite. In this particular example the matrix **A** is *diagonal* while **C** is *tridiagonal*, i.e., its only non-zero elements are c_{ii} on the principal diagonal, and $c_{i,i+1}$ $(=c_{i+1,i})$ on the two neighbouring diagonals.

Equation (2.2.10) can also be constructed by introducing $\theta = \{\theta_1, \theta_2, \ldots, \theta_N\}$ and noting that

$$\begin{bmatrix} \theta_1 \\ \theta_2 \\ \cdot \\ \cdot \\ \theta_N \end{bmatrix} = \begin{bmatrix} k_1 & & & \\ & k_2 & & \\ & & \cdot & \\ & & & k_N \end{bmatrix} \begin{bmatrix} 1 & 0 & \cdots & 0 \\ -1 & 1 & \cdots & 0 \\ \cdot & \cdot & \cdots & \cdot \\ 0 & \cdot & \cdots & 0 \\ 0 & \cdot & \cdots & 1 \end{bmatrix} \begin{bmatrix} u_1 \\ u_2 \\ \cdot \\ \cdot \\ u_N \end{bmatrix} \tag{2.2.12}$$

which will be written

$$\theta = \mathbf{K}\mathbf{E}^T\mathbf{u} \ , \tag{2.2.13}$$

where

$$\mathbf{E}=\begin{bmatrix}1 & -1 & 0 & \cdots & 0\\ 0 & 1 & -1 & \cdots & 0\\ . & . & . & . & .\\ 0 & \cdots & & 1 & -1\\ 0 & \cdots & & 0 & 1\end{bmatrix},\ E^{-1}=\begin{bmatrix}1 & 1 & 1 & \cdots & 1\\ 0 & 1 & 1 & \cdots & 1\\ . & . & . & . & .\\ 0 & 0 & 0 & \cdots & 1\\ 0 & 0 & 0 & \cdots & 1\end{bmatrix} \quad (2.2.14)$$

and $\mathbf{K}=diag(k_1,k_2,...,k_N)$.

Using the matrix $\mathbf{E}$ we may write equation (2.2.1) - (2.2.2) in the form

$$\mathbf{M\ddot{u}}=-\mathbf{E}\theta+\mathbf{F}\ , \quad (2.2.15)$$

so that on using (2.2.13) we find

$$\mathbf{M\ddot{u}}+\mathbf{EKE}^T\mathbf{u}=\mathbf{F}\ , \quad (2.2.16)$$

and

$$\mathbf{A}=\mathbf{M}\ \ ,\ \ \mathbf{C}=\mathbf{EKE}^T\ . \quad (2.2.17)$$

For free vibration analysis there are two important end conditions. The right hand end may be *free*, in which case there is *no* restriction on the $(u_i)_1^N$, or it may be *fixed*, in which case $u_N=0$.

Examples 2.2

1. Verify that the stiffness matrix in equation (2.2.10) satisfies the conditions of Theorem 1.4.2. Obtain a proof that applies to principal minors of any order i, such that $1\le i\le N$.

2. Consider the multiple pendulum of Figure 2.2.5. Show that the kinetic and potential energies of the system for small oscillations are given by

$$2T=m_1\dot{y}_1^2+m_2\dot{y}_2^2+...+m_N\dot{y}_N^2,$$

$$2V=\sigma_1\frac{y_1^2}{\ell_1}+\sigma_2\frac{(y_2-y_1)^2}{\ell_2}+...+\sigma_N\frac{(y_N-y_{N-1})^2}{\ell_N}$$

where $\sigma_r=g\sum_{s=1}^{r}m_s$.

2.3 Transverse vibration of a beam

Figure 2.3.1 shows a simple discrete model for the transverse vibration of a beam; it consists of $N+2$ masses $(m_r)_{-1}^N$ linked by massless rigid rods of lengths $(\ell_r)_o^N$ which are themselves connected by N rotational springs of stiffnesses $(k_r)_1^N$. The mass and stiffness of the beam, which are actually distributed along the length, have been lumped at $N+2$ points. A beam with distributed mass is governed by a partial differential equation which is fourth order in x, second in t, namely

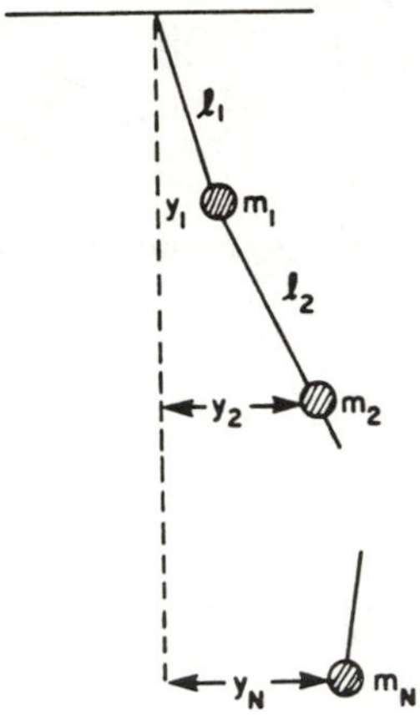

Figure 2.2.5 – A Compound Pendulum Made up of N Inextensible Strings

$$A\rho\frac{\partial^2 u}{\partial t^2}+\frac{\partial^2}{\partial x^2}(EI\frac{\partial^2 u}{\partial x^2})=F(x,t) \ . \tag{2.3.1}$$

Here $A\rho$, I, which may both vary with distance x along the beam, are respectively the linear density of the beam and second moment of area of the cross-section; E is Young's modulus. The discrete system is governed by a set of four first-order difference equations, which may be deduced from Figure 2.3.2.

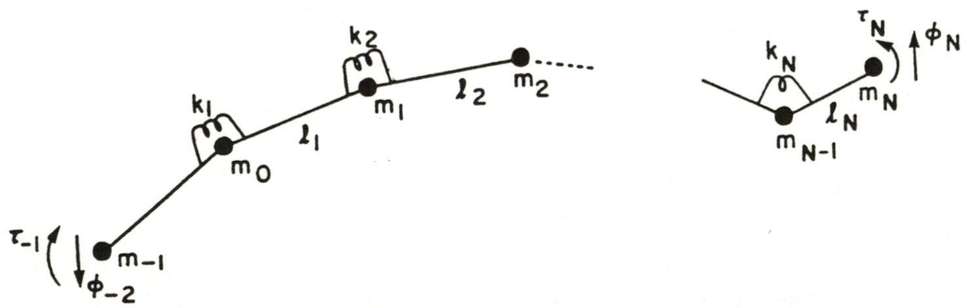

Figure 2.3.1 - A Discrete Model of a Vibrating Beam

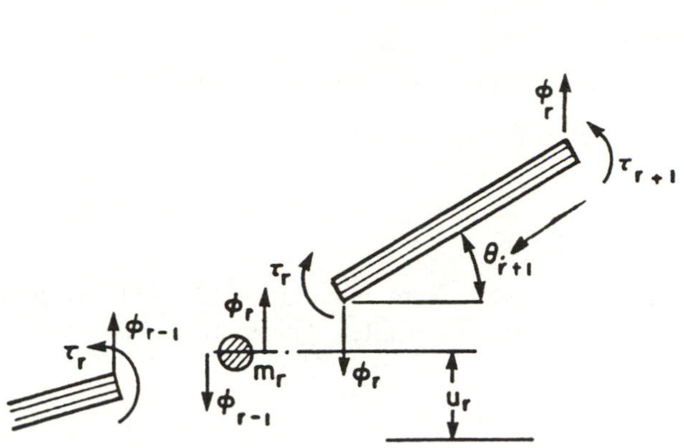

Figure 2.3.2 - The Configuration Around m_r

For small displacements, the rotations are

$$\theta_r=(u_r-u_{r-1})/\ell_r\ ,\quad r=0,1,...,N\ . \tag{2.3.2}$$

If the rth spring has rotational stiffness k_r, then the moment τ_r needed to produce a relative rotation $\theta_{r+1}-\theta_r$ of the two rigid rods on either side of m_r is

$$\tau_r=k_{r+1}(\theta_{r+1}-\theta_r)\ ,\quad r=0,1,...,N-1\ . \tag{2.3.3}$$

Equilibrium of the rod linking m_r and m_{r+1} yields the shearing forces

$$\phi_r=(\tau_r-\tau_{r+1})/\ell_{r+1}\ ,\quad r=-1,0,..,N-1\ , \tag{2.3.4}$$

while Newton's equation of motion for mass m_r is

$$m_r\ddot{u}_r=\phi_r-\phi_{r-1}\ ,\quad r=-1,0,...,N\ . \tag{2.3.5}$$

Here ϕ_{-2}, ϕ_N and τ_{-1}, τ_N denote external shearing forces and bending moments, respectively, applied to the ends.

Suppose that the left hand end is *clamped* so that

$$u_{-1}=0=u_o\ , \tag{2.3.6}$$

then only the masses $(m_r)_1^N$ move, and the governing equations may be written

$$\theta=\mathbf{L}^{-1}\mathbf{E}^T\mathbf{u}\ , \tag{2.3.7}$$

$$\tau=\mathbf{K}\mathbf{E}^T\theta\ , \tag{2.3.8}$$

$$\phi=\mathbf{L}^{-1}\mathbf{E}\tau-\ell_N^{-1}\tau_N\mathbf{e}_N\ , \tag{2.3.9}$$

$$\mathbf{M}\ddot{\mathbf{u}}=-\mathbf{E}\phi+\phi_N\mathbf{e}_N\ , \tag{2.3.10}$$

where $\mathbf{u}=\{u_1,u_2,...,u_N\}$, $\theta=\{\theta_1,\theta_2,\ldots,\theta_N\}$, $\tau=\{\tau_o,\tau_1,\ldots,\tau_{N-1}\}$, $\phi=\{\phi_o,\phi_1,\ldots,\phi_{N-1}\}$, $\mathbf{K}=diag(k_r)$, $\mathbf{L}=diag(\ell_r)$, $\mathbf{M}=diag(m_r)$, $\mathbf{e}_N=\{0,0,...,0,1\}$

and $\mathbf{E}$ is given in equation (2.2.14).

Equations (2.3.7) - (2.3.10) may be combined to give

$$\mathbf{M\ddot{u}}+\mathbf{Cu}=\phi_N\mathbf{e}_N+\ell_N^{-1}\tau_N\mathbf{E}\mathbf{e}_N\ , \tag{2.3.11}$$

where

$$\mathbf{C}=\mathbf{E}\mathbf{L}^{-1}\mathbf{E}\mathbf{K}\mathbf{E}^T\mathbf{L}^{-1}\mathbf{E}^T\ . \tag{2.3.12}$$

This equation has the same general form as equation (2.2.11). We note that $\mathbf{M}$ and $\mathbf{C}$ are again symmetric and positive definite. The matrix $\mathbf{C}$ is now a *pentadiagonal* matrix with the form

$$\mathbf{C}=\begin{bmatrix} a_1 & -b_1 & c_1 & 0 & 0 & \cdots & 0 & 0 \\ -b_1 & a_2 & -b_2 & c_2 & 0 & \cdots & 0 & 0 \\ c_1 & -b_2 & a_3 & -b_3 & c_3 & \cdots & 0 & 0 \\ 0 & c_2 & -b_3 & a_4 & -b_4 & \cdot & 0 & 0 \\ \cdot & \cdot & \cdot & \cdot & \cdot & \cdots & \cdot & \cdot \\ 0 & 0 & 0 & 0 & 0 & \cdots & a_{N-1} & -b_{N-1} \\ 0 & 0 & 0 & 0 & 0 & \cdots & -b_{N-1} & a_N \end{bmatrix} \tag{2.3.13}$$

Examples 2.3

1. Show that the coefficients in the matrix (2.3.13) are given by

$$a_n=p_{n,n}+2p_{n,n+1}+p_{n+1,n+1}\ ,\quad b_n=p_{n,n+1}+p_{n+1,n+1}+p_{n+1,n+2}\ ,$$

$$c_n=p_{n+1,n+2}$$

where

$$p_{n,n}=(k_n+k_{n+1})/\ell_n^2\ ,\quad p_{n,n+1}=k_{n+1}/(\ell_n\ell_{n+1})\ ,$$

and $k_n \doteq 0$ when $n>N$.

2.4 Generalized coordinates and Lagrange's equations

The idea that a discrete system is one composed of a finite number of masses connected by springs is unnecessarily restrictive. The general concept is that of a system whose motion is specified by N *generalized coordinates* $(q_r)_1^N$ that are functions of time t alone. The systems considered in Sections 2.2, 2.3 are indeed discrete in this sense and the generalized coordinates corresponding to the system in Figure 2.2.1 are $(u_r)_1^N$. However, the more general concept would also cover, for instance, a model of a non-uniform longitudinally vibrating rod constructed by using the *finite element method* (see for example, Zienkiewicz (1971)).

In such a model, shown in Figure 2.4.1, the rod is first divided into $N+1$ elements. In the rth element, shown in Figure 2.4.2, the longitudinal displacement $y(x,t)$ is taken to have a simple approximate form

$$y(x,t)=y_r(t)(1-\xi)+y_{r+1}(t)\xi\ , \quad x_r \leq x \leq x_{r+1}\ , \tag{2.4.1}$$

where

$$\xi=(x-x_r)/\ell_r\ , \tag{2.4.2}$$

runs from 0 at the left hand end of the element to 1 at the right. Equations (2.4.1) with $r=0,1,...,N$ express the displacement at every point of the rod in terms of the $N+2$ generalized coordinates $(y_r)_o^{N+1}$. When the end conditions are imposed there will be, as before, only N coordinates $(y_r)_1^N$.

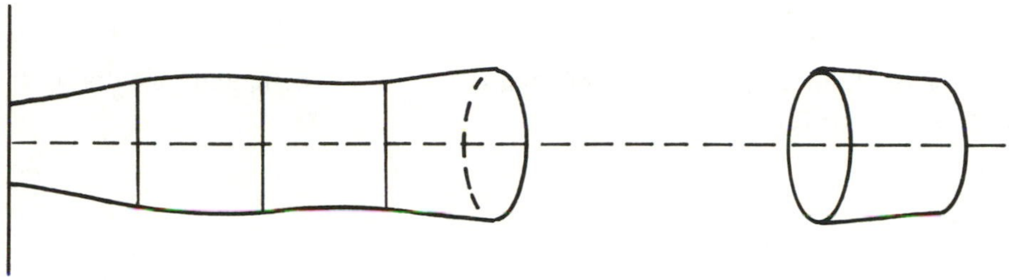

Figure 2.4.1 - A Rod Divided into Elements

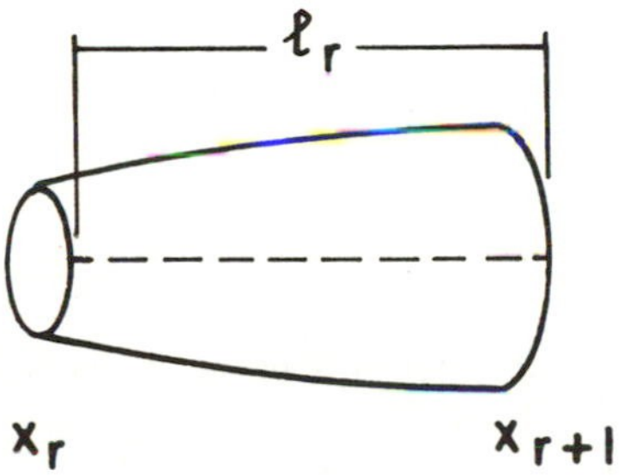

Figure 2.4.2 - One Element of the Rod

When the finite element method is used, it is not possible to set up the equations of motion by using Newton's equations of motion, for there is no actual 'mass' to which forces are applied. Instead we may use Lagrange's equations. For a conservative system with *kinetic energy* T and *potential* or *strain energy* V, which are functions of N coordinates $(q_r)_1^N$, Lagrange's equations state that

$$\frac{d}{dt}\left(\frac{\partial T}{\partial \dot{q}_r}\right)-\frac{\partial T}{\partial q_r}+\frac{\partial V}{\partial q_r}=Q_r\ , \quad (r=1,2,...,N)\ . \tag{2.4.3}$$

Here Q_r is the generalized force corresponding to q_r in the sense that the work done by external forces acting on the system when the system is displaced from a configuration specified by $(q_r)_1^N$ to one specified by $(q_r+\delta q_r)_1^N$, is

$$\delta W_e=\sum_{r=1}^{N} Q_r \delta q_r \,. \tag{2.4.4}$$

For the system shown in Figure 2.2.1 the kinetic and potential energies are

$$T=\frac{1}{2}\sum_{i=1}^{N} m_r \dot{y}_r^2 \,, \qquad V=\frac{1}{2}\sum_{i=1}^{N} k_r(y_{r+1}-y_r)^2 \,, \tag{2.4.5}$$

and $Q_r=F_r(t)$. Thus

$$\frac{\partial T}{\partial \dot{y}_r}=m_r \dot{y}_r \,, \quad \frac{\partial V}{\partial y_r}=-k_r(y_{r+1}-y_r)+k_{r-1}(y_r-y_{r-1}) \,, \tag{2.4.6}$$

and equation (2.4.3) yields (2.2.1).

For the finite element model of Figure 2.4.1, the kinetic and potential energies of the system will be

$$T=\frac{1}{2}\int_o^{\ell} S\rho[\dot{y}(x,t)]^2 dx \,, \tag{2.4.7}$$

$$V=\frac{1}{2}\int_o^{\ell} SE\,[\frac{\partial y}{\partial x}(x,t)]^2 dx \,. \tag{2.4.8}$$

where $S(x)$, $\rho(x)$, $E(x)$ are the (possibly variable) cross-sectional area, density and Young's modulus of the rod. On inserting the assumed form of $y(x,t)$ given in (2.4.1) we find

$$T=\frac{1}{2}\sum_{r=o}^{N}\int_o^1 S(x_r+\ell_r\xi)\rho(x_r+\ell_r\xi)[\dot{y}_r(1-\xi)+\dot{y}_{r+1}\xi]^2 \ell_r d\xi, \tag{2.4.9}$$

$$V=\frac{1}{2}\sum_{r=o}^{N}\int_o^1 S(x_r+\ell_r\xi)E(x_r+\ell_r\xi)[y_{r+1}-y_r]^2 \ell_r^{-1} d\xi \,. \tag{2.4.10}$$

On carrying out the integrations, perhaps numerically if $S(x)$, $\rho(x)$, $E(x)$ are variable, we may write

$$T=\frac{1}{2}\sum_{r=o}^{N+1}\sum_{s=o}^{N+1} a_{rs}\dot{y}_r \dot{y}_s \,, \tag{2.4.11}$$

$$V=\frac{1}{2}\sum_{r=o}^{N+1}\sum_{s=o}^{N+1} c_{rs} y_r y_s \,. \tag{2.4.12}$$

In this example, since the product terms which appear in (2.4.9), (2.4.10) involve only immediately neighbouring values of r, the coefficients a_{rs}, c_{rs} will be zero unless $|\,r-s\,|\le 1$. If the rod is fixed at both ends, then

$$y_o=0=y_{N+1} \,, \tag{2.4.13}$$

so that all the sums in (2.4.11), (2.4.12) run from 1 to N. In this case

$$\frac{\partial T}{\partial \dot{y}_r} = \sum_{s=1}^{N} a_{rs}\dot{y}_s \; , \qquad \frac{\partial V}{\partial y_r} = \sum_{s=1}^{N} c_{rs}y_s \; , \tag{2.4.14}$$

and equation (2.4.3) yields the following equation for free vibration:

$$\sum_{s=1}^{N} a_{rs}\ddot{y}_s + \sum_{s=1}^{N} c_{rs}y_s = 0 \; , \; (r=1,2,...,N) \; . \tag{2.4.15}$$

This equation may, as before, be condensed into the matrix equation

$$\mathbf{A}\ddot{\mathbf{y}} + \mathbf{C}\mathbf{y} = \mathbf{0} \; . \tag{2.4.16}$$

On the basis of these examples we now pass to the general case. For a conservative system with generalized coordinates $(q_r)_1^N$ which specify small displacements from a position of stable equilibrium, the kinetic and potential energies will have the form

$$T = \frac{1}{2}\sum_{r=1}^{N}\sum_{s=1}^{N} a_{rs}\dot{q}_r\dot{q}_s \; , \tag{2.4.17}$$

$$V = \frac{1}{2}\sum_{r=1}^{N}\sum_{s=1}^{N} c_{rs}q_r q_s \; , \tag{2.4.18}$$

where the matrices $\mathbf{A}=(a_{rs})$ and $\mathbf{C}=(c_{rs})$ are *symmetric*, in that

$$a_{rs} = a_{sr} \; , \quad c_{rs} = c_{sr} \; . \tag{2.4.19}$$

The equations governing free vibration may be written

$$\mathbf{A}\ddot{\mathbf{q}} + \mathbf{C}\mathbf{q} = \mathbf{0} \; . \tag{2.4.20}$$

We note that equations (2.4.17), (2.4.18) may be written

$$T = \frac{1}{2}\dot{\mathbf{q}}^T\mathbf{A}\dot{\mathbf{q}} \; , \qquad V = \frac{1}{2}\mathbf{q}^T\mathbf{C}\mathbf{q} \; . \tag{2.4.21}$$

It is not possible for any arbitrarily chosen symmetric matrix **A** to be an inertia matrix, because the kinetic energy T is an essentially positive quantity, i.e., it is always positive except when each of the $\dot{q}_r$ is zero, in which case it is zero. Thus **A** must be *positive definite* (see Section 1.4).

The restrictions on the matrix **C** are slightly less severe since, although the strain energy will always be positive or zero, it will actually be zero if the system has a rigid-body displacement. Notice, for example, that the V of (2.4.5) will be zero if **y** is the rigid body displacement

$$y_o = y_1 = \cdots = y_N = y_{N+1} \; . \tag{2.4.22}$$

This will be a possible displacement of the system in Figure 2.2.1 only if both ends are *free*. We conclude that if the system is not constrained so that one point is fixed, then **C** is *positive semi-definite*.

Example 2.4

1. Use equations (2.4.9), (2.4.10) to evaluate the mass and stiffness matrices for a *uniform* rod in longitudinal vibration subject to the end conditions (2.4.13).

2.5 Natural frequencies and normal modes

The matrix equation (2.4.20) represents a set of second order equations with constant coefficients. Following usual practice we seek the solution in the form

$$\mathbf{q} = \begin{bmatrix} q_1 \\ q_2 \\ \cdot \\ q_N \end{bmatrix} = \begin{bmatrix} x_1 \\ x_2 \\ \cdot \\ x_N \end{bmatrix} \sin(\omega t + \phi) , \tag{2.5.1}$$

where the constants x_r, frequency ω and phase angle ϕ are to be determined. When $\mathbf{q}$ has the form (2.5.1), then

$$\ddot{\mathbf{q}} = -\omega^2 \mathbf{q} = -\omega^2 \mathbf{x} \sin(\omega t + \phi) , \tag{2.5.2}$$

so that equation (2.4.20) demands that

$$(\mathbf{C} - \lambda \mathbf{A})\mathbf{x} = \mathbf{0} , \quad \lambda = \omega^2 . \tag{2.5.3}$$

This is the eigenvalue equation (1.4.1) and, since $\mathbf{A}$ is positive-definite and $\mathbf{C}$ is either positive semi-definite or positive-definite, the whole of the analysis developed in Section 1.4 can be used here. Thus the equation has N eigenvalues $(\lambda_i)_1^N$ satisfying

$$0 \leq \lambda_1 \leq \lambda_2 < \cdots \leq \lambda_N , \tag{2.5.4}$$

and N corresponding eigenvectors $(\mathbf{x}^{(i)})_1^N$ satisfying

$$(\mathbf{C} - \lambda_i \mathbf{A})\mathbf{x}^{(i)} = \mathbf{0} . \tag{2.5.5}$$

The frequencies $\omega_i = (\lambda_i)^{1/2}$ are called the *natural* frequencies of the system, and the eigenvectors are called the *normal* or *principal* modes.

In order to become acquainted with the properties of natural frequencies and normal modes we shall consider the system specified by equation (2.2.10) and, to simplify the algebra, shall assume that

$$(m_r)_1^N = m , \quad (k_r)_1^N = k . \tag{2.5.6}$$

In this case the eigenvalue equation may be written

$$\begin{bmatrix} 2-\lambda & -1 & 0 & \cdots & 0 \\ -1 & 2-\lambda & -1 & \cdots & 0 \\ \cdot & \cdot & \cdot & \cdots & \cdot \\ 0 & \cdots & & 2-\lambda & -1 \\ 0 & \cdots & & -1 & 1-\lambda \end{bmatrix} \begin{bmatrix} x_1 \\ x_2 \\ \cdot \\ x_{N-1} \\ x_N \end{bmatrix} = 0 , \tag{2.5.7}$$

where

$$\lambda = m\omega^2/k . \tag{2.5.8}$$

To solve for the x_r we use the idea suggested in Ex. 1.4.3, namely to write (2.5.7) as the recurrence relation

$$-x_{r-1}+(2-\lambda)x_r-x_{r+1}=0 , \quad (r=0,1,...,N) . \tag{2.5.9}$$

The first of equations (2.5.7) may be written in this form if x_o is taken to be zero; this may be interpreted as staling that the left hand mass (m_o) is fixed. On the other hand, the last of equations (2.5.7) may be written in the form (2.5.9) if x_{N+1} is taken to be equal to x_N. Thus the end conditions for the recurrence (2.5.9) are

$$x_o=0=x_{N+1}-x_N . \tag{2.5.10}$$

The recurrence relation has the general solution

$$x_r = A\cos r\theta + B\sin r\theta , \tag{2.5.11}$$

where, on substitution into (2.5.9) we find that θ must satisfy

$$\begin{aligned} \cos(r-1)\theta+\cos(r+1)\theta &= 2\cos\theta\cos r\theta=(2-\lambda)\cos r\theta , \\ \sin(r-1)\theta+\sin(r+1)\theta &= 2\cos\theta\sin r\theta=(2-\lambda)\sin r\theta , \end{aligned} \tag{2.5.12}$$

i.e.,

$$2\cos\theta = 2-\lambda . \tag{2.5.13}$$

The end conditions will be satisfied if and only if

$$A=0=\sin(N+1)\theta-\sin N\theta=2\cos[(N+1/2)\theta]\sin\theta/2 , \tag{2.5.14}$$

so that the possible values of θ are

$$\theta=\theta_i=\frac{(2i-1)\pi}{2N+1} , \quad (i=1,2,...,N) , \tag{2.5.15}$$

while the corresponding values of λ are

$$\lambda_i = 2-2\cos\theta_i = 4\sin^2[\frac{(2i-1)\pi}{2(2N+1)}] . \tag{2.5.16}$$

Thus, in the ith mode, the displacement amplitude of the rth mass is

$$x_r = \sin r\theta_i = \sin[\frac{(2i-1)r\pi}{(2N+1)}] . \tag{2.5.17}$$

The modes for the case $N=4$, which are shown in Figure 2.5.1, exhibit properties that are held by all eigenvectors of a *tri-diagonal* matrix (such as that in (2.5.7)), namely

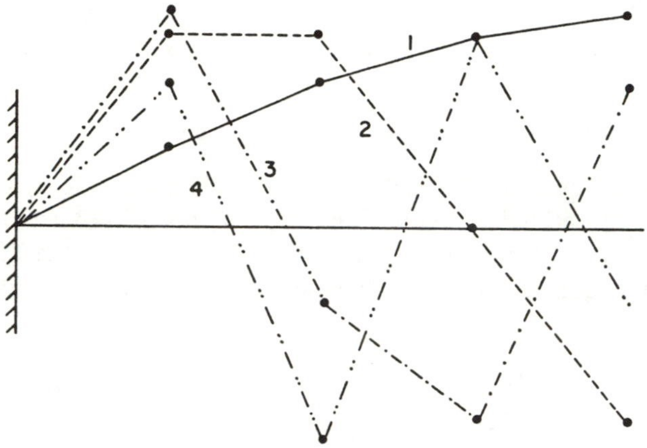

Figure 2.5.1 - The Modes of the Spring-Mass System for N=4

(a) the ith mode crosses the axis (i-1) times – the zeros at the ends are not counted;

(b) the *nodes* (points where the mode crosses the axis) of the ith mode interlace those of the neighbouring $((i-1)$th and $(i+1)$th) modes.

If instead of being free at the right hand end, the system were pinned there, then the analysis would be unchanged except that the end conditions would be

$$x_o = 0 = x_N . \tag{2.5.18}$$

In this case θ would have to satisfy

$$\sin N\theta = 0 \tag{2.5.19}$$

so that

$$\theta = \phi_i = \frac{i\pi}{N} , \quad i = 1,2,...,N-1 \tag{2.5.20}$$

and the corresponding eigenvalues, which we will label $(\lambda_i^o)_1^{N-1}$, would be

$$\lambda_i^o = 4\sin^2(\frac{i\pi}{2N}) \ . \tag{2.5.21}$$

In the ith mode, the rth displacement amplitude is

$$y_r = \sin(r\phi_i) = \sin[\frac{ri\pi}{N}] \ . \tag{2.5.22}$$

The two sets of eigenvalues $(\lambda_i)_1^N$ and $(\lambda_i^o)_1^{N-1}$ are related in a way which will be found to be general for problems of this type (see equation (2.8.14)), namely

$$0 < \lambda_1 < \lambda_1^o < \cdots < \lambda_{N-1} < \lambda_{N-1}^o < \lambda_N \ . \tag{2.5.23}$$

Examples 2.5

1. Consider the beam system of Figure 2.3.1 in the case when $(m_i)_{-1}^N = m$, $(k_i)_1^N = k$, $(\ell_i)_o^N = \ell$. Show that the recurrence relation linking the $(u_r)_o^{N+2}$ may be written

   $$u_{r-2} - 4u_{r-1} + (4-\lambda)u_r - 4u_{r+1} + u_{r+2} = 0$$

   where $\lambda = m\omega^2\ell^2/k$. Seek a solution of the recurrence relation of the form

   $$u_r = A\cos r\theta + B\sin r\theta + C\cosh r\phi + D\sinh r\phi$$

   and find θ, ϕ so that the end conditions $u_{-1} = 0 = u_o = u_{N-1} = u_N$ are satisfied. Hence find the natural frequencies and normal modes of the system; i.e. a clamped-clamped beam. A physically more acceptable discrete approximation of a beam is considered in detail by Gladwell (1962) and Lindberg (1963).

2.6 Principal coordinates and receptances

Theorem 1.4.5 states that the vectors $(\mathbf{x}^{(i)})_1^N$ span the space of N-vectors, so that any arbitrary vector $\mathbf{q}(t)$ may be written

$$\mathbf{q}(t) = p_1\mathbf{x}^{(1)} + p_2\mathbf{x}^{(2)} + \cdots + p_N\mathbf{x}^{(N)} \ . \tag{2.6.1}$$

This may be condensed into the matrix equation

$$\mathbf{q} = \mathbf{X}\mathbf{p} \ , \tag{2.6.2}$$

where $\mathbf{X}$ is the $N \times N$ matrix with the $\mathbf{x}^{(i)}$ as its columns. The coordinates $p_1, p_2, \ldots, p_N$, called the *principal coordinates*, will in general be functions of t; they indicate the extent to which the various eigenvectors $\mathbf{x}^{(i)}$ participate in the vector $\mathbf{q}$. The energies T, V take particularly simple forms when $\mathbf{q}$ is expressed in terms of the principal coordinates. For equation (2.6.2) implies

$$\dot{\mathbf{q}} = \mathbf{X}\dot{\mathbf{p}} \ , \tag{2.6.3}$$

so that

$$T = \frac{1}{2}(\mathbf{X}\dot{\mathbf{p}})^T\mathbf{A}(\mathbf{X}\dot{\mathbf{p}}) = \frac{1}{2}\dot{\mathbf{p}}^T(\mathbf{X}^T\mathbf{A}\mathbf{X})\dot{\mathbf{p}} \,. \tag{2.6.4}$$

But the element in row i, column j of the matrix $\mathbf{X}^T\mathbf{A}\mathbf{X}$ is simply $\mathbf{x}^{(i)T}\mathbf{A}\mathbf{x}^{(j)}$ and this is zero if $i \neq j$, a_i if $i = j$. Thus

$$\mathbf{X}^T\mathbf{A}\mathbf{X} = diag(a_1, a_2, \ldots, a_N) \,, \tag{2.6.5}$$

so that

$$T = \frac{1}{2}\{a_1\dot{p}_1^2 + a_2\dot{p}_2^2 + \cdots \; a_N\dot{p}_N^2\}. \tag{2.6.6}$$

Similarly

$$V = \frac{1}{2}\mathbf{p}^T(\mathbf{X}^T\mathbf{C}\mathbf{X})\mathbf{p} \tag{2.6.7}$$

and

$$\mathbf{X}^T\mathbf{C}\mathbf{X} = diag(\lambda_1 a_1, \lambda_2 a_2, \ldots \lambda_N a_N) \tag{2.6.8}$$

so that

$$V = \frac{1}{2}\{\lambda_1 a_1 p_1^2 + \lambda_2 a_2 p_2^2 + \cdots \; \lambda_N a_N p_N^2\}. \tag{2.6.9}$$

Equations (2.6.6), (2.6.9) show that the search for eigenvalues and eigenvectors for a symmetric matrix pair $(\mathbf{C},\mathbf{A})$ is equivalent to the search for a coordinate transformation $\mathbf{q} \rightarrow \mathbf{p}$ which will simultaneously convert two quadratic forms $\mathbf{q}^T\mathbf{A}\mathbf{q}$ and $\mathbf{q}^T\mathbf{C}\mathbf{q}$ to sums of squares.

We shall now use the principal coordinates to obtain the response of a system to sinusoidal forces. Equations (2.4.3) and (2.4.21) show that the equation governing the response to generalized forces $(Q_r)_1^N$ is

$$\mathbf{A}\ddot{\mathbf{q}} + \mathbf{C}\mathbf{q} = \mathbf{Q} \tag{2.6.10}$$

where $\mathbf{Q} = \{Q_1, Q_2, \ldots, Q_N\}$. If the forces have frequency ω and are all in phase, then $\mathbf{Q}$ may be written

$$\mathbf{Q} = \boldsymbol{\Phi}\sin(\omega t + \phi) \,. \tag{2.6.11}$$

In this case equations (2.5.1) - (2.5.2) yield

$$(\mathbf{C} - \lambda\mathbf{A})\mathbf{x} = \boldsymbol{\Phi} \,. \tag{2.6.12}$$

To solve this equation we express $\mathbf{x}$ in terms of the eigenvectors $\mathbf{x}^{(i)}$, so that

$$\mathbf{x} = \pi_1\mathbf{x}^{(1)} + \pi_2\mathbf{x}^{(2)} + \cdots \; \pi_N\mathbf{x}^{(n)} = \mathbf{X}\boldsymbol{\pi} \,, \tag{2.6.13}$$

where $\pi_1, \pi_2, \ldots, \pi_N$ are the amplitudes of the principal coordinates $p_1, p_2, \ldots, p_N$. Substitute (2.6.13) into (2.6.12) and multiply the resulting equation by $\mathbf{X}^T$; the result is

$$\mathbf{X}^T(\mathbf{C}-\lambda\mathbf{A})\mathbf{X}\pi=\mathbf{X}^T\Phi=\Xi\,. \tag{2.6.14}$$

But now the matrix of coefficients of the set of N equations for the unknowns $\pi_1,\pi_2,\ldots,\pi_N$ is *diagonal*, and the ith equation is simply

$$(\lambda_i-\lambda)a_i\pi_i=\Xi_i\;, \tag{2.6.15}$$

so that

$$\pi_i=\frac{\Xi_i}{a_i(\lambda_i-\lambda)}\,. \tag{2.6.16}$$

In order to interpret this result we consider the response to a single generalized force Q_r. In this case

$$\Phi=\{0,0,\ldots,\Phi_r,0,\ldots,0\}\;,\quad \Xi=\Phi_r\{x_r^{(1)},x_r^{(2)},\ldots,x_r^{(N)}\}\,, \tag{2.6.17}$$

$$\pi_1=\frac{\Phi_r x_r^{(i)}}{a_i(\lambda_i-\lambda)}\;, \tag{2.6.18}$$

and the sth displacement amplitude is

$$x_s=\alpha_{rs}\Phi_r\;, \tag{2.6.19}$$

where

$$\alpha_{rs}=\sum_{i=1}^{N}\frac{x_r^{(i)}x_s^{(i)}}{a_i(\lambda_i-\lambda)}\,. \tag{2.6.20}$$

The quantity α_{rs} is the *receptance* giving the amplitude of response of q_s to a unit amplitude generalized force Q_r. The fact that α_{rs} is symmetric, i.e.,

$$\alpha_{rs}=\alpha_{sr} \tag{2.6.21}$$

is a reflection of the *reciprocal* theorem which holds for forced harmonic excitation.

Example 2.6

1. Use the orthogonality of the $(\mathbf{x}^{(i)})_1^N$ w.r.t. the inertia matrix to show that

$$\mathbf{x}^{(i)}\mathbf{A}\mathbf{q}=p_i a_i \ .$$

2.7 Rayleigh's Principle

Consider a conservative system with generalized coordinates $(q_r)_1^N$ vibrating with harmonic motion given by (2.5.1). Its kinetic and potential energies will be

$$T=\frac{1}{2}\dot{\mathbf{q}}^T\mathbf{A}\dot{\mathbf{q}}=\omega^2\cos^2\omega t T_o \ , \tag{2.7.1}$$

$$V=\frac{1}{2}\mathbf{q}^T\mathbf{C}\mathbf{q}=\sin^2\omega t V_o \ , \tag{2.7.2}$$

where

$$T_o=\frac{1}{2}\mathbf{x}^T\mathbf{A}\mathbf{x} \ , \ \ V_o=\frac{1}{2}\mathbf{x}^T\mathbf{C}\mathbf{x} \ . \tag{2.7.3}$$

Since the system is conservative,

$$T+V=const. , \tag{2.7.4}$$

so that

$$\omega^2\cos^2\omega t T_o+(1-\cos^2\omega t)V_o=const. , \tag{2.7.5}$$

and therefore

$$\omega^2 T_o=V_o \ . \tag{2.7.6}$$

This we may write as

$$\lambda=\frac{V_o}{T_o}=\frac{\mathbf{x}^T\mathbf{C}\mathbf{x}}{\mathbf{x}^T\mathbf{A}\mathbf{x}} \ . \tag{2.7.7}$$

If the system is vibrating freely at frequency ω, then ω must be one of the natural frequencies and $\mathbf{x}$ the corresponding eigenvector. If $\omega=\omega_i$, then $\lambda=\lambda_i$ and $\mathbf{x}=\mathbf{x}^{(i)}$ and equation (2.7.7) yields

$$\lambda_i=\frac{\mathbf{x}^{(i)T}\mathbf{C}\mathbf{x}^{(i)}}{\mathbf{x}^{(i)T}\mathbf{A}\mathbf{x}^{(i)}} \ , \tag{2.7.8}$$

which agrees with equation (1.4.7).

Rayleigh's Principle states that *the stationary values of the Rayleigh Quotient*

$$\lambda_R=\frac{\mathbf{x}^T\mathbf{C}\mathbf{x}}{\mathbf{x}^T\mathbf{A}\mathbf{x}} \ , \tag{2.7.9}$$

viewed as a function of the components $(x_r)_1^N$, occur when $\mathbf{x}$ is an eigenvector $\mathbf{x}^{(i)}$. The corresponding stationary value of λ_R is λ_i.

Proof

Rayleigh's Principle has a long history – see for example Temple and Bickley (1933) or Washizu (1982). We shall state the proof in a number of ways because each is instructive. First consider λ_R as the ratio of V_o and T_o and write down the partial derivative of this quotient w.r.t. x_r. We have

$$\frac{\partial T_o}{\partial x_r}=a_{r1}x_1+a_{r2}x_2+\cdots+a_{rn}x_N \ , \tag{2.7.10}$$

$$\frac{\partial V_o}{\partial x_r}=c_{r1}x_1+c_{r2}x_2+\cdots+c_{rn}x_N \ , \tag{2.7.11}$$

and

$$\frac{\partial}{\partial x_r}\left(\frac{V_o}{T_o}\right)=\frac{1}{T_o}\frac{\partial V_o}{\partial x_r}-\frac{V_o}{T_o^2}\frac{\partial T_o}{\partial x_r}=\frac{1}{T_o}\left\{\frac{\partial V_o}{\partial x_r}-\lambda_R\frac{\partial T_o}{\partial x_r}\right\}, \tag{2.7.12}$$

so that, on inserting the expressions for $\partial V_o/\partial x_r$ and $\partial T_o/\partial x_r$ we obtain just the rth row of the matrix equation (2.5.3) with λ_R in place of λ. The complete set of N equations which state that V_o/T_o is stationary w.r.t. all the $(x_r)_1^N$ is the matrix equation (2.5.3) which is satisfied when $\mathbf{x}$ is an eigenvector $\mathbf{x}^{(i)}$ and λ is the corresponding eigenvalue λ_i.

Now express the energies in terms of the principal coordinates. If

$$p_i=\pi_i\sin(\omega t+\phi) \ , \tag{2.7.13}$$

then equations (2.6.6), (2.6.9) show that

$$T_o=\frac{1}{2}\{a_1\pi_1^2+a_2\pi_2^2+\cdots+a_N\pi_N^2\} \ , \tag{2.7.14}$$

$$V_o=\frac{1}{2}\{\lambda_1a_1\pi_1^2+\lambda_2a_2\pi_2^2+\cdots+\lambda_Na_N\pi_N^2\} \ . \tag{2.7.15}$$

Since $\mathbf{A}$ is assumed to be positive definite, there is no loss in generality in taking each $a_i=1$, then

$$\lambda_R=\frac{\lambda_1\pi_1^2+\lambda_2\pi_2^2+\cdots+\lambda_N\pi_N^2}{\pi_1^2+\pi_2^2+\cdots+\pi_N^2} \ , \tag{2.7.16}$$

so that, in particular,

$$\lambda_R-\lambda_1=\frac{(\lambda_2-\lambda_1)\pi_2^2+\cdots+(\lambda_N-\lambda_1)\pi_N^2}{\pi_1^2+\pi_2^2+\cdots+\pi_N^2} \ . \tag{2.7.17}$$

Since the quantities $\lambda_2-\lambda_1$, $\lambda_3-\lambda_1, \ldots, \lambda_N-\lambda_1$ are non-negative on account of (2.5.4), equation (2.7.17) states that

$$\lambda_R \geq \lambda_1 \,. \tag{2.7.18}$$

If λ_1 is strictly less than λ_2 then equality occurs only when $\pi_2=0=...=\pi_N$, i.e., when the system is vibrating in its first principal mode. Equation (2.7.18) states the important property that whatever values are taken for $(x_r)_1^N$, the values of the Rayleigh quotient will always be greater than λ_1 and (when $\lambda_1<\lambda_2$) will be equal to λ_1 only if the ratios $x_1{:}x_2{:}...x_N$ correspond to those of the first eigenvector $x_1^{(1)}{:}x_2^{(1)}{:}...x_N^{(1)}$. Equation (2.7.18) shows that λ_1 is the *global minimum* of λ_R, and it may be proved in an exactly similar way that

$$\lambda_R \leq \lambda_N \,, \tag{2.7.19}$$

so that λ_N is the global *maximum* of λ_R.

If λ_i is an intermediate eigenvalue, so that $\lambda_1<\lambda_i<\lambda_N$, then

$$\lambda_R-\lambda_i=\frac{-\sum_{j=1}^{i-1}(\lambda_i-\lambda_j)\pi_j^2+\sum_{j=i+1}^{N}(\lambda_j-\lambda_i)\pi_j^2}{\pi_1^2+\pi_2^2+\cdots+\pi_N^2}\,. \tag{2.7.20}$$

In this case λ_R will not be strictly less nor strictly greater than λ_i for variations of the π_j; λ_R has a saddle point in the ith mode($\pi_j=0$, $j\neq 1$). However, for computational purposes it is important that the difference between λ_R and λ_i depends on the *squares* of the quantities π_j. This means that if $\mathbf{x}$ is 'nearly' in the ith mode, so that the π_j with $j\neq i$ are much smaller than π_i, i.e., $\pi_i\approx 1$, $\pi_j=0(\varepsilon)$, then $\lambda_R-\lambda_i=0(\varepsilon^2)$.

Since $\mathbf{A}$ is positive definite, $\mathbf{x}^T\mathbf{A}\mathbf{x}>0$, and the problem of finding the stationary values of the Rayleigh quotient λ_R given by equation (2.7.9) is equivalent to finding the stationary values of $\mathbf{x}^T\mathbf{C}\mathbf{x}$ subject to the restriction that $\mathbf{x}^T\mathbf{A}\mathbf{x}=1$. This in turn is equivalent to finding the stationary values of

$$F\equiv\mathbf{x}^T\mathbf{C}\mathbf{x}-\lambda\mathbf{x}^T\mathbf{A}\mathbf{x}\,, \tag{2.7.21}$$

subject to $\mathbf{x}^T\mathbf{A}\mathbf{x}=1$. Here λ acts as a Lagrange parameter. Note that

$$\frac{\partial F}{\partial x_r}=2\sum_{s=1}^{n}c_{rs}x_s-2\lambda\sum_{s=1}^{n}a_{rs}x_s\,,$$

so that the set of equations $\partial F/\partial x_r=0$ yields equation (2.5.3), viz.

$$(\mathbf{C}-\lambda\mathbf{A})\mathbf{x}=0\,. \;\blacksquare \tag{2.7.22}$$

2.8 Vibration under constraint

The concept of a system vibrating under constraint is important in the solution of inverse problems. Suppose a system has generalized coordinates $(q_r)_1^N$, but they are constrained to satisfy a relation

$$f(q_1,q_2,...,q_N)=0 \ . \tag{2.8.1}$$

For small vibrations about $q_1=0=...=q_N$, this relation may be replaced by

$$\mathbf{q}^T\mathbf{d}=d_1q_1+d_2q_2+...+d_Nq_N=0 \ , \tag{2.8.2}$$

where

$$d_r=\frac{\partial f}{\partial q_r}(q_1,q_2,...,q_N)\,|_{\,q_1=0=q_2=...=q_N} \ . \tag{2.8.3}$$

Two of the most important constraints will correspond to a certain q_r being zero, or two, q_r and q_s, being equal. Now suppose that the system is vibrating with frequency ω, where $\omega^2=\lambda$, and

$$\mathbf{q}=\mathbf{x}\sin\omega t \ . \tag{2.8.4}$$

Rayleigh's Principle states that the (natural frequencies)^2 will be the stationary values of F, given in equation (2.7.21), but now subject to the further constraint

$$\mathbf{x}^T\mathbf{d}=0 \ . \tag{2.8.5}$$

Thus we must find the stationary values of

$$\mathbf{F}=\mathbf{x}^T\mathbf{C}\mathbf{x}-\lambda\mathbf{x}^T\mathbf{A}\mathbf{x}-2\nu\mathbf{x}^T\mathbf{d} \ , \tag{2.8.6}$$

where ν is another Lagrange parameter (the 2 is inserted purely for convenience). The equations $\partial F/\partial x_r=0$ now yield

$$\mathbf{C}\mathbf{x}-\lambda\mathbf{A}\mathbf{x}-\nu\mathbf{d}=0 \ . \tag{2.8.7}$$

By comparing this with equation (2.6.12) we see that $\nu\mathbf{d}$ is a generalized force; it is the force required to maintain the constraint (2.8.5).

In order to analyse equation (2.8.7) we express $\mathbf{x}$ in terms of principal coordinates, using equation (2.6.13). Then

$$\mathbf{C}\mathbf{X}\pi-\lambda\mathbf{A}\mathbf{X}\pi-\nu\mathbf{d}=0 \ . \tag{2.8.8}$$

Multiply throughout by $\mathbf{X}^T$ and use equations (2.6.5) and (2.6.8) which show that both $\mathbf{X}^T\mathbf{A}\mathbf{X}$ and $\mathbf{X}^T\mathbf{C}\mathbf{X}$ will be diagonal matrices; the rth row of the resulting equation is

$$\lambda_r a_r \pi_r-\lambda a_r \pi_r-\nu b_r=0 \ , \quad r=1,2,...,N \ , \tag{2.8.9}$$

where

$$\mathbf{b}=\mathbf{X}^T\mathbf{d} \ . \tag{2.8.10}$$

Equations (2.8.9) yield

$$\pi_r=\frac{\nu b_r}{a_r(\lambda_r-\lambda)} \ , \tag{2.8.11}$$

which, when substituted in the constraint (2.8.5); i.e.,

$$\mathbf{x}^T\mathbf{d} \equiv \pi^T \mathbf{X}^T \mathbf{d} \equiv \pi^T \mathbf{b} = 0 , \tag{2.8.12}$$

yields the frequency equation

$$B(\lambda) \equiv \sum_{i=1}^{N} \frac{b_i^2}{a_i(\lambda_i - \lambda)} = 0 . \tag{2.8.13}$$

The form of this equation has important consequences. Consider first the case in which none of the b_i is zero. The coefficients $(b_i^2/a_i)_1^N$ will all be positive and the graph of $B(\lambda)$ against λ will have the form shown in Figure 2.8.1. Since $B(\lambda_i+0)$ is very large negative, $B(\lambda_{i+1}-0)$ is very large positive, and $B(\lambda)$ is steadily increasing between λ_i and λ_{i+1}, $B(\lambda)$ will have just $N-1$ zeros, $(\lambda_i^{\bullet})_i^{N-1}$, which interlace the λ_i in the sense that

$$\lambda_i < \lambda_i^{\bullet} < \lambda_{i+1} , \quad (i=1,2,...,N-1) . \tag{2.8.14}$$

This inequality may be interpreted as follows: *If a linear constraint is applied to a system, each natural frequency increases, but does not exceed the next natural frequency of the original system.*

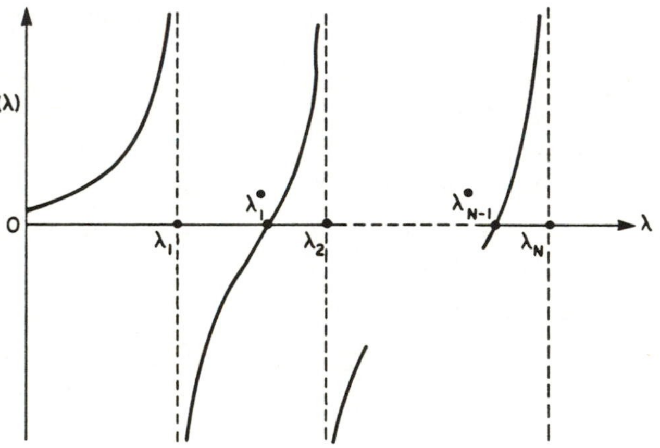

Figure 2.8.1 – The Eigenvalues of a Constrained System Interlace the Original Eigenvalues

If all the b_i are non-zero then the inequalities in (2.8.14) are strictly obeyed. Now, however, suppose some of the b_i are zero; in particular consider the constraint

$$\pi_1 = 0 , \tag{2.8.15}$$

for which $(b_i)_2^N = 0$. In this case $(\pi_i)_2^N$ are the principal coordinates of the system and the corresponding eigenvalues are

$$\lambda_i^{\bullet} = \lambda_{i=1} , \quad (i = 1,2,...,N-1) . \tag{2.8.16}$$

If the constraint is

$$\pi_j = 0 , \quad 1 < j \leq N , \tag{2.8.17}$$

then the principal coordinates are $\pi_1, \pi_2, \ldots, \pi_{j-1}, \pi_{j+1}, \ldots, \pi_N$, so that

$$\lambda_i^{\bullet} = \lambda_i , \quad i = 1,2,...,j-1 ; \quad \lambda_i^{\bullet} = \lambda_{i+1} , \quad i = j,j+1,...,N-1 . \tag{2.8.18}$$

If the constraint is (2.8.12) and some particular b_j is zero, then equation (2.8.9) shows that

$$\pi_i = \delta_{i,j} \equiv \begin{cases} 1 & i = j \\ 0 & i \neq j \end{cases} \tag{2.8.19}$$

is a solution corresponding to $\lambda = \lambda_j$. This means that a constraint (2.8.12) with $b_j = 0$ does not affect the jth principal mode. Figure 2.8.2 shows the form of $B(\lambda)$ when $b_2 = 0$. The graph may pass to the left of λ_2 and cut the axes at $\lambda_1^{\bullet} < \lambda_2$, in which case $\lambda_2^{\bullet} = \lambda_2$, or it may pass to the right, in which case $\lambda_1^{\bullet} = \lambda_2$, $\lambda_2^{\bullet} > \lambda_2$.

If *two* constraints are applied, then the constrained system will have $N-2$ eigenvalues $(\lambda_i^{\bullet\bullet})_1^{N-2}$ satisfying

$$\lambda_i^{\bullet} \leq \lambda_i^{\bullet\bullet} \leq \lambda_{i+1}^{\bullet} , \quad (i = 1,2,...,N-2) , \tag{2.8.20}$$

where $\lambda_i^{\bullet}$ are the eigenvalues of the system subject to one of the constraints. Thus

$$\lambda_i \leq \lambda_i^{\bullet} \leq \lambda_i^{\bullet\bullet} \leq \lambda_{i+1}^{\bullet} \leq \lambda_{i+2} , \quad (i = 1,2,...,N-2) , \tag{2.8.21}$$

or

$$\lambda_i \leq \lambda_i^{\bullet\bullet} \leq \lambda_{i+2} , \quad (i = 1,2,...,N-2) . \tag{2.8.22}$$

2.9 Iterative and independent definitions of eigenvalues

The results of Sections 2.7 and 2.8 may be used to yield some important properties of eigenvalues in relation to the Rayleigh quotient.

In the following discussion we shall assume, without loss of generality, that all vectors $\mathbf{x}$ have been normalized so that

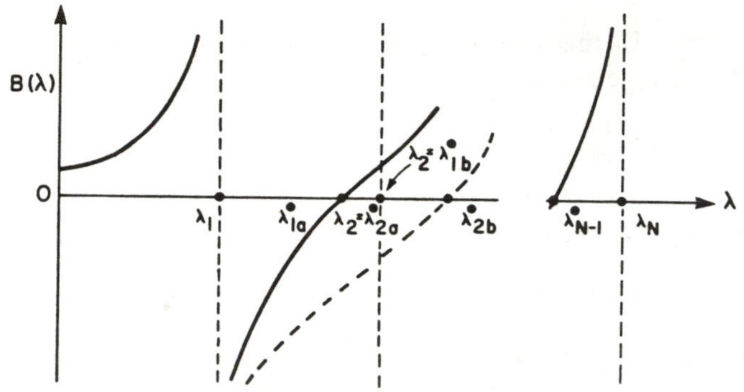

Figure 2.8.2 – *The form of* $B(\lambda)$ *when* $b_2=0$; *either a)* $\lambda_1^*<\lambda_2$, $\lambda_2^*=\lambda_2$ *or b)* $\lambda_1^*=\lambda_2$, $\lambda_2^*>\lambda_2$.

$$\mathbf{x}^T\mathbf{A}\mathbf{x}=1\ . \tag{2.9.1}$$

This means, in particular that the principal modes $\mathbf{x}^{(i)}$ satisfy

$$\mathbf{x}^{(i)T}\mathbf{A}\mathbf{x}^{(i)}=a_i=1\ ,\quad i=1,2,...,N\ . \tag{2.9.2}$$

First, the inequality (2.7.18) states that λ_1 is the (global) minimum of the Rayleigh quotient, and that the minimum is attained in the first mode. With the normalization of equations (2.9.1), (2.9.2) this means

$$\lambda_1=\min\lambda_R=\min\mathbf{x}^T\mathbf{C}\mathbf{x}=\mathbf{x}^{(1)T}\mathbf{C}\mathbf{x}^{(1)}\ . \tag{2.9.3}$$

Equation (2.8.16) with $i=1$ shows that λ_2 is the first eigenvalue λ_1^* of the system constrained so that $\pi_1=0$. Thus,

$$\lambda_2=\min_{(\pi_1=0)}\lambda_R=\min_{(\pi_1=0)}\mathbf{x}^T\mathbf{C}\mathbf{x}=\mathbf{x}^{(2)T}\mathbf{C}\mathbf{x}^{(2)}\ , \tag{2.9.4}$$

and using Ex. 2.6.1 we may write the constraint $\pi_1=0$ as

$$\mathbf{x}^{(1)T}\mathbf{A}\mathbf{x}=0\ , \tag{2.9.5}$$

which we may express in the form '**x** is orthogonal to $\mathbf{x}^{(1)}$ w.r.t. **A**'. Equation (2.9.4) may be used to evaluate λ_2 only when $\mathbf{x}^{(1)}$ has been found – for if $\mathbf{x}^{(1)}$ is unknown, the constraint $\pi_1=0$ may not be computed; the equation may thus be viewed as an *iterative* definition of λ_2. The corresponding definition of $\lambda_j(2\leq j\leq N)$ is

$$\lambda_j = \min_{(\pi_1=0=\cdots\pi_{j-1})} \lambda_R = \min_{(\pi_1=0=\cdots\pi_{j-1})} \mathbf{x}^T\mathbf{C}\mathbf{x} = \mathbf{x}^{(j)T}\mathbf{C}\mathbf{x}^{(j)} . \tag{2.9.6}$$

This defines λ_j as the lowest eigenvalue of the system subject to the $j-1$ constraints $\pi_1=0=...=\pi_{j-1}$, i.e.,

$$\mathbf{x}^{(i)T}\mathbf{A}\mathbf{x}=0 , \quad (i=1,2,...,j-1) . \tag{2.9.7}$$

To obtain an *independent* classification of eigenvalues we first note that the inequality

$$\lambda_1^{\bullet} \leq \lambda_2 , \tag{2.9.8}$$

may be interpreted as stating that λ_2 is the maximum value that can be attained by the first eigenvalue of the system subject to one constraint. Since the typical constraint (2.8.2) is specified by the vector **d**, we may write

$$\lambda_2 = \max_{\mathbf{d}} \min_{\mathbf{d}^T\mathbf{x}=0} \mathbf{x}^T\mathbf{C}\mathbf{x} . \tag{2.9.9}$$

This definition of λ_2 does not depend on $\mathbf{x}^{(1)}$ first being found; it is an *independent* or *minimax* definition of λ_2. The corresponding definition of λ_j is that it is the maximum value of the lowest eigenvalue of the system subject to $j-1$ constraints, i.e.,

$$\lambda_j = \max_{(\mathbf{d}^{(1)},\ldots,\mathbf{d}^{(j-1)})} \min_{\mathbf{d}^{(i)T}\mathbf{x}=0;\, i=1,2,...,j-1} \mathbf{x}^T\mathbf{C}\mathbf{x} . \tag{2.9.10}$$

The minimax definition of eigenvalues seems to have been noted first by Fischer (1905); see also Courant and Hilbert (1953, Section 1.4) and Gould (1966).

CHAPTER 3

JACOBIAN MATRICES

Let no one say that I have said nothing new;
the arrangement of the subject is new.
Pascal's Pensées

3.1 Sturm sequences

Before we will be in a position to solve the inverse problem for the systems considered in Section 2.2 we must complete certain mathematical preliminaries. These are related to the special, tridiagonal form of the matrix **C** in equation (2.2.10). Because of this special form, the eigenvalues and eigenvectors of equation (2.5.3) have special properties, in addition to those described in Chapter 2; the latter are common to all systems with positive definite inertia matrix **A** and positive semi-definite stiffness matrix **C**. The most important property of the eigenvalues of such matrices is that they are real and simple, i.e., distinct (Theorem 3.1.3). Thus $\lambda_1<\lambda_2<\cdots<\lambda_N$. If $\mathbf{x}^{(r)}$ is the rth eigenvector, then, as r increases, the eigenvector oscillates more and more (Theorem 3.3.1) in such a way that the zeros of $\mathbf{x}^{(r)}$ interlace those of the neighbouring $\mathbf{x}^{(r-1)}$ and $\mathbf{x}^{(r+1)}$ (Theorem 3.3.4). We shall now establish these and other results. Throughout Chapters 3 - 5 we make extensive use of the analysis developed by Gantmakher and Krein (1950). Before embarking on the analysis we first reduce the eigenvalue equation

$$(\mathbf{C}-\lambda\mathbf{A})\mathbf{x}=0 \tag{3.1.1}$$

to standard form. The matrix **A** is

$$\mathbf{A}=diag(m_1,m_2,...,m_N)\ . \tag{3.1.2}$$

Write

$$\mathbf{A}=\mathbf{L}\mathbf{L}^T\ , \tag{3.1.3}$$

where

$$\mathbf{L}=\mathbf{L}^T=diag(m_1^{1/2},m_2^{1/2},...,m_N^{1/2})\ , \tag{3.1.4}$$

introduce the vector $\mathbf{u}$ related to $\mathbf{x}$ by

$$\mathbf{u}=\mathbf{L}^T\mathbf{x}\ ,\quad \mathbf{x}=(\mathbf{L}^T)^{-1}\mathbf{u}\ , \tag{3.1.5}$$

and multiply equation (3.1.1) on the left by $\mathbf{L}^{-1}$ to obtain

$$\mathbf{L}^{-1}(\mathbf{C}-\lambda\mathbf{L}\mathbf{L}^T)(\mathbf{L}^T)^{-1}\mathbf{u}=\mathbf{0}\ , \tag{3.1.6}$$

i.e.,

$$(\mathbf{B}-\lambda\mathbf{I})\mathbf{u}=0\ , \tag{3.1.7}$$

where

$$\mathbf{B}=\mathbf{L}^{-1}\mathbf{C}(\mathbf{L}^T)^{-1}\ . \tag{3.1.8}$$

The matrix $\mathbf{B}$, like $\mathbf{C}$, is symmetric, tridiagonal and positive definite (or positive semi-definite if $\mathbf{C}$ is positive semi-definite), and has the same eigenvalues as $\mathbf{C}$. A symmetric tridiagonal matrix is termed a *Jacobian* matrix. In its standard form we shall assume that $\mathbf{B}$ has *negative* off-diagonal elements (Ex. 3.1.1). We write

$$\mathbf{B}=\begin{bmatrix} a_1 & -b_1 & 0 & 0 & 0 \\ -b_1 & a_2 & -b_2 & 0 & 0 \\ . & . & . & . & . \\ 0 & 0 & 0 & a_{N-1} & -b_{N-1} \\ 0 & 0 & 0 & -b_{N-1} & a_N \end{bmatrix}\ . \tag{3.1.9}$$

The analysis now centres on the *principal minors* of the matrix $\mathbf{B}-\lambda\mathbf{I}$. We define

$$P_0=1\ ,\ P_1(\lambda)=a_1-\lambda\ ,\ P_2(\lambda)=\begin{vmatrix} a_1-\lambda & -b_1 \\ -b_1 & a_2-\lambda \end{vmatrix}\ ,\quad \text{etc.,} \tag{3.1.10}$$

so that finally

$$P_N(\lambda)=|\,\mathbf{B}-\lambda\mathbf{I}\,|\ . \tag{3.1.11}$$

The minors satisfy the recurrence relation

$$P_{r+1}(\lambda)=(a_{r+1}-\lambda)P_r(\lambda)-b_r^2P_{r-1}(\lambda)\ ,\ (r=1,2,...,N-1), \tag{3.1.12}$$

which enables us to calculate $P_2,P_3,...,P_N$ successively from P_0, P_1.

We now prove:

Theorem 3.1.1. If $b_r^2>0$ (r=1,2,...,N-1), then the $(P_r(\lambda))_1^N$ form a *Sturm* sequence, the defining properties of which are

1. $P_0(\lambda)$ has everywhere the same sign ($P_0(\lambda)\equiv 1$).
2. When $P_r(\lambda)$ vanishes, $P_{r+1}(\lambda)$ and $P_{r-1}(\lambda)$ are non-zero and have opposite signs.

Proof

In order to establish property 2 we note first that two successive P_r cannot be *simultaneously* zero – i.e., for the same $\lambda=\lambda^o$. For if $P_{s+1}(\lambda^o)=0=P_s(\lambda^o)$ then equation (3.1.12) shows that $P_{s-1}(\lambda^o)=0$, so that finally we must have P_1 and P_0 zero; but $P_0(\lambda^o)=1$, which yields a contradiction.

The latter part of property 2 now follows directly from (3.1.12). ■

Before proceeding further we must define $s_r(\lambda)$. This is the integer-valued function equal to the cumulative number of sign-changes in the sequence P_0, $P_1(\lambda)$, $P_2(\lambda)$,..., $P_r(\lambda)$. Thus if

$$\mathbf{B}=\begin{bmatrix} 2 & -1 & 0 \\ -1 & 3 & -2 \\ 0 & -2 & 4 \end{bmatrix}, \tag{3.1.13}$$

then,

$$P_0=1\ ,\ \ P_1(\lambda)=-\lambda+2\ ,$$
$$P_2(\lambda)=\lambda^2+5\lambda+5\ ,\ \ P_3(\lambda)=-\lambda^3+9\lambda^2-21\lambda+12. \tag{3.1.14}$$

For $\lambda=0$ the sequence of values is 1, 2, 5, 12. Since there is no change of sign in the sequence, each $s_r(0)=0$. For $\lambda=3$ the sequence is 1, -1, -1, 3, so that $s_1(3)=s_2(3)=1$, $s_3(3)=2$.

We now prove:

Theorem 3.1.2. $s_r(\lambda)$ changes only when λ passes through a zero of the last polynomial, $P_r(\lambda)$.

Proof:

Clearly, $s_r(\lambda)$ can change only when λ passes through a zero of one of the $P_s(\lambda)$, $(s\le r)$; it therefore suffices to prove that $s_r(\lambda)$ does not change at all when λ passes through a zero of an *intermediate* $P_s(\lambda)$, $(s<r)$. Suppose $P_s(\lambda^o)=0$, where $1\le s<r$, then $P_{s-1}(\lambda^o)$ and $P_{s+1}(\lambda^o)$ will be both non-zero and have opposite signs. The signs of the triad $P_{s-1}(\lambda^o)$, $P_s(\lambda^o)$, $P_{s+1}(\lambda^o)$ are therefore $+0-$ or $-0+$. Suppose the first to be the case and that $P_s(\lambda)$ increases as λ passes through λ^o, (the other possibilities may be handled similarly). Then for values of λ sufficiently close to λ^o and less than λ^o the signs are $+--$, while for values of λ sufficiently close

and greater than λ^o the signs are + + -. Thus, whether λ is greater than or less than λ^o there is just one change of sign in the triad of values of $P_{s-1}(\lambda)$, $P_s(\lambda)$, $P_{s+1}(\lambda)$. In other words the triad of polynomials $P_{s-1}(\lambda)$, $P_s(\lambda)$, $P_{s+1}(\lambda)$ will not contribute any change to $s_r(\lambda)$ as λ passes through λ^o. But no other members of the sequence will contribute any change to $s_r(\lambda)$ as λ passes through λ^o (unless λ^o is a zero of another $P_t(\lambda)$, $|t-s|\geq 2$, in which case again there will be no change in $s_r(\lambda)$) so that $s_r(\lambda)$ will not change at all. ■

Theorem 3.1.3. The zeros of $P_r(\lambda)$ are real and simple. In addition if $P_r(\lambda^o)\neq 0$ and $s_r(\lambda^o)=k$, then $P_r(\lambda)$ has k zeros less than λ^o.

Proof:

Since $P_s(\lambda)=(-)^s\lambda^s+\cdots$, all $P_s(\lambda)$ will be positive for sufficiently large negative λ, i.e., $\lambda\leq\alpha$, so that $s_r(\alpha)=0$: α may be taken to be *zero* if **B** is positive definite. On the other hand, for sufficiently large positive λ, i.e., $\lambda\geq\beta$, the $P_s(\lambda)$ will alternate in sign, so that $s_r(\beta)=r$. Now since $s_r(\lambda)$ can increase only when λ passes through a zero of $P_r(\lambda)$, all the zeros of $P_r(\lambda)$ must be distinct. For if λ^o were a zero of even multiplicity then $s_r(\lambda)$ would not increase at all as λ passed through λ^o, while $s_r(\lambda)$ would increase only by unity if λ^o were a zero of odd multiplicity. The second part of the theorem now follows immediately. ■

Corollary 1. The eigenvalues of a Jacobian matrix are *distinct*.

Corollary 2. The number of zeros of $P_r(\lambda)$ satisfying $\alpha<\lambda<\beta$ is equal to $s_r(\beta)-s_r(\alpha)$.

Corollary 3. If λ^o is a zero of $P_r(\lambda)$ then, as λ passes from λ^o- to λ^o+ the sign of $P_{r-1}(\lambda)P_r(\lambda)$ changes from + to -, and $s_r(\lambda)$ increases by unity.

Theorem 3.1.4. Between any two neighbouring zeros of $P_r(\lambda)$ there lies one and only one zero of $P_{r-1}(\lambda)$, and one and only one zero of $P_{r+1}(\lambda)$.

Proof:

Let μ_1, μ_2 be the two neighbouring zeros. Suppose, for the sake of argument that $P_r(\mu_1-)>0$, then $P_r(\mu_1+)<0$ and $P_r(\mu_2-)<0$. By Corollary 3, $P_{r-1}(\mu_1+)>0$ and $P_{r-1}(\mu_2-)<0$, so that $P_{r-1}(\lambda)$ changes sign between μ_1+ and μ_2-, and therefore has at least one zero in (μ_1,μ_2).

Now property 2 of Sturm's sequences shows that $P_{r+1}(\mu_i)$ and $P_{r-1}(\mu_i)$, (i=1,2) have opposite signs. Thus $P_{r+1}(\mu_1+)<0$, $P_{r+1}(\mu_2-)>0$ so that $P_{r+1}(\lambda)$ has *at least one* zero in (μ_1,μ_2). Now suppose, if possible, that $P_{r-1}(\lambda)$ (or $P_{r+1}(\lambda)$) had two (or more) zeros in (μ_1,μ_2) then $P_r(\lambda)$ would have a zero in (μ_1,μ_2), contrary to the hypothesis that μ_1,μ_2 are neighbouring zeros. ■

This theorem is usually stated in the form: the eigenvalues of successive principal minors interlace each other.

3.2 Orthogonal polynomials

There is an intimate connection between Jacobian matrices and orthogonal polynomials. In this section we outline some of the basic properties of orthogonal polynomials.

Two polynomials $p(x)$, $q(x)$ are said to be *orthogonal* w.r.t. the weight function $w(x)>0$ over an interval (a,b) if

$$<p,q>\equiv\int_a^b w(x)p(x)q(x)dx=0 \ . \tag{3.2.1}$$

A familiar example is provided by the Laguerre polynomials $(L_n(x))_o^\infty$, i.e.,

$$L_o(x)=1\ ,\ \ L_1(x)=x-1\ ,\ \ L_2(x)=x^2-4x+2\ ,... \tag{3.2.2}$$

which are orthogonal w.r.t. the weight function e^{-x} over (o,∞), i.e.,

$$\int_o^\infty e^{-x}L_n(x)L_m(x)dx=0\ ,\quad m\neq n\ . \tag{3.2.3}$$

One of the important properties of such polynomials is that they satisfy a *three-term recurrence relation*. The relation for the $L_n(x)$, for example, is

$$L_{n+1}(x)=(x-2n-1)L_n(x)-n^2L_{n-1}(x)\ . \tag{3.2.4}$$

In this section we shall be concerned, not with a *continuous* orthogonality relation of the form (3.2.1), but with a *discrete* orthogonality relation

$$<p,q>\equiv\sum_{i=1}^{N}w_ip(\xi_i)q(\xi_i)=0\ ;\ \ (w_i)_1^N>0\ . \tag{3.2.5}$$

where $(\xi_i)_1^N$ are N points, satisfying $\xi_1<\xi_2<\cdots<\xi_N$.

To introduce the concept formally we let $\mathbf{P}_N$ denote the linear space of polynomials of order N, i.e., the set of all polynomials $p(x)$ with degree $k<N$, with real coefficients. On this space $<,>$ acts as an *inner product* since it is *positive definite*, *bilinear* and *symmetric*, i.e.,

1)

$$<p,q>\equiv||p||^2>0 \text{ if } p(x)\neq0$$

2)

$$<\alpha p,q>=\alpha<p,q>\ ,\ <p+q,r>=<p,r>+<q,r>$$

3)

$$<p,q>=<q,p>\ .$$

In addition

4)

$$<xp,q>=<p,xq> .$$

We now prove

Theorem 3.2.1. There is a unique sequence of monic polynomials, i.e., $(q_i(x))_o^N$ such that $q_i(x)$ has degree i and leading coefficient (of x^i) unity, which are orthogonal with respect to the inner product $<,>$, i.e., for which

$$<q_i,q_j>=0 , \quad i \neq j . \tag{3.2.6}$$

Proof:

The $q_i(x)$ may be constructed by applying the familiar Gram-Schmidt orthogonalization procedure to the linearly independent polynomials $(x^i)_o^{N-1}$. Thus

$$q_o=1 , \quad q_i(x)=x^i-\sum_{j=0}^{i-1}\alpha_{i,j}q_j(x) , \quad (i=1,2,...,N-1), \tag{3.2.7}$$

where

$$<q_i,q_j>=<x^i,q_j>-\alpha_{i,j}<q_j,q_j>=0 \tag{3.2.8}$$

so that

$$\alpha_{i,j}=<x^i,q_j>/||q_j||^2 , \quad j=0,1,...,i-1 . \quad \blacksquare \tag{3.2.9}$$

We note that the polynomial

$$q_N(x)=\sum_{i=1}^{N}(x-\xi_i) \tag{3.2.10}$$

is the monic polynomial of degree N in the sequence. It is orthogonal to $(q_i)_o^{N-1}$ – in fact it is orthogonal to all functions, since $q_N(\xi_i)=0$, $i=1,2,...,N$.

The Gram-Schmidt procedure does not provide a computationally convenient means for computing the q_i; instead we use Forsythe (1957).

Theorem 3.2.2. The monic polynomials $(q_i)_o^N$ satisfy a three-term recurrence relation of the form

$$q_i(x)=(x-\alpha_i)q_{i-1}(x)-\beta_{i-1}^2q_{i-2}(x) , \quad i-1,2,...,N , \tag{3.2.11}$$

with the initial values

$$q_{-1}(x)=0 , \quad q_o(x)=1 . \tag{3.2.12}$$

Proof:

$q_i(x)-xq_{i-1}(x)$ is a polynomial of degree $(i-1)$. It may therefore be expressed in terms of (the linearly independent – see Ex. 3.2.1) $q_o,q_1,...,q_{i-1}$. Thus

$$q_i(x)-xq_{i-1}(x)=c_oq_o+c_1q_1+\cdots+c_{i-1}q_{i-1}\,. \tag{3.2.13}$$

Take the inner product of this equation with $q_j(x)$, $(j=0,1,\ldots,i\text{-}1)$; thus

$$<q_i,q_j>-<q_{i-1},xq_j>=\sum_{k=o}^{i-1}c_k<q_k,q_j>=c_j\,||q_j||^2\,, \tag{3.2.14}$$

where the second term on the left has been rewritten by using property 4, above. But if $j=0,1,\ldots,i-1$, then the first term on the left is zero, and if $j=0,1,\ldots,i-3$, then xq_j has degree at most $i-2$ and so is orthogonal to q_{i-1}. Thus $c_j=o$ if $j=0,1,2,\ldots,i-3$ and there only *two* terms c_{i-1} and c_{i-2} on the right of (3.2.13), i.e.,

$$q_i(x)-xq_{i-1}(x)=c_{i-2}q_{i-2}(x)+c_{i-1}q_{i-1}(x). \tag{3.2.15}$$

Moreover equation (3.2.14) gives

$$\alpha_i=-c_{i-1}=<q_{i-1},xq_{i-1}>/\,||q_{i-1}||^2 \tag{3.2.16}$$

$$c_{i-2}=-<q_{i-1},xq_{i-2}>/\,||q_{i-2}||^2\,. \tag{3.2.17}$$

But xq_{i-2} is a monic polynomial of degree $i-1$; it may therefore be expressed in the form

$$xq_{i-2}(x)=q_{i-1}(x)+\sum_{j=o}^{i-2}d_jq_j(x) \tag{3.2.18}$$

so that

$$<q_{i-1},xq_{i-2}>=||q_{i-1}||^2 \tag{3.2.19}$$

and thus c_{i-2} is negative and equal to $-\beta_{i-1}^2$, where

$$\beta_i=||q_i||/||q_{i-1}||\,. \quad \blacksquare \tag{3.2.20}$$

Equations (3.2.11), (3.2.12) with (3.2.16), (3.2.20) enable us to compute the polynomials $\{q_i\}_1^{N-1}$ step by step. Thus with $q_{-1}=0$, $q_o=1$ we first compute α_1 from (3.2.16); this substituted into (3.2.11) gives q_1. Now we compute α_2, β_1 and find q_2, etc.

In the inverse problem we will need to express the weights w_i in terms of the polynomials q_{N-1} and q_N. For this we note that if $f(x)$ is any polynomial in $\mathbf{P}_{N-1}$, i.e., of degree $N-2$ or less, then

$$<q_{N-1},f>\equiv\sum_{i=1}^{N}w_iq_{N-1}(\xi_i)f(\xi_i)=0\,. \tag{3.2.21}$$

But if such a combination

$$\sum_{i=1}^{N}m_if(\xi_i)\,,\quad m_i=w_iq_{N-1}(\xi_i) \tag{3.2.22}$$

is zero for any $f(x)$ in $\mathbf{P}_{N-1}$ then

$$\sum_{i=1}^{N} m_i \xi_i^k = 0 \ , \quad (k=0,1,\ldots,N-2) \ , \tag{3.2.23}$$

since each x^k is in $\mathbf{P}_{N-1}$, i.e.,

$$\mathbf{Am} = \begin{bmatrix} 1 & 1 & 1 & \cdots & 1 \\ \xi_1 & \xi_2 & \xi_3 & \cdots & \xi_N \\ \xi_1^2 & \xi_2^2 & \xi_3^2 & \cdots & \xi_N^2 \\ \cdot & \cdot & \cdot & \cdot & \cdot \\ \xi_1^{N-2} & \xi_2^{N-2} & \xi_3^{N-2} & \cdots & \xi_N^{N-2} \end{bmatrix} \begin{bmatrix} m_1 \\ m_2 \\ m_3 \\ \cdots \\ \cdot \\ m_N \end{bmatrix} = 0. \tag{3.2.24}$$

It is shown in Ex. 3.2.2 that this equation has the solution

$$m_i = \gamma / \prod_{j=1}^{N}{}' (\xi_i - \xi_j). \tag{3.2.25}$$

Apart from the arbitrariness of the factor γ, this is the unique solution. The prime means that the term $j=i$ is omitted. Now since

$$q_N(\xi) = \prod_{j=1}^{N} (\xi - \xi_j) \ , \tag{3.2.26}$$

we have

$$q_N'(\xi_i) = \prod_{j=1}^{N}{}' (\xi_i - \xi_j) \ , \tag{3.2.27}$$

where the prime on the left denotes differentiation!

Returning to equation (3.2.25) we can deduce that, for some γ,

$$w_i q_{N-1}(\xi_i) = \gamma / q'_N(\xi_i) \ , i=1,2,\ldots,N \ . \tag{3.2.28}$$

Since the $\{q_i\}_o^N$ satisfy the three-term recurrence relation (3.2.11) it follows, by the arguments used in Section 3.1, that the zeros of $q_N(x)$ and $q_{N-1}(x)$ must interlace and therefore (Ex. 3.2.3) $q_{N-1}(\xi_i) q'_N(\xi_i) > 0$. This means that the weights

$$w_i = \gamma / \{q_{N-1}(\xi_i) q'_N(\xi_i)\} \tag{3.2.29}$$

are positive.

This equation is important: it means that if the monic polynomials $q_N(\xi)$, $q_{N-1}(\xi)$ are given, and if their zeros interlace, then they may be viewed as the Nth and $(N-1)$th members, respectively, of a sequence of monic polynomials orthogonal w.r.t. the weights w_i given by (3.2.29), and the points $\{\xi_i\}_1^N$.

Examples 3.2

1. Show that if the polynomials $\{q_i\}_o^k$, $k<N$, are orthogonal w.r.t. the inner-product (3.2.5), then they are linearly independent. Hence deduce that any polynomial $p(x)$ of degree $k-1$ may be expressed uniquely in the form

$$p(x)=\sum_{j=o}^{k-1} c_j q_j(x) \ ,$$

and that $q_k(x)$ is orthogonal to each polynomial of degree $k-1$.

2. Show that if

$$\Delta=\begin{vmatrix} 1 & 1 & \cdots & 1 \\ \xi_1 & \xi_2 & \cdots & \xi_{N-1} \\ \xi_1^2 & \xi_2^2 & \cdots & \xi_{N-1}^2 \\ \cdots & \cdots & \cdots & \cdots \\ \xi_1^{N-2} & \xi_2^{N-2} & \cdots & \xi_{N-1}^{N-2} \end{vmatrix},$$

then

$$\Delta=\prod_{j=2}^{N-1}\ \prod_{k=1}^{j-1}(\xi_j-\xi_k)=\Gamma/\prod_{j=1}^{N-1}(\xi_N-\xi_j) \ ,$$

where

$$\Gamma=\prod_{j=2}^{N}\ \prod_{k=1}^{j-1}(\xi_j-\xi_k) \ .$$

Hence deduce (3.2.25).

3. The zeros $\{\xi_i\}_1^N$ of $q_n(x)$, and $\{\xi_i^*\}_1^{N-1}$ of $q_{N-1}(x)$ must satisfy $\xi_1<\xi_1^*<\xi_2< \cdots <\xi_{N-1}^* \quad <\xi_N$. Show that $(-)^{N-i}q_N'(\xi_i)>0$, $(-)^{N-i}q_{N-1}(\xi_i)>0$ and hence $q_N'(\xi_i)q_{N-1}(\xi_i)>0$.

3.3 Eigenvectors of Jacobian matrices

In this section we establish some properties of the eigenvectors of Jacobian matrices, in preparation for the solution of 'inverse mode problems' in Section 6.1. We return to the analysis of Section 3.1 and prove

Theorem 3.3.1. The sequence $(u_r^{(j)})_{r=1}^N$ for the jth eigenvector has exactly $j-1$ sign reversals.

Proof:

The $u_r^{(j)}$ are determined from equation (3.1.7) which may be written

$$-b_{r-1}u_{r-1}^{(j)}+(a_r-\lambda_j)u_r^{(j)}-b_r u_{r+1}^{(j)}=0 \ , \quad (r=1,2,...,N) \tag{3.3.1}$$

where $u_o^{(j)}$, $u_{N+1}^{(j)}$ are interpreted as zero, i.e.,

$$u_o^{(j)}=0=u_{N+1}^{(j)} \ . \tag{3.3.2}$$

Choose an arbitrary $b_N>0$ and put

$$v_1 = u_1^{(j)}\ ,\ \ v_2 = b_1 u_2^{(j)}\ , \ldots ,\ v_{N+1} = b_1 b_2 \cdots b_N u_{N+1}^{(j)} \tag{3.3.3}$$

and multiply equation (3.3.1) by $b_1 b_2 \cdots b_{r-1}$ to obtain

$$-b_{r-1}^2 v_{r-1} + (a_r - \lambda_j) v_r - v_{r+1} = 0\ ,\ \ (r = 1,2,...,N)\ . \tag{3.3.4}$$

On comparing this equation with (3.1.12) we see that it has the solution

$$v_o = 0\ ,\ \ v_r = v_1 P_{r-1}(\lambda_j)\ ,\ \ (r = 1,2,...,N+1) \tag{3.3.5}$$

which, because of $P_N(\lambda_j) = 0$, satisfies the end-condition $v_{N+1} = 0$.

Thus,

$$u_r^{(j)} = u_1^{(j)} (b_1 b_2 \cdots b_{r-1})^{-1} P_{r-1}(\lambda_j)\ , \tag{3.3.6}$$

and since λ_j lies between the $(j-1)$th and jth zeros of $P_{N-1}(\lambda)$, $s_{N-1}(\lambda_j) = j-1$ ■

Before establishing further properties of the eigenvector we introduce the concept of a u-line.

Definition: Let $\mathbf{u} = \{u_1, u_2, \ldots, u_{N+1}\}$ be a vector. We shall define the u-line as the broken line in the plane joining the points with coordinates

$$x_r = r\ ,\ \ y_r = u_r\ ,\ \ (r = 1,2,...,N+1)\ . \tag{3.3.7}$$

Thus, between (x_r, y_r) and (x_{r+1}, y_{r+1}), $y(x)$ is defined by

$$y(x) = (r+1-x) u_r + (x-r) u_{r+1}\ ,\ \ (r = 1,2,...,N) \tag{3.3.8}$$

as shown in Figure 3.3.1.

Now return to Theorem 3.3.1. For arbitrary (real) λ the sequence given by

$$u_o = 0\ ,\ \ u_r(\lambda) = u_1 (b_1 b_2 \cdots b_{r-1})^{-1} P_{r-1}(\lambda)\ ,\ \ (r = 1,2,...,N+1), \tag{3.3.9}$$

satisfies the recurrence (3.3.1) for $r = 1,2,...,N-1$. (It will satisfy the last equation with $u_{N+1} = 0$ if and only if $P_N(\lambda) = 0$). For arbitrary λ, the vector $\mathbf{u}(\lambda)$ = $\{u_1(\lambda), ..., u_{N+1}(\lambda)\}$ defines a $u(\lambda)$ - line. We now investigate the *nodes* of this line, i.e., the points x at which $y(x) = 0$. First we note that if $u_r(\lambda) = 0$, i.e., $P_{r-1}(\lambda) = 0$, then $P_r(\lambda)$ and $P_{r-2}(\lambda)$, i.e., u_{r+1} and u_{r-1}, will have opposite signs, so that the $u(\lambda)$ - line will cross the x-axis at $x = r$. Secondly, if u_r and u_{r+1} have opposite signs, then $y(x)$ has a node between r and $r+1$. This implies that the $u(\lambda_j)$ - line has exactly j nodes, *excluding* the left hand end where $u_o = 0$, but *including* the right hand end. Moreover, if $\lambda_j \le \lambda < \lambda_{j+1}$, then the $u(\lambda)$ - line will have exactly j nodes, again excluding the left hand end where $u_o = 0$. Table 3.3.1 shows the signs of y_r for the whole range of λ-values, for the case $N = 3$. The last line in the table shows the number of nodes in the $u(\lambda)$. Figure 3.3.1 shows the form of the $u(\lambda)$ for the starred values of λ. We now establish an identity which will enable us to prove further results concerning the eigenvectors.

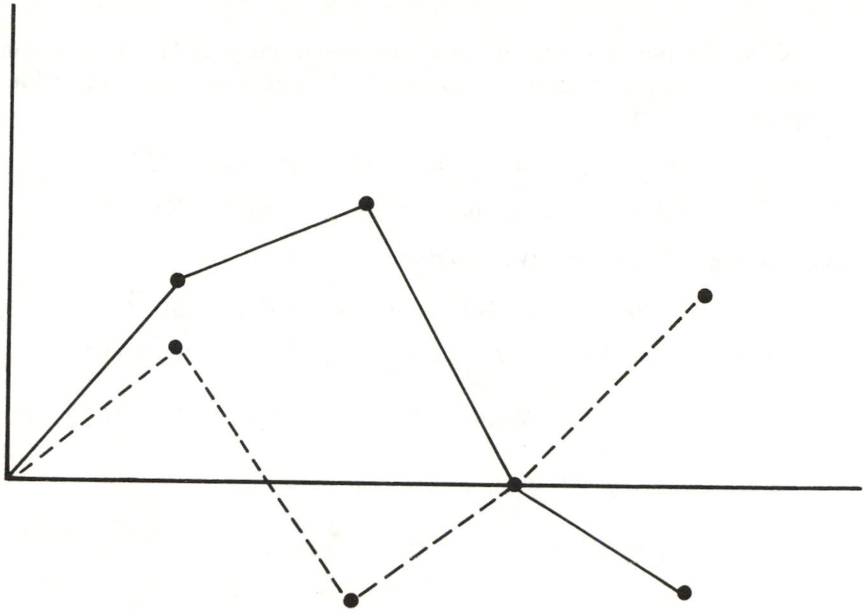

Figure 3.3.1 – *Figure 3.3.1 - The $u(\lambda)$ - lines for λ^* – , and λ^{**} ----.*

Table 3.3.1 – *The Signs of u_r for Different Values of λ (The broken lines show the positions of the nodes)*

λ	0	λ_1	λ^*	λ	λ_2	λ^{**}	λ_3
U_1	+	+	+	+	+	+	+
U_2	+	+	+	0	−	−	−
U_3	+	+	0	−	−	0	+
U_4	+	0	−	−	0	+	0
	0	1	1	1	2	2	3

Consider the solutions **u**, **v** of the equations (3.3.1) corresponding to λ, μ respectively. Suppose that $u_o=0=v_o$ and that some positive value has been assigned to b_N. Then

$$-b_{r-1}u_{r-1}+a_r u_r-b_r u_{r+1}=\lambda u_r\ ,\quad (r=1,2,...,N)$$

$$-b_{r-1}v_{r-1}+a_r v_r-b_r v_{r+1}=\mu v_r\ ,\quad (r=1,2,...,N)\ . \tag{3.3.10}$$

Eliminating a_r from these two equations, we find

$$b_{r-1}(u_{r-1}v_r-v_{r-1}u_r)-b_r(u_r v_{r+1}-v_r u_{r+1})=(\mu-\lambda)u_r v_r \tag{3.3.11}$$

so that on summing for $r=p,p+1,...,q$, $(1\le p\le q\le N)$, we obtain

$$b_{p-1}(u_{p-1}v_p-v_{p-1}u_p)-b_q(u_q v_{q+1}-v_q u_{q+1})=(\mu-\lambda)\sum_{r=p}^{q} u_r v_r\ . \tag{3.3.12}$$

In particular, when $p=1$, we have, since $u_o=0=v_o$,

$$-b_q(u_q v_{q+1}-v_q u_{q+1})=(\mu-\lambda)\sum_{r=1}^{q} u_r v_r\ . \tag{3.3.13}$$

We now prove

Theorem 3.3.2. If $\lambda<\mu$, then between any two nodes of the $u(\lambda)$ - line there is at least one node of the $u(\mu)$ - line.

Proof:

Let α, $\beta(\alpha<\beta)$ be two neighbouring nodes of the u-line and suppose that

$$p-1\le\alpha<p\ ,\quad q<\beta\le q+1\ ,\quad (p\le q)\ ,$$

so that

$$y(\alpha,\lambda)\equiv(p-\alpha)u_{p-1}+(\alpha-p+1)u_p=0\ , \tag{3.3.14}$$

$$y(\beta,\lambda)\equiv(q+1-\beta)u_q+(\beta-q)u_{q+1}=0 \tag{3.3.15}$$

and $y(x,\lambda)\neq0$ for $\alpha<x<\beta$. For the sake of definiteness suppose that $y(x,\lambda)>0$ for $\alpha<x<\beta$, then $u_p,u_{p+1},...,u_q$ are all positive. We now need to prove that $y(x,\mu)$ has a zero between α and β. Suppose $y(x,\mu)$ has no such zero, that is, it has the same sign for $\alpha<x<\beta$. Without loss of generality we can assume that

$$y(x,\mu)>0 \text{ for } \alpha<x<\beta\ .$$

that is $y(\alpha,\mu)\ge0$, $y(\beta,\mu)\ge0$ and v_p, $v_{p+1},...,v_q$ are all positive. Thus,

$$(p-\alpha)v_{p-1}+(\alpha-p+1)v_p\ge0\ , \tag{3.3.16}$$

$$(q+1-\beta)v_q+(\beta-q)v_{q+1}\ge0\ , \tag{3.3.17}$$

and on eliminating α between (3.3.14), (3.3.16), and β between (3.3.15), (3.3.17) we deduce that

$$u_{p-1}v_p - v_{p-1}u_p \leq 0 \ , \tag{3.3.18}$$

$$u_{q+1}v_q - v_{q+1}u_q \geq 0 \ . \tag{3.3.19}$$

But this means that the left hand side of (3.3.12) is negative while the right hand side is positive, providing a contradiction. If we had assumed that $y(x,\mu)<0$ for $\alpha<x<\beta$ then we would have found the left hand side positive and the right negative. ■

Theorem 3.3.3. If λ increases continuously, then the nodes of the $u(\lambda)$ - line shift continuously to the left.

Proof:

Let $\alpha_1(\lambda)$, $\alpha_2(\lambda)$,... be the nodes of the $u(\lambda)$ - line, and suppose $0<\alpha_1(\mu)$, $\alpha_2(\mu)$,... are the nodes of the $u(\mu)$ - line. We need to prove that

$$\alpha_r(\mu)<\alpha_r(\lambda) \tag{3.3.20}$$

for all those values of r corresponding to the $u(\lambda)$ - line. Since, by Theorem 3.3.2, there is a least one of the $\alpha_r(\mu)$ between any two of the $\alpha_r(\lambda)$, it is sufficient to prove

$$\alpha_1(\mu)<\alpha_1(\lambda) \ . \tag{3.3.21}$$

Suppose, if possible that $\alpha_1(\mu)\geq\alpha_1(\lambda)$ and that

$$q<\alpha_1(\lambda)\leq q+1 \quad (1\leq q\leq N) \ , \tag{3.3.22}$$

then all $u_1,u_2,...,u_q$ and $v_1,v_2,...,v_q$ are positive while

$$(q+1-\alpha_1)u_q+(\alpha_1-q)u_{q+1}=0 \ , \tag{3.3.23}$$

$$(q+1-\alpha_1)v_q+(\alpha_1-q)v_{q+1}\geq 0 \tag{3.3.24}$$

so that

$$u_q v_{q+1} - v_q u_{q+1} \geq 0 \tag{3.3.25}$$

which, when used in (3.3.13), provides a contradiction. ■

Table 3.3.1 shows how the first node of $u(\lambda)$ appears at the right hand end $(N+1)$ when $\lambda=\lambda_1$ and gradually shifts to the left, how the second zero appears when $\lambda=\lambda_2$, etc.

Theorem 3.3.4. The nodes of two successive eigenvectors alternate.

Proof:

Let the eigenvectors correspond to λ_j and λ_{j+1}. The nodes of the $u(\lambda_j)$ and $u(\lambda_{j+1})$ – lines are $(\alpha_r(\lambda_j))_{r=1}^{j}$ and $(\alpha_r(\lambda_{j+1}))_{r=1}^{j+1}$ respectively; and $\alpha_j(\lambda_j) = \alpha_{j+1}(\lambda_{j+1})=N+1$. Theorem 3.3.3 shows that $\alpha_1(\lambda_{j+1})<\alpha_1(\lambda_j)$, while Theorem 3.3.2 applied to the two zeros $\alpha_{j-1}(\lambda_j)$ and $\alpha_j(\lambda_j) \equiv N+1$ shows that

$\alpha_j(\lambda_{j+1}) > \alpha_{j-1}(\lambda_j)$. These two inequalities imply that the only possible ordering of the nodes is

$$0 < \alpha_1(\lambda_{j+1}) < \alpha_1(\lambda_j) < \alpha_2(\lambda_{j+1}) < \cdots < \alpha_{j-1}(\lambda_j) < \alpha_j(\lambda_{j+1})$$
$$< \alpha_j(\lambda_j) = \alpha_{j+1}(\lambda_{j+1}) = N+1 \ . \ \blacksquare \qquad (3.3.26)$$

The derivation of certain other important properties of the eigenmodes will be deferred until Section 5.7, where properties of an oscillatory matrix will be used.

Examples 3.3

1. Show that the last term, $u_N^{(j)}$, of any eigenmode of (3.3.1) must be non-zero, i.e., $u_N^{(j)} \neq 0$.
2. Show that if the matrix **B** of (3.1.9), with negative off-diagonal elements has an eigenpair λ_i, $u^{(i)}$, then the corresponding matrix $\mathbf{B}^*$ with positive off-diagonal elements, has eigenpairs λ_i, $\mathbf{S}u^{(i)}$ where $\mathbf{S}$ is given by $\mathbf{S} = diag(1, -1, 1, \cdots (-)^{N-1})$. This means that the eigenvector corresponding to the smallest eigenvalue, λ_1, has $N-1$ sign changes, while that corresponding to λ_N has none. Show that if the eigenvalues of $\mathbf{B}^*$ are numbered *in reverse*, i.e., $\lambda_1 > \lambda_2 > \cdots > \lambda_N > 0$, then Theorem 3.3.1 remains valid.

CHAPTER 4

INVERSION OF DISCRETE SECOND-ORDER SYSTEMS

People are generally better persuaded by the reasons which they themselves discovered than by those which have come into the mind of others.

Pascal's Pensées

4.1 Introduction

Research on these inverse problems began in the Soviet Union, with the work of M.G. Krein. It appears that his primary interest was in the qualitative properties of the solution of, and the inverse problems for, the Sturm-Liouville equation (see Chapter 8), and the discrete problems were studied because such problems were met in any approximate analysis of Sturm-Liouville problems. Krein's early papers (1933, 1934) concern the theory of Sturm sequences, while the Supplement to Gantmakher and Krein (1950), and Krein (1952) make use of the theory of continued fractions developed by Stieltjes (1918). Krein sees his results as giving mechanical interpretations of Stieltjes' analysis.

First consider the very simplest inverse problem shown in Figure 4.1.1. If m_1, k_1,k_2 were given, then the analysis of Chapter 2 shows us how we could find the single natural frequency of each of the two systems. In the inverse problem, data and results are interchanged. Thus certain natural frequencies are assumed to be known, and it is required to find m_1:k_1:k_2. Clearly, since three quantities are to be found, it will be necessary to know more than one natural frequency. That is why *two* possible end conditions are shown in Figure 4.1.1. It is also clear that if values

Figure 4.1.1 – The simplest spring-mass system

of m_1, k_1, k_2 have been found which yield a set of specified natural frequencies, then am_1, ak_1, ak_2 will be another solution for any positive constant a. This lack of uniqueness may be eliminated by specifying either the total mass, m_1, or the stiffness k_1, or even the total stiffness k of system of Figure 4.1.1b given by

$$\frac{1}{k}=\frac{1}{k_1}+\frac{1}{k_2} . \tag{4.1.1}$$

We shall now show that, with one of these totals and the natural frequencies of systems (a) and (b), m_1, k_1, k_2 may be found uniquely.

Suppose that the system is vibrating freely with frequency ω, and $\omega^2=\lambda$. The equation governing system (a) is

$$-m_1\lambda y_1=-k_1y_1 , \tag{4.1.2}$$

so that the single (natural frequency)^2 is

$$\omega_1^2=\lambda_1=k_1/m_1 . \tag{4.1.3}$$

The equation governing (b) is similarly

$$-m_1\lambda y_1=-k_1y_1-k_2y_1 , \tag{4.1.4}$$

so that its (natural frequency)^2 is

$$\omega_1^{*2}=\lambda_1^*=(k_1+k_2)/m_1 . \tag{4.1.5}$$

Thus if ω_1,ω_1^* are given and satisfy $0<\omega_1<\omega_1^*$, (i.e., $0<\lambda_1<\lambda_1^*$) then equations (4.1.3), (4.1.5) determine $m_1:k_1:k_2$ uniquely, in fact

$$m_1 : k_1 : k_2=1 : \lambda_1 : \lambda_1^*-\lambda_1 . \tag{4.1.6}$$

In this chapter we shall not follow the historical development of the subject. Instead we shall consider an inverse problem for a Jacobian matrix, and then show that, once it is solved, all the important inverse problems for discrete second-order systems can also be solved.

4.2 An inverse problem for a Jacobian matrix

It was shown in Section 2.2 that the (natural frequencies)^2 of a lumped mass system may be obtained as the eigenvalues of a Jacobian matrix

$$\mathbf{B} = \begin{bmatrix} a_1 & -b_1 & 0 & \cdots & 0 & 0 \\ -b_1 & a_2 & -b_2 & \cdots & 0 & 0 \\ \cdot & \cdot & \cdot & \cdot & \cdot & \cdot \\ 0 & 0 & 0 & \cdots & a_{N-1} & -b_{N-1} \\ 0 & 0 & 0 & \cdots & -b_{N-1} & a_N \end{bmatrix}. \tag{4.2.1}$$

It will be assumed that the co-diagonal elements, $-b_1$ are *all of the same sign* which, for the sake of argument, will be taken to be *strictly negative.*

The basic theorem is

Theorem 4.2.1. There is a unique Jacobian matrix **B** having specified eigenvalues $(\lambda_i)_1^N$, where

$$\lambda_1 < \lambda_2 < \cdots < \lambda_N \ ,$$

and with normalized eigenvectors $\mathbf{u}^{(i)}$ having specified values of $(u_1^{(i)})_1^N$ or of $(u_N^{(i)})_1^N$.

Proof

This theorem is at once an existence (there *is* ...) and a uniqueness theorem (...a unique...). We shall prove existence by actually constructing a matrix, and will do so by using the so-called Lanczos algorithm. This algorithm has the advantage that numerically it is well conditioned. An independent proof that the matrix is unique will be left to the reader.

The proof will be presented for the case in which the $(u_1^{(i)})_1^N$ are specified.

The eigenvectors $\mathbf{u}^{(i)}$ satisfy

$$\mathbf{B}\mathbf{u}^{(i)} = \lambda_i \mathbf{u}^{(i)} \ . \tag{4.2.2}$$

Put $\mathbf{U} = [\mathbf{u}^{(1)}, \mathbf{u}^{(2)}, \ldots, \mathbf{u}^{(N)}]$, then the orthogonality condition

$$\mathbf{u}^{(i)T}\mathbf{u}^{(j)} = \delta_{ij} \ , \tag{4.2.3}$$

yields

$$\mathbf{U}^T\mathbf{U} = \mathbf{I} \ . \tag{4.2.4}$$

But, therefore

$$\mathbf{U}\mathbf{U}^T = \mathbf{I} \ , \tag{4.2.5}$$

so that

$$\sum_{i=1}^{N} u_r^{(i)} u_s^{(i)} = \delta_{rs} . \tag{4.2.6}$$

This is conveniently written by introducing

$$\mathbf{x}^{(r)} = \{u_r^{(1)}, u_r^{(2)}, \ldots, u_r^{N}\} , \tag{4.2.7}$$

for then

$$\mathbf{x}^{(r)T} \mathbf{x}^{(s)} = \delta_{rs} . \tag{4.2.8}$$

The set of equations (4.2.2) for $i = 1,2,...,N$ may be written

$$\mathbf{BU} = \mathbf{U}\Lambda , \tag{4.2.9}$$

or equivalently

$$\mathbf{XB} = \Lambda\mathbf{X} , \tag{4.2.10}$$

where $\mathbf{X} = [\mathbf{x}^{(1)}, \mathbf{x}^{(2)}, \ldots, \mathbf{x}^{(N)}] = \mathbf{U}^T$. Thus

$$[\mathbf{x}^{(1)} \mathbf{x}^{(2)} \cdots \mathbf{x}^{(N)}] \begin{bmatrix} a_1 & -b_1 & & & \\ -b_1 & a_2 & -b_2 & & \\ \cdot & \cdot & \cdot & \cdot & \cdot \\ & & & a_{N-1} & -b_{N-1} \\ & & & -b_{N-1} & a_N \end{bmatrix} = \Lambda [\mathbf{x}^{(1)} \mathbf{x}^{(2)} \cdots \mathbf{x}^{(N)}] . \tag{4.2.11}$$

The first column of this equation is

$$a_1 \mathbf{x}^{(1)} - b_1 \mathbf{x}^{(2)} = \Lambda \mathbf{x}^{(1)}, \tag{4.2.12}$$

so that if $\mathbf{x}^{(1)T} \mathbf{x}^{(2)}$ is to be zero, then

$$a_1 = \mathbf{x}^{(1)T} \Lambda \mathbf{x}^{(1)} . \tag{4.2.13}$$

Put

$$\mathbf{d}^{(2)} = a_1 \mathbf{x}^{(1)} - \Lambda \mathbf{x}^{(1)} , \tag{4.2.14}$$

and use the notation of equation (1.2.31). If $||\mathbf{x}^{(2)}||_2 = 1$, b_1 must be chosen so that

$$b_1 = ||\mathbf{d}^{(2)}||_2 , \tag{4.2.15}$$

and then

$$\mathbf{x}^{(2)} = \mathbf{d}^{(2)} / b_1 . \tag{4.2.16}$$

Having found a_1, b_1, we consider the ith column of equation (4.2.11); it is

$$-b_{i-1} \mathbf{x}^{(i-1)} + a_i \mathbf{x}^{(i)} - b_i \mathbf{x}^{(i+1)} = \Lambda \mathbf{x}^{(i)} . \tag{4.2.17}$$

At the previous stage, $\mathbf{x}^{(i)}$ was found so that $\mathbf{x}^{(i)T} \mathbf{x}^{(i-1)} = 0$.

Thus a_i must be chosen so that

$$a_i = \mathbf{x}^{(i)T}\Lambda\mathbf{x}^{(i)} , \tag{4.2.18}$$

then

$$\mathbf{d}^{(i+1)} = -b_{i-1}\mathbf{x}^{(i-1)} + a_i\mathbf{x}^{(i)} - \Lambda\mathbf{x}^{(i)} , \tag{4.2.19}$$

and

$$b_i = ||\mathbf{d}^{(i+1)}||_2 , \ \mathbf{x}^{(i+1)} = \mathbf{d}^{(i+1)}/b_i . \tag{4.2.20}$$

This procedure, carried out for $i = 2,...,N$ is called the *Lanczos* algorithm; see Lanczos (1950), Golub (1973), Golub and Van Loan (1983). It produces a matrix **B** and, at the same time, constructs the eigenvectors of **B**.

Examples 4.2

1. Show that the vectors $\mathbf{x}^{(i)}$ constructed in the Lanczos algorithm satisfy

$$\mathbf{x}^{(i)T}\mathbf{x}^{(j)} = \delta_{ij} \quad , \quad i,j = 1,2,...,N ,$$

even though this orthogonality is apparently established only for $|i-j| \leq 1$.

4.3 Variants of the inverse problem for a Jacobian matrix

A matrix obtained by deleting the first (or last) row and column of **B** will be called a truncated matrix, and will be denoted by $\mathbf{B}^o$. Let $(\lambda_i)_1^N$ and $(\lambda_i^o)_1^{N-1}$ be the eigenvalues of **B** and $\mathbf{B}^o$ respectively. Since **B** and $\mathbf{B}^o$ are Jacobian, the eigenvalues satisfy

$$\lambda_1 < \lambda_2 < \cdots < \lambda_N , \tag{4.3.1}$$

$$\lambda_1^o < \lambda_2^o < \cdots < \lambda_{N-1}^o . \tag{4.3.2}$$

Since the eigenvalues λ_i^o are obtained by imposing the constraint $q_1 = 0 (q_N = 0)$ on the eigenproblem for **B**, the theory of Section 2.8 shows that they will interlace the λ_i, i.e.,

$$\lambda_1 < \lambda_1^o < \lambda_2 < \lambda_2^o \cdots < \lambda_{N-1} < \lambda_{N-1}^o < \lambda_N . \tag{4.3.3}$$

One of the early inverse problems to be studied was the reconstruction of the matrix **B** such that **B** and $\mathbf{B}^o$ had eigenvalues $(\lambda_i)_1^N$ and $(\lambda_1^o)_1^{N-1}$ respectively. The problem seems to have been studied first by Hochstadt (1967). He proved that there is *at most* one matrix **B** with the required property. Hochstadt (1974) attempted to construct this unique matrix, but did not show that his method would always lead to *real* values of the co-diagonal elements b_i. Hald (1976) presented another construction and showed that, in theory, it would always work provided that the eigenvalues satisfy the interlacing condition (4.3.3). In practice the construction was found to break down due to loss of significant figures. Hald also showed that Hochstadt's construction *will* lead to real b_i provided that (4.3.3) is satisfied. Gray and Wilson (1976) presented an alternative, inductive construction of the required matrix **B**. An independent uniqueness proof was given by Hald (1976).

In this section we shall present two methods for constructing **B**. The first relies on the theory of orthogonal polynomials described in Section 3.2, while the second, which can be generalized to inverse problems for truncated band matrices, reduces the problem to the inverse problem described in Section 4.2.

The first method is best described by supposing that $\mathbf{B}^o$ is obtained from **B** by deleting the *last* row and column, not the first. Two spectra $(\lambda_i)_1^N$ and $(\lambda_i^o)_1^{N-1}$ are given, and satisfy the conditions (4.3.1) - (4.3.3). Two polynomials

$$p_N(\lambda)=\prod_{i=1}^{N}(\lambda-\lambda_i)\ ,\ \ p_{N-1}(\lambda)=\prod_{i=1}^{N-1}(\lambda-\lambda_i^o) \tag{4.3.4}$$

are formed. These polynomials are the Nth and $(N-1)$th monic polynomials of the sequence of orthogonal polynomials with weights given by equation (3.2.29), i.e.,

$$w_i=\gamma/\{p_{N-1}(\lambda_i)p_N'(\lambda_i)\}\ , \tag{4.3.5}$$

and points $(\lambda_i)_1^N$. In addition they are the N-th and $(N-1)$th principal minors of the required Jacobian matrix **B**. The polynomials $p_r(\lambda)$ therefore satisfy

$$p_r(\lambda)=(\lambda-a_r)p_{r-1}(\lambda)-b_{r-1}^2p_{r-2}(\lambda)\ . \tag{4.3.6}$$

Hald's method of reconstructing **B** is as follows: he starts from $p_N(\lambda)$, $p_{N-1}(\lambda)$, constructs $p_{N-2}(\lambda)$ and in the process finds a_N, b_{N-1}. Then from $p_{N-1}(\lambda)$, $p_{N-2}(\lambda)$ he constructs $p_{N-3}(\lambda)$ and finds a_{N-1}, b_{N-2} and so on. This process is inherently unstable because the polynomials $p_{N-2}, p_{N-3}, \ldots, p_1$ are found by successively cancelling the leading terms in the preceding pair of polynomials; the process becomes unstable because of cancellation of leading digits.

de Boor and Golub (1978) proceed quite differently. Having found the weights w_i by using (4.3.5), they construct the polynomials in the natural order by using the analysis of Section 3.2, i.e.,

$$p_{-1}(\lambda)\equiv 0\ \ ,\ \ p_o(\lambda)=1\ , \tag{4.3.7}$$

$$p_r(\lambda)=(\lambda-a_r)p_{r-1}(\lambda)-b_{r-1}^2p_{r-2}(\lambda)\ ,\ \ r=1,2,\ldots,N\ , \tag{4.3.8}$$

with the numbers a_r, b_r computed by

$$a_r=\frac{<p_{r-1},\lambda p_{r-1}>}{||p_{r-1}||^2}\ ,\ b_r=\frac{||p_r||}{||p_{r-1}||}\ ,\ r=1,2,..,N\ . \tag{4.3.9}$$

This process is numerically stable.

The only major difficulty encountered by de Boor and Golub lay in the computation of the weights w_i. In seeking to overcome this difficulty they used the reflection of **B** about its second diagonal. The matrix

$$\mathbf{T} = \begin{bmatrix} 0 & 0 & \cdots & 0 & 1 \\ 0 & 0 & \cdots & 1 & 0 \\ \cdot & \cdot & \cdot & \cdot & \cdot \\ 1 & 0 & \cdots & 0 & 0 \end{bmatrix} \tag{4.3.10}$$

is orthogonal and symmetric, so that $\mathbf{T}^2=\mathbf{I}$. It transforms $\mathbf{B}$ into

$$\overline{\mathbf{B}}=\mathbf{TBT}=\begin{bmatrix} a_N & -b_{N-1} & 0 & \cdots & & 0 \\ -b_{N-1} & a_{N-1} & -b_{N-2} & \cdots & & 0 \\ \cdots & \cdots & \cdots & \cdots & \cdots & \cdots \\ 0 & \cdots & & -b_2 & a_2 & -b_1 \\ 0 & \cdots & & & -b_1 & a_1 \end{bmatrix}. \tag{4.3.11}$$

If therefore the elements of $\overline{\mathbf{B}}$ are denoted by $\bar{a}_r$, $\bar{b}_r$, then

$$\bar{a}_r=a_{N+1-r}\ , \quad \bar{b}_r=b_{N-r}\ . \tag{4.3.12}$$

The matrix $\mathbf{B}$ is said to be persymmetric if

$$\mathbf{B}=\overline{\mathbf{B}}\ , \tag{4.3.13}$$

i.e.,

$$a_r=a_{N+1-r}\ , \quad b_r=b_{N-r}\ . \tag{4.3.14}$$

The $(N-1)$th order minor of $\lambda\mathbf{I}-\mathbf{B}$ is $p_{N-1}(\lambda)$: Let the corresponding minor of $\lambda\mathbf{I}-\overline{\mathbf{B}}$ be denoted by $\bar{p}_{N-1}(\lambda)$. We prove:

Theorem 4.2.1. For $i=1,2,\ldots,N$

$$p_{N-1}(\lambda_i)\bar{p}_{N-1}(\lambda_i)=(b_1b_2\cdots b_{N-1})^2=b^2\ . \tag{4.3.15}$$

Proof:

$p_{N-1}(\lambda_i)\bar{p}_{N-1}(\lambda_i)$ is the product of the $(N-1)$th order left principal minor with the $(N-1)$th right principal minor of the singular matrix

$$\mathbf{A}=\lambda_i\mathbf{I}-\mathbf{B}\ , \tag{4.3.16}$$

i.e.,

$$p_{N-1}(\lambda_i)=A\begin{pmatrix}1,2,\ldots,N-1\\1,2,\ldots,N-1\end{pmatrix}, \quad \bar{p}_{N-1}(\lambda_i)=A\begin{pmatrix}2,3,\ldots,N\\2,3,\ldots,N\end{pmatrix}. \tag{4.3.17}$$

Now using Sylvester's identity (Corollary 2 of Theorem 5.2.2), with

$$A\begin{pmatrix}2,3,\ldots,N-1\\2,3,\ldots,N-1\end{pmatrix}$$

as pivotal block, we obtain

$$0=\left|A\begin{pmatrix}2,3,\ldots,N-1\\2,3,\ldots,N-1\end{pmatrix}\right|\cdot|\mathbf{A}|=\begin{vmatrix}A\begin{matrix}1,2,\ldots,N-1\\1,2,\ldots,N-1\end{matrix} & A\begin{matrix}1,2,\ldots,N-1\\2,3,\ldots,N\end{matrix}\\ A\begin{matrix}2,3,\ldots,N\\1,2,\ldots N-1\end{matrix} & A\begin{matrix}2,3,\ldots,N\\2,3,\ldots,N\end{matrix}\end{vmatrix} \tag{4.3.18}$$

i.e.,

$$0=p_{N-1}(\lambda_i)\bar{p}_{N-1}(\lambda_i)-(b_1b_2\cdots b_{N-1})^2\ . \ \blacksquare \tag{4.3.19}$$

This result means that the polynomials $\bar{p}_N(\lambda)$ $(=p_N(\lambda))$, $\bar{p}_{N-1}(\lambda),\ldots,$ $\bar{p}_1(\lambda),1$ are the monic polynomials related to the weights

$$w_i=\frac{b^2}{\bar{p}_{N-1}(\lambda_i)\bar{p}_N'(\lambda_i)}=\frac{p_{N-1}(\lambda_i)}{p_N'(\lambda_i)}\ . \tag{4.3.20}$$

These weights may be more easily constructed than those in (4.3.5). In this procedure the terms of the matrix **B** are computed in the order $\bar{a}_1,\bar{b}_1,\ldots,\bar{b}_{N-1},\bar{a}_N$, i.e., a_N, $b_{N-1},\ldots,b_1,a_1$.

The second method of constructing **B** is due to Golub and Boley (1977). It relies on the fact that, once we know the eigenvalues $(\lambda_i)_1^N$ and $(\lambda_i^o)_1^{N-1}$, we may compute the quantities $u_N^{(i)}$, i.e., the vector $\mathbf{x}^{(N)}$, needed for the inversion procedure of Section 4.2. (Barcilon (1978) concentrated on the eigen *vectors* corresponding to λ_i and λ_i^o, rather than using the λ_i^o to find the quantities $u_N^{(i)}$; his subsequent analysis did not lend itself to computation.)

The eigenvalues of $\mathbf{B}^o$ are the stationary values of $\mathbf{u}^T\mathbf{B}\mathbf{u}$ subject to $\mathbf{u}^T\mathbf{u}=1$ and the constraint $u_N=0$. Thus they are the stationary values of

$$f=\mathbf{u}^T\mathbf{B}\mathbf{u}-\lambda\mathbf{u}^T\mathbf{u}-2\nu\mathbf{u}^T\mathbf{e}_N\ , \tag{4.3.21}$$

where $\mathbf{e}_N=\{0,0,\ldots,1\}$, and λ, ν are Lagrange parameters. The condition that f be stationary yields

$$\mathbf{B}\mathbf{u}-\lambda\mathbf{u}-\nu\mathbf{e}_N=0\ . \tag{4.3.22}$$

Since the eigenvectors $\mathbf{u}^{(i)}$ span the space of N-vector we may write

$$\mathbf{u}=\sum_{i=1}^{N}\alpha_i\mathbf{u}^{(i)}\ , \tag{4.3.23}$$

then

$$\mathbf{B}\mathbf{u}=\sum_{i=1}^{N}\lambda_i\alpha_i\mathbf{u}^{(i)}\ , \tag{4.3.24}$$

so that

$$\sum_{i=1}^{N}(\lambda_i-\lambda)\alpha_i\mathbf{u}^{(i)}-\nu\mathbf{e}_N=0\ , \tag{4.3.25}$$

and the orthogonality condition $\mathbf{u}^{(i)T}\mathbf{u}^{(j)}=\delta_{ij}$ gives

$$(\lambda_i-\lambda)\alpha_i=\nu\mathbf{u}^{(i)T}\mathbf{e}_N=\nu u_N^{(i)}\ , \tag{4.3.26}$$

so that

$$\mathbf{u}=\nu\sum_{i=1}^{N}\frac{u_N^{(i)}\mathbf{u}^{(i)}}{\lambda_i-\lambda}\ , \tag{4.3.27}$$

and the condition $u_N=0$ yields the eigenvalue equation

$$\sum_{i=1}^{N}\frac{[u_N^{(i)}]^2}{\lambda_i-\lambda}=0. \tag{4.3.28}$$

Since the matrix **B** is Jacobian, none of the coefficients $u_N^{(i)}$ will be zero. The analysis of Section 2.8 shows that the roots λ_i^o of this equation will satisfy

$$\lambda_1<\lambda_1^o<\lambda_2<\cdots<\lambda_{N-1}<\lambda_{N-1}^o<\lambda_N\ . \tag{4.3.29}$$

Since $\sum_{i=1}^{N}[u_N^{(i)}]^2=1$, we have the identity

$$\sum_{i=1}^{N}\frac{[u_N^{(i)}]^2}{\lambda_i-\lambda}=\frac{\prod_{i=1}^{N-1}(\lambda_i^o-\lambda)}{\prod_{i=1}^{N}(\lambda_i-\lambda)}\ , \tag{4.3.30}$$

and hence

$$[u_N^{(i)}]^2=\prod_{j=1}^{N-1}(\lambda_j^o-\lambda_i)/\prod_{j=1}^{N}{}'(\lambda_j-\lambda_i)\ , \tag{4.3.31}$$

where $'$ indicates that the term $j=i$ is omitted. This equation yields the values of $x_i^{(N)}=u_N^{(i)}$.

There is a third inverse problem which appears in a number of contexts. Given two strictly increasing sequences $(\lambda_i)_1^N$ and $(\lambda_i^{\bullet})_1^N$ with

$$\lambda_1<\lambda_1^{\bullet}<\lambda_2<\lambda_2^{\bullet}\cdots<\lambda_N<\lambda_N^{\bullet}\ , \tag{4.3.32}$$

determine an Nth order Jacobian matrix **B** which has $(\lambda_i)_1^N$ as its eigenvalues, and for which the matrix $\mathbf{B}^{\bullet}$ obtained by changing a_N to $a_N^{\bullet}$ has $(\lambda_i^{\bullet})_1^N$ as its eigenvalues. We shall show how the given data allows the determination of the elements $u_N^{(i)}$ for matrix **B**.

The quantities $(\lambda_i^{\bullet})_1^N$ are the eigenvalues of

$$\mathbf{B}^{\bullet}\mathbf{u}=\lambda\mathbf{u}\ . \tag{4.3.33}$$

Write

$$\mathbf{u}=\sum_{i=1}^{N}\alpha_i\mathbf{u}^{(i)}\ , \tag{4.3.34}$$

then

$$\mathbf{B}^{\bullet}\mathbf{u}=\mathbf{B}\mathbf{u}+(a_N^{\bullet}-a_N)u_N\mathbf{e}^{(N)}\ , \tag{4.3.35}$$

so that equation (4.3.33) becomes

$$\sum_{i=1}^{N}\lambda_i\alpha_i\mathbf{u}^{(i)}+(a_N^{\bullet}-a_N)u_N\mathbf{e}^{(N)}=\lambda\sum_{i=1}^{N}\alpha_i\mathbf{u}^{(i)}\ , \tag{4.3.36}$$

and therefore

$$(\lambda-\lambda_i)\alpha_i=(a_N^{\bullet}-a_N)u_Nu_N^{(i)}\ , \tag{4.3.37}$$

which when substituted into (4.3.34) yields

$$\mathbf{u}=(a_N^{\bullet}-a_N)u_N\sum_{i=1}^{N}\frac{u_N^{(i)}\mathbf{u}^{(i)}}{\lambda-\lambda_i}\ . \tag{4.3.38}$$

Equating the last components on each side of this equation, we obtain the eigenvalue equation

$$1=(a_N^{\bullet}-a_N)\sum_{i=1}^{N}\frac{[u_N^{(i)}]^2}{\lambda-\lambda_i}\ . \tag{4.3.39}$$

The roots of this equation are $(\lambda_i^{\bullet})_1^N$, so that

$$1-(a_N^{\bullet}-a_N)\sum_{i=1}^{N}\frac{[u_N^{(i)}]^2}{\lambda-\lambda_i}=\prod_{i=1}^{N}\left(\frac{\lambda-\lambda_i^{\bullet}}{\lambda-\lambda_i}\right)\ , \tag{4.3.40}$$

and therefore

$$(a_N^{\bullet}-a_N)[u_N^{(i)}]^2=(\lambda_i^{\bullet}-\lambda_i)\prod_{j=1}^{N}{}'\left(\frac{\lambda_i-\lambda_j^{\bullet}}{\lambda_i-\lambda_j}\right)\ . \tag{4.3.41}$$

Now by comparing the traces of **B** and $\mathbf{B}^{\bullet}$ we see that

$$(a_N^{\bullet}-a_N)=\sum_{j=1}^{N}(\lambda_j^{\bullet}-\lambda_j)>0\ . \tag{4.3.42}$$

Thus the $[u_N^{(i)}]^2$ are expressed in terms of the data, and the interlacing condition (4.3.32) ensures that $[u_N^{(i)}]^2$ is positive. This problem may also be solved by using the theory of orthogonal polynomials.

The final inversion problem concerns the reconstruction of a persymmetric matrix **B**, i.e., one satisfying equations (4.3.13), (4.3.14). We prove

Theorem 4.3.2. There is a unique persymmetric Jacobian matrix **B** with given eigenvalues $(\lambda_i)_1^N$ satisfying $\lambda_1<\lambda_2<\cdots<\lambda_N$.

Proof. The simplest method of proof is perhaps to show that if the eigenvalues $(\lambda_i)_1^N$ are known, then it is possible to find the weights for the construction of the orthogonal polynomials $p_r(\lambda)$. Indeed if **B** is persymmetric then the minor $p_r(\lambda)$ is equal to $\bar{p}_r(\lambda)$. But then Theorem 4.3.1 shows that

$$[p_{N-1}(\lambda_i)]^2 = b^2 \ , \ \ i.e., p_{N-1}(\lambda_i) = \pm b \ , \tag{4.3.43}$$

so that equation (4.3.20) yields

$$w_i = \pm b / p_N'(\lambda_i) \ . \tag{4.3.44}$$

Since the signs of $p_N'(\lambda_i)$ will alternate, then so must the signs of b if the w_i are to be positive. The magnitude of b is irrelevant to the construction of the $p_r(\lambda)$. See Hochstadt (1979) for another variant of the inverse eigenvalue problem.

Examples 4.3

1. Verify that the $q_N^{(i)}$ computed from equation (4.3.41) do satisfy

$$\sum_{i=1}^{N} [q_N^{(i)}]^2 = 1 \ .$$

4.4 Inverse eigenvalue problems for spring-mass systems

We now return to physical problems and will show how the analysis of Sections 4.2-4.3 enables us to reconstruct spring-mass systems from natural frequency data.

First consider the system shown in Figure 4.4.1. This is the one considered by Gantmakher and Krein (1950). The matrix equations for systems (a) and (b) are respectively

$$\begin{bmatrix} k_1+k_2 & -k_2 & & \\ -k_2 & k_2+k_3 & -k_3 & \\ \cdot & \cdot & \cdot & \cdot \\ & & k_{N-1}+k_N & -k_N \\ & & -k_N & k_N \end{bmatrix} \begin{bmatrix} x_1 \\ x_2 \\ \cdot \\ x_{N-1} \\ x_N \end{bmatrix} = \lambda \begin{bmatrix} m_1 & & & \\ & m_2 & & \\ & & \cdot & \\ & & & \\ & & & m_N \end{bmatrix} \begin{bmatrix} x_1 \\ x_2 \\ \cdot \\ x_{N-1} \\ x_N \end{bmatrix} \tag{4.4.1}$$

and

$$\begin{bmatrix} k_1+k_2 & -k_2 & & \\ -k_2 & k_2+k_3 & -k_3 & \\ \cdot & \cdot & \cdot & \cdot \\ & & k_{N-1}+k_N & -k_N \\ & & -k_N & k_N+k_{N+1} \end{bmatrix} \begin{bmatrix} x_1 \\ x_2 \\ \cdot \\ x_{N-1} \\ x_N \end{bmatrix} = \lambda \begin{bmatrix} m_1 & & & \\ & m_2 & & \\ & & \cdot & \\ & & & \\ & & & m_N \end{bmatrix} \begin{bmatrix} x_1 \\ x_2 \\ \cdot \\ x_{N-1} \\ x_N \end{bmatrix} \tag{4.4.2}$$

These may be written

$$(\mathbf{C}-\lambda\mathbf{A})\mathbf{x}=0 \text{ and } (\mathbf{C}^{\bullet}-\lambda A)\mathbf{x}=0 . \tag{4.4.3}$$

As in Section 3.1, write

$$\mathbf{A}=\mathbf{M}^{1/2}\mathbf{M}^{1/2} , \ \mathbf{x}=\mathbf{M}^{-1/2}\mathbf{u} , \tag{4.4.4}$$

to obtain

$$(\mathbf{B}-\lambda\mathbf{I})\mathbf{u}=0 , \ (\mathbf{B}^{\bullet}-\lambda\mathbf{I})\mathbf{u}=0 , \tag{4.4.5}$$

where

$$\mathbf{B}=\mathbf{M}^{-1/2}\mathbf{C}\mathbf{M}^{-1/2} , \ \mathbf{B}^{\bullet}=\mathbf{M}^{-1/2}\mathbf{C}^{\bullet}\mathbf{M}^{-1/2} . \tag{4.4.6}$$

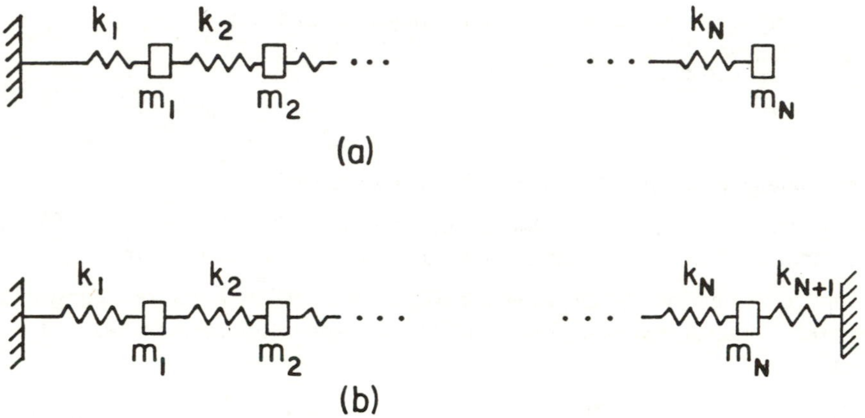

Figure 4.4.1 – The end of the system is a) free and b) attached by a spring to a fixed point.

We note that $\mathbf{B}^{\bullet}$ differs from $\mathbf{B}$ only in its last diagonal element, i.e.,

$$a_N=\frac{k_N}{m_N} \quad , \ a_N^{\bullet}=\frac{k_N+k_{N+1}}{m_N} . \tag{4.4.7}$$

This is an example of problem III of Section 4.3. The necessary and sufficient conditions for there to exist an actual physical model (with positive stiffnesses) is that given natural frequencies satisfy (4.3.33), i.e.,

$$0<\lambda_1<\lambda_1^{\bullet}<\lambda_2< \cdots <\lambda_{N-1}^{\bullet}<\lambda_N<\lambda_N^{\bullet} . \tag{4.4.8}$$

The steps in the solution are then:

(1) Use equation (4.3.42), to find k_{N+1}/m_N.

(2) Use equations (4.3.41), (4.3.42) to find $u_N^{(i)}$, $i=1,2,\ldots,N$.

(3) Use the Lanczos algorithm to find **B**.

(4) Unravel **B** to find the masses and stiffnesses. There remains only 4 to be discussed.

Consider the static behaviour of system (a). If a concentrated force $f_1 = k_1$ is applied to mass m_1 then only spring k_1 will be stretched, and that by unity. Since the right hand end is free, all the displacements of the masses will be unity, i.e.,

$$\mathbf{x} = \{1,1,1,...,1\} . \tag{4.4.9}$$

Equation (2.2.10) shows that the displacements are given by

$$\mathbf{Cx} = \{k_1, 0, ..., 0, 0\} . \tag{4.4.10}$$

But this equation is equivalent to

$$\mathbf{Bu} = \{m_1^{-1/2} k_1, \ldots, 0, 0\} , \tag{4.4.11}$$

and, according to (4.4.9) its solution must be

$$\mathbf{u} = \mathbf{M}^{1/2}\mathbf{x} = \{m_1^{1/2}, m_2^{1/2}, \ldots, m_N^{1/2}\} . \tag{4.4.12}$$

But starting from any assumed value $u_N = m_N^{1/2}$ we may solve the tridiagional set of equations (4.4.11), without knowing the value of $m_1^{-1/2}k_1$, and so deduce the remaining terms $m_r^{1/2}$ $(r = 1,2,...,N-1)$. Now knowing the total mass m, we may compute all the m_r.

The tridiagonal matrix

$$C = \mathbf{M}^{1/2}\mathbf{B}\mathbf{M}^{1/2} , \tag{4.4.13}$$

now satisfies

$$\mathbf{C}\begin{bmatrix} 1 \\ 1 \\ \cdot \\ 1 \end{bmatrix} = \begin{bmatrix} k_1 \\ 0 \\ \cdot \\ 0 \end{bmatrix} . \tag{4.4.14}$$

But (Ex. 4.4.1), this implies that, according to (2.2.17),

$$\mathbf{E}^{-1}\mathbf{C}\mathbf{E}^{-T} = \mathbf{K} \tag{4.4.15}$$

is the required diagonal matrix of stiffnesses. This complete the reconstruction of the system.

Now consider the following problem. With the right hand end *free*, the system of Figure 4.4.2(a) has (natural frequencies)^2 $(\lambda_i)_1^N$, while with the right hand end fixed, as in Figure 4.4.2(b), the (natural frequencies)^2 are $(\lambda_i^o)_1^{N-1}$. Reconstruct the system.

With the end free, the governing equation is (4.4.5), i.e.,

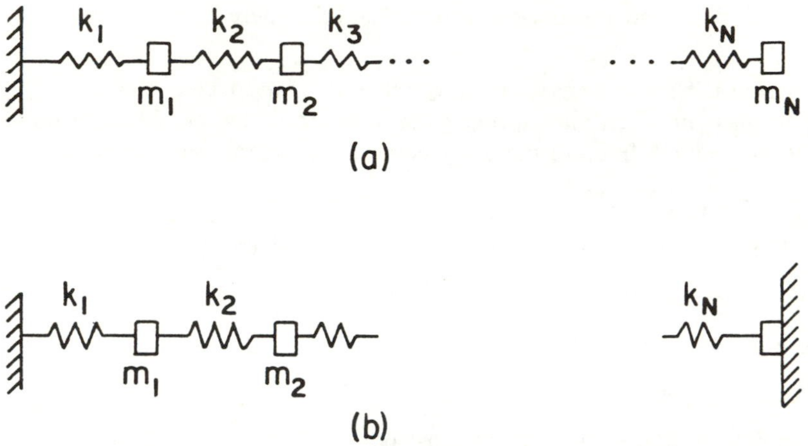

Figure 4.4.2 – The end of the system is either a) free or b) fixed.

$$(\mathbf{B}-\lambda\mathbf{I})\mathbf{u}=0 \ , \tag{4.4.16}$$

while with the end fixed, so that $u_N=0$, it is

$$(\mathbf{B}^o-\lambda\mathbf{I})\mathbf{u}^o=0 \ , \tag{4.4.17}$$

where $\mathbf{B}^o$ is obtained from $\mathbf{B}$ by deleting the last row and column. The necessary and sufficient condition for the existence of a real system, with positive sitffnesses, having the given eigenvalues, is, according to equations (4.3.1) - (4.3.3),

$$0<\lambda_1<\lambda_1^o<\lambda_2< \cdots <\lambda_{N-1}<\lambda_{N-1}^o<\lambda_N \ . \tag{4.4.18}$$

The steps in the solution are:

(1) Use equation (4.3.31) to find $u_N^{(i)}$, $i=1,2,...,N$.

(2) Use the Lanczos algorithm to find $\mathbf{B}$.

(3) Use the analysis of equations (4.4.9) - (4.4.15) to find the m_r and k_r.

It was shown in Section 4.3 that a persymmetric Jacobian matrix $\mathbf{B}$ can be reconstructed uniquely from its eigenvalues. We shall now consider a physical problem relating to a persymmetric matrix. Figure 4.4.3 shows a system of $2N$ masses connected by $(2N+1)$ springs and fixed at each end. Suppose that the system is *symmetrical* about the mid point, so that

$$m_r=m_{2N+1-r} \ , \quad k_r=k_{2N+1-r} \ , \quad (r=1,2,...,N) \ . \tag{4.4.18}$$

The odd numbered principal modes of the system will be symmetrical about the mid-point; they will thus be the principal modes of one half (say the left-hand half) of the system with the mid-point of the system free, as in Figure 4.4.4(a). Thus the odd numbered eigenvalues $\Lambda_1,\Lambda_3, \ldots ,\Lambda_{2N-1}$ of the complete system will be the eigenvalues of the left-hand half under the conditions fixed-free, i.e.,

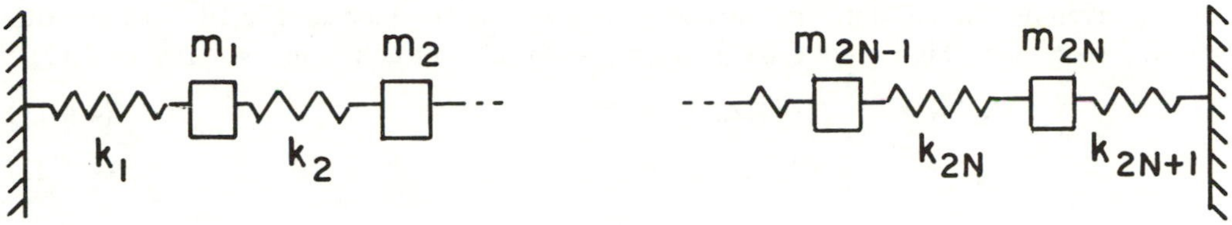

Figure 4.4.3 – A symmetrical system with 2N masses

$$\Lambda_{2i-1}=\lambda_i \ , \quad i=1,2,...,N \ . \tag{4.4.19}$$

On the other hand, the even-numbered principal modes of the system will be antisymmetrical about the mid-point so that the even-numbered eigenvalues $\Lambda_2,\Lambda_4,\ldots,\Lambda_{2N}$ will be the eigenvalues of the left-hand half under the conditions fixed-fixed, as in Figure 4.4.4(b).

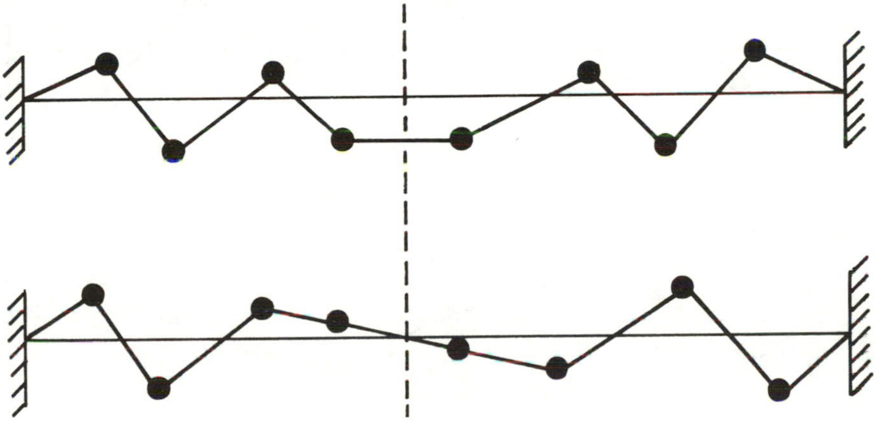

Figure 4.4.4 – The odd numbered modes are symmetrical, while the even numbered ones are antisymmetrical.

Thus

$$\Lambda_{2i}=\lambda_i^* \ , \quad i=1,2,...,N \ . \tag{4.4.20}$$

This means that the whole system may be uniquely constructed, using the analysis of equations (4.4.1) - (4.4.15) from the eigenvalues $\Lambda_1,\ldots,\Lambda_{2N}$ and the total mass.

Figure 4.4.5 shows a symmetrical system with $2N-1$ masses and $2N$ springs. Now the odd-numbered symmetrical modes will be the modes of the left-hand half with $(m_N/2)$ at the end and free there, as in Figure 4.4.6(a). On the other hand, the even-numbered, antisymmetrical modes will be the modes of left-hand half with m_N fixed as in Figure 4.4.6(b). Thus in the notation of equations (4.4.16), (4.4.17)

$$\Lambda_{2i-1}=\lambda_i\ , \quad i=1,2,\ldots,N\ , \tag{4.4.21}$$

$$\Lambda_{2i}=\lambda_i^o\ , \quad i=1,2,\ldots,N-1\ . \tag{4.4.22}$$

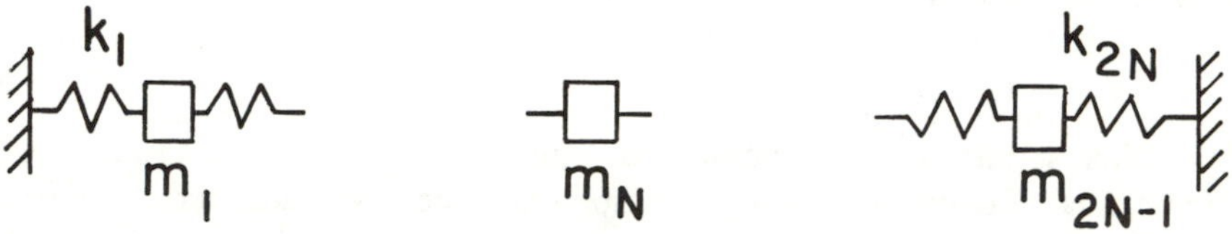

Figure 4.4.5 – A symmetrical system with $2N-1$ masses.

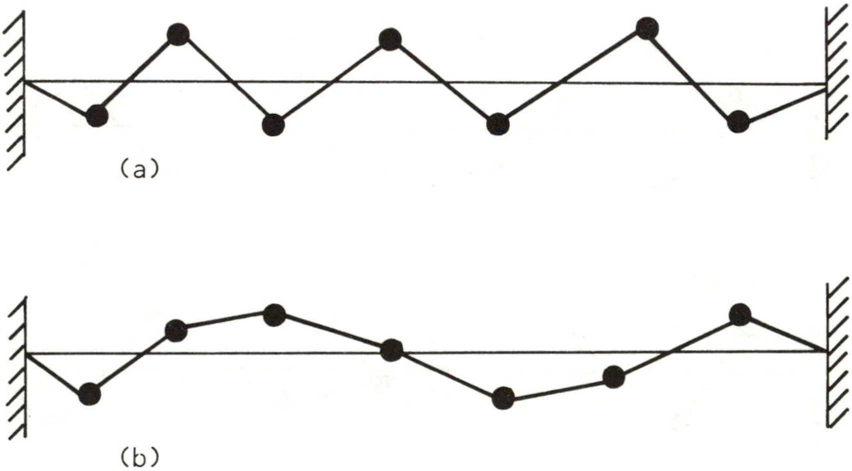

Figure 4.4.6 – a) The odd numbered modes are symmetrical. b) The even numbered modes are antisymmetrical.

As a final inversion problem, suppose that an oscillating force $F\sin\omega t$ is applied to the free end of the spring-mass system of Figure 4.4.1. The matrix equation governing the response $x\sin\omega t$ is

$$(C-\lambda\mathbf{A})\mathbf{x}=F\mathbf{e}_N\ . \tag{4.4.23}$$

Write

$$\mathbf{x}=\sum_{i=1}^{N}\alpha_i \mathbf{x}^{(i)} , \tag{4.4.24}$$

where $\mathbf{x}^{(i)}$ is the ith principal mode, normalized so that

$$\mathbf{x}^{(i)T}\mathbf{A}\mathbf{x}^{(j)}=\delta_{ij} . \tag{4.4.25}$$

We obtain

$$(\lambda_i-\lambda)\alpha_i=Fx_N^{(i)} , \tag{4.4.26}$$

and hence

$$\mathbf{x}=F\sum_{i=1}^{N}\frac{x_N^{(i)}\mathbf{x}^{(i)}}{(\lambda_i-\lambda)} , \tag{4.4.27}$$

so that

$$x_N=F\sum_{i=1}^{N}\frac{[x_N^{(i)}]^2}{\lambda_i-\lambda} . \tag{4.4.28}$$

This means that if the response $x_N(\lambda)/F$ is known as a function of λ, in particular if its poles $(\lambda_i)_1^N$ and zeros $(\mu_i)_1^{N-1}$ are known, then $x_N^{(i)}$ can be found and the system may be reconstructed.

An alternative procedure for inversion may be found in Gladwell and Gbadeyan (1985).

Examples 4.4

1. Write

$$\mathbf{C}=\begin{bmatrix} c_{11} & c_{12} & 0 & \cdots & 0 \\ c_{12} & c_{22} & c_{23} & \cdots & 0 \\ . & . & . & . & . \\ 0 & \cdot & \cdot & c_{N-1,N} & c_{NN} \end{bmatrix} .$$

   Show that if $\mathbf{C}$ satisfies equation (4.4.14) then $\mathbf{C}\mathbf{E}^{-T}$ is bi-diagonal and $\mathbf{E}^{-1}\mathbf{C}\mathbf{E}^{-T}$ is diagonal.

2. By evaluating the potential and kinetic energies of the systems in Figure 4.4.4(a) and 4.4.4(b), verify the assertions made in equations (4.4.19)-(4.4.22).

3. Suppose that the eigenvalues $(\lambda_n)_1^N$ of the system of Figure 4.4.2(a) are known, as are the eigenvalues $(\lambda_n^*)_1^N$ when the mass m_N replaced by some unknown mass m_N^*. Reconstruct the system and show that m_N^* may be found as a proportion of the total mass.

4. Suppose that the eigenvalues $(\lambda_n)_1^N$ of the system of Figure 4.4.1(b) are known, as are the eigenvalues $(\lambda_n^*)_1^N$ when the stiffness k_{N+1} is replaced by some unknown stiffness k_{N+1}^*. Show that there is a one-parameter family of systems, each member of which has the stated eigenvalues.

CHAPTER 5

FURTHER PROPERTIES OF MATRICES

There are then two kinds of intellect: the one able to penetrate acutely and deeply into the conclusions of given premises, and this is the precise intellect; the other able to comprehend a great number of premises without confusing them, and this is the mathematical intellect.

Pascal's Pensées

5.1 Introduction

The basic eigenvalue analysis of real symmetric matrices was discussed in Chapter 1. The eigenvalue properties described there are shared by all positive-definite (or semi-definite) matrices. This chapter, which may be missed on a first reading, provides proofs of some of the results which were used in Chapter 1. Foremost among these are Theorem 5.3.1, that a real symmetric matrix has N real eigenvectors which are orthonormal, and thus span the vector space; and Theorem 5.3.7 which provides necessary and sufficient conditions for the matrix **A** to be positive definite.

In Chapter 3 attention was focused on a narrower class, Jacobian (or tridiagonal) matrices, and it was found that they had additional eigenvalue properties, e.g., they had distinct eigenvalues and, with increasing i, the eigenvector $\mathbf{u}^{(i)}$ became increasingly oscillatory, meaning that there was an increasing number of sign changes among the elements $u_1^{(i)}$, $u_2^{(i)}$,...$u_N^{(i)}$. It will be shown in this chapter that many of the eigen-properties of such matrices are shared by a wider class of so-called oscillatory (or oscillation) matrices. Actually there are twin classes of matrices, *oscillatory* and *sign-oscillatory* (Section 5.5). If **A** is oscillatory then $\mathbf{A}^* = \mathbf{SAS}$ (see Ex. 3.3.2) is sign-oscillatory, and *vice versa*. The Jacobian matrix **B** of equation (3.1.9) is actually sign-oscillatory. These matrices were introduced and exhaustively studied by Gantmakher and Krein (1950). It will be fouond in Chapters 6 and 7 that the matrices appearing in lumped-mass or finite element models of strings, rods and beams are all oscillatory or sign-oscillatory: this chapter serves as reference material for the study of oscillatory matrices.

The theorem upon which the whole of the analysis of oscillatory matrices depends is Perron's theorem (Theorem 5.4.1). This relates to a *positive* matrix, one which has all positive coefficients, and states that such a matrix has one eigenvalue, the greatest in magnitude, which is real, simple and positive; the corresponding eigenvector has all its coefficients positive.

The matrices appearing in mechanics are usually not positive (such matrices appear in Economics and Operational Analysis), but *oscillatory* (Section 5.5). In order to apply Perron's theorem to such matrices, we need two essential steps. First (Ex. 5.5.3), if **A** is oscillatory, then $\mathbf{B}\equiv\mathbf{A}^{N-1}$ is *completely positive*. This term, which is introduced in Section 5.5, means that not only all the coefficients of **B**, but also all the *minors* (Section 5.1) of **B** are positive. Note that the eigenvalues of **B** are the $(N-1)$th powers, λ_i^{N-1}, of the eigenvalues of **A**, while its eigenvectors are the same as those of **A**. The other step which is needed is the introduction of the concept of an associated matrix (Section 5.4). The associated matrix $\check{\mathbf{A}}_p$ is formed from all the

$$\check{N}\equiv\binom{N}{p}$$

pth-order minors of **A**. The Binet Cauchy Theorem, Theorem 5.2.3, shows (Theorem 5.4.2) that the eigenvalues of $\check{\mathbf{A}}_p$ are simply products of p eigenvalues of **A**. The argument then runs as follows. Suppose **A** is oscillatory, then $\mathbf{B}\equiv\mathbf{A}^{N-1}$ is completely positive, and hence for each $p=1,2,\ldots,N$, $\check{\mathbf{B}}_p$ is positive. Now apply Perron's theorem to $\check{\mathbf{B}}_p$. The first conclusion (Theorem 5.7.1) is that the eigenvalues of **A** are positive and distinct. The second (Theorem 5.7.2) is that the eigenvectors, like the eigenvectors of the Jacobian matrices appearing in Chapter 3, become increasingly oscillatory as the eigenvalue number increases.

Before beginning the analysis proper, we point out a notational matter which must be understood if confusion is to be avoided. In Chapter 3, in dealing with the tridiagonal matrix **B** having *negative* off-diagonal elements (equation (3.1.9)) the eigenvalues were labelled in increasing order, i.e., $0<\lambda_1<\lambda_2<\cdots<\lambda_N$. The eigenvectors then satisfied Theorem 3.3.1. In Ex. 3.3.2 it was pointed out that if the eigenvalues of **B***, a tridiagonal matrix with positive off-diagonal elements (i.e. an osillatory matrix), are numbered in the order $\lambda_1>\lambda_2>\cdots>\lambda_N>0$, then the eigenvectors still satisfy Theorem 3.3.1. In this chapter, in dealing with oscillatory matrices, and **B*** is one, we shall keep this same ordering, i.e., $\lambda_1>\lambda_2>\cdots>\lambda_N>0$. Theorem 5.7.2 is a generalization of Theorem 3.3.1.

5.2 Minors

Let **A** be a square matrix of order N, with elements a_{ij}. The determinant constructed from the elements in p different rows and p different columns of **A** is called a *minor* of **A**. Thus if $N=3$, then

$$a_{13}=A\begin{pmatrix}1\\3\end{pmatrix}\ ,\quad \begin{vmatrix}a_{11} & a_{13}\\ a_{21} & a_{23}\end{vmatrix}=A\begin{pmatrix}1\ 2\\1\ 3\end{pmatrix}\ ,\quad |\mathbf{A}|=A\begin{pmatrix}1\ 2\ 3\\1\ 2\ 3\end{pmatrix}, \tag{5.2.1}$$

are minors of **A**. A general minor is denoted by

$$A\begin{pmatrix}i_1 i_2 \cdots i_p\\ k_1 k_2 \cdots k_p\end{pmatrix} = \begin{vmatrix} a_{i_1k_1} & a_{i_1k_2} & \cdots & a_{i_1k_p}\\ a_{i_2k_1} & a_{i_2k_2} & \cdots & a_{i_2k_p}\\ \cdot & \cdot & \cdot & \cdot\\ a_{i_pk_1} & a_{i_pk_2} & \cdots & a_{i_pk_p}\end{vmatrix}, \tag{5.2.2}$$

where $1\le i_1<i_2 \cdots <i_p\le N$, $1\le k_1<k_2< \cdots <k_p\le N$.

The *cofactor* of a_{ij}, introduced in Section 1.3, may be written

$$A_{ij}=(-)^{i+j}\hat{a}_{ij}\ ,\quad \hat{a}_{ij}=A\begin{pmatrix}1 & 2 \cdots & i-1 & i+1 & \cdots N\\ 2 & 2 \cdots & j-1 & j+1 & \cdots N\end{pmatrix} \tag{5.2.3}$$

$\hat{a}_{ij}$ will be called *the minor of* a_{ij}.

Theorem 5.2.1. Let $\hat{\mathbf{A}}=(a_{ij})_1^N$ then the minors of $\hat{\mathbf{A}}$ are given by

$$\hat{\mathbf{A}}\begin{pmatrix}i_1 & i_2 \cdots & i_p\\ k_1 & k_2 \cdots & k_p\end{pmatrix} = |\mathbf{A}|^{p-1}A\begin{pmatrix}j_1 & j_2 \cdots & j_{N-p}\\ \ell_1 & \ell_2 \cdots & \ell_{N-p}\end{pmatrix} \tag{5.2.4}$$

where $j_1<j_2 \cdots <j_{N-p}$ is the set of indices complementary to $i_1<i_2< \cdots i_p$ and $\ell_1<\ell_2< \cdots \ell_{N-p}$ to $k_1<k_2 \cdots <k_p$.

Proof:

Consider the theorem for

$$A\begin{pmatrix}1 & 2 \cdots p\\ 1 & 2 \cdots p\end{pmatrix};$$

the general case may be obtained by a suitable rearrangement of the rows and columns. Since $\hat{a}_{ij}=(-)^{i+j}A_{ij}$, we may write

$$A\begin{pmatrix}1 & 2 \cdots p\\ 1 & 2 \cdots p\end{pmatrix} = \begin{vmatrix}A_{11} & A_{12} & \cdots & A_{1p}\\ A_{21} & A_{22} & \cdots & A_{2p}\\ A_{p1} & A_{p2} & \cdots & A_{pp}\end{vmatrix}. \tag{5.2.5}$$

Multiplying this result by $|\mathbf{A}|$, and writing the determinant in (5.2.5) as that of an $N\times N$ matrix, we find

$$\hat{\mathbf{A}}\begin{pmatrix}1 & 2 & \cdots & p\\ 1 & 2 & \cdots & p\end{pmatrix}|\mathbf{A}| = \begin{vmatrix} A_{11} & \cdots & A_{1p} & A_{1p+1} & \cdots & A_{1N}\\ A_{21} & \cdots & A_{2p} & A_{2p+1} & \cdots & A_{2N}\\ \cdot & \cdot & \cdot & \cdot & \cdot & \cdot\\ A_{p1} & \cdots & A_{pp} & A_{pp+1} & \cdots & A_{pN}\\ 0 & \cdots & 0 & 1 & \cdots & 0\\ \cdot & \cdot & \cdot & \cdot & \cdot & \cdot\\ 0 & \cdots & 0 & 0 & \cdots & 1 \end{vmatrix} \cdot \begin{vmatrix} a_{11} & \cdots & a_{N1}\\ a_{12} & \cdots & a_{N2}\\ & & \\ & & \\ & & \\ & & \\ a_{1N} & \cdots & a_{NN} \end{vmatrix}$$

so that, on using (1.3.10), we obtain

$$\hat{A}\begin{pmatrix}1 & 2 & \cdots & p\\ 1 & 2 & \cdots & p\end{pmatrix}|\mathbf{A}| = \begin{vmatrix} |A| & 0 & \cdots & 0 & \cdots & 0\\ 0 & |A| & \cdots & 0 & \cdots & 0\\ \cdot & \cdot & \cdot & \cdot & \cdot & \cdot\\ 0 & 0 & |A| & 0 & & 0\\ a_{1,p+1} & a_{2,p+1} & \cdots & & & a_{N,p+1}\\ \cdot & \cdot & \cdot & \cdot & \cdot & \cdot\\ a_{1,n} & a_{2,n} & \cdots & & & a_{N,N} \end{vmatrix}$$

$$= |\mathbf{A}|^{p} A\begin{pmatrix}p+1 & p+2 & \cdots & N\\ p+1 & p+2 & \cdots & N\end{pmatrix} \tag{5.2.7}$$

so that the theorem holds when $|\mathbf{A}| \neq 0$. Continuity considerations show that the theorem holds also when $|\mathbf{A}| = 0$; this means that, when $|\mathbf{A}| = 0$, the rank of $\hat{\mathbf{A}}$ is at most 1. ■

Corollary. When $p = N$

$$\hat{A}\begin{pmatrix}1 & 2 & \cdots & N\\ 1 & 2 & \cdots & N\end{pmatrix} = |\hat{\mathbf{A}}| = |\mathbf{A}|^{N-1}. \tag{5.2.8}$$

Theorem 5.2.2. Let $\mathbf{A}=(a_{ij})_1^N$ be an arbitrary square matrix, and let $1 \leq p < N$. Put

$$b_{ik} = A\begin{pmatrix}1 & 2 & \cdots & p & i\\ 1 & 2 & \cdots & p & k\end{pmatrix}, \quad (i,k = p+1,\ldots,N)$$

and

$$\mathbf{B} = (b_{ik})_{p+1}^{N}$$

then

$$|\mathbf{B}| = B\begin{pmatrix}p+1 & \cdots & N\\ p+1 & \cdots & N\end{pmatrix} = A\begin{pmatrix}1 & 2 & \cdots & p\\ 1 & 2 & \cdots & p\end{pmatrix}^{N-p-1} A\begin{pmatrix}1 & 2 & \cdots & N\\ 1 & 2 & \cdots & N\end{pmatrix}. \tag{5.2.9}$$

Proof:

Theorem 5.2.1, with p replaced by $N-p-1$, shows that

$$c_{rs} = \hat{A}\begin{pmatrix} p+1 & \cdots & r-1 & r+1 & \cdots & N \\ p+1 & \cdots & s-1 & s+1 & \cdots & N \end{pmatrix} = |\mathbf{A}|^{N-p-2} A\begin{pmatrix} 1 & 2 & \cdots & p & r \\ 1 & 2 & \cdots & p & s \end{pmatrix}$$

$$= |\mathbf{A}|^{N-p-2} b_{rs} \,. \tag{5.2.10}$$

The corollary to Theorem 5.2.1 shows that if $\mathbf{C}=(c_{rs})_{p+1}^{N}$ then

$$|\mathbf{C}| = \left\{ \hat{A}\begin{pmatrix} p+1 & \cdots & N \\ p+1 & \cdots & N \end{pmatrix} \right\}^{N-p-1} . \tag{5.2.11}$$

But, according to (5.2.10),

$$|\mathbf{C}| = B\begin{pmatrix} p+1 & \cdots & N \\ p+1 & \cdots & N \end{pmatrix} |\mathbf{A}|^{(N-p-2)(N-p)} \tag{5.2.12}$$

and, from Theorem 5.2.1,

$$\hat{A}\begin{pmatrix} p+1 & \cdots & N \\ p+1 & \cdots & N \end{pmatrix} = |\mathbf{A}|^{N-p-1} A\begin{pmatrix} 1 & 2 & \cdots & p \\ 1 & 2 & \cdots & p \end{pmatrix} \tag{5.2.13}$$

so that, on substituting (5.2.13) into (5.2.11) we find

$$|\mathbf{C}| = |\mathbf{A}|^{(N-p-1)^2} A\begin{pmatrix} 1 & 2 & \cdots & p \\ 1 & 2 & \cdots & p \end{pmatrix}^{N-p-1} \tag{5.2.14}$$

which, on comparison with (5.2.12), yields the required result when $|\mathbf{A}| \neq 0$. Continuity considerations show that the theorem still holds when $|\mathbf{A}| = 0$. ■

Corollary 1

$$B\begin{pmatrix} i_1 & i_2 & \cdots & i_g \\ k_1 & k_2 & \cdots & k_g \end{pmatrix} = \left[A\begin{pmatrix} 1 & 2 & \cdots & p \\ 1 & 2 & \cdots & p \end{pmatrix} \right]^{g-1} A\begin{pmatrix} 1 & 2 & \cdots & p & i_1 & \cdots & i_g \\ 1 & 2 & \cdots & p & k_1 & \cdots & k_g \end{pmatrix}, \tag{5.2.15}$$

where $p < i_1 < i_2 \cdots i_g \le N$, $p < k_1 < k_2 < \cdots k_g \le N$.

Corollary 2

Suppose $1 \le i_1 < i_2 \cdots < i_p \le N$, $1 \le k_1 < k_2 \cdots < k_p \le N$,

$$b_{ik} = A\begin{pmatrix} i_1 & i_2 & \cdots & i_p & i \\ k_1 & k_2 & \cdots & k_p & k \end{pmatrix} \quad \begin{matrix} i = i_p+1, & \cdots & N, \\ k = k_p+1 & \cdots & N, \end{matrix}$$

and

$$i_p+1\leq \ell_1<\ell_2\cdots<\ell_q\leq N\ ,\ \ k_p+1\leq m_1<m_2\cdots<m_q\leq N\ ,$$

then

$$B\begin{pmatrix}\ell_1 & \ell_2\cdots\ell_q\\ m_1 & m_2\cdots m_q\end{pmatrix}=A\begin{pmatrix}i_1\cdots i_p\\ k_1\cdots k_p\end{pmatrix}^{q-1}A\begin{pmatrix}i_1 & i_2\cdots i_p & \ell_1\cdots\ell_q\\ k_1 & k_2\cdots k_p & m_1\cdots m_q\end{pmatrix}. \quad (5.2.16)$$

This is the general form of *Sylvester's Identity* (Gantmakher (1959), Vol. I, p.32).

Theorem 5.2.3. (Binet-Cauchy) (Gantmakher (1959), Vol. I, p.9).

If **A**, **B**, **C** are matrices of orders $M\times K$, $K\times N$ and $M\times N$ respectively and $\mathbf{C}=\mathbf{AB}$, then the minors of **C** are given by

$$C\begin{pmatrix}i_1 & i_2\cdots i_p\\ k_1 & k_2\cdots k_p\end{pmatrix}=\Sigma A\begin{pmatrix}i_1 & i_2\cdots i_p\\ j_1 & j_2\cdots j_p\end{pmatrix}B\begin{pmatrix}j_1 & j_2\cdots j_p\\ k_1 & k_2\cdots k_p\end{pmatrix}$$

where the sum extends over all $j_1,j_2,...,j_p$ satisfying $1\leq j_1<j_2<\cdots<j_p\leq K$.

Theorem 5.2.4. Let **A** be a square matrix of order N. If all minors

$$A\begin{pmatrix}i_1, & i_2, & \cdots i_p\\ 1, & 2, & \cdots p\end{pmatrix}\quad 1\leq i_1<i_2<\cdots<i_p\leq N$$

of order $p(<N)$ are positive, and all minors

$$A\begin{pmatrix}i, & i+1, & \cdots i+p\\ 1, & 2, & \cdots p+1\end{pmatrix}\quad 1\leq i,\ i+p\leq N$$

for *consecutive* sets of $p+1$ indices $i_1,i_2,...,i_{p+1}$ are positive, then *all* minors

$$A\begin{pmatrix}i_1, & i_2, & \cdots i_{p+1}\\ 1, & 2, & \cdots p+1\end{pmatrix}\quad 1<i_1<i_2<\cdots<i_{p+1}\leq N$$

are positive.

Proof:

Suppose that $p=2$ and $N=3$, and consider the zero determinant

$$\begin{vmatrix}a_{i1} & a_{i2} & a_{i3} & a_{i1} & a_{i2}\\ a_{j1} & a_{j2} & a_{j3} & a_{j1} & a_{j2}\\ a_{k1} & a_{k2} & a_{k3} & a_{k1} & a_{k2}\\ a_{m1} & a_{m2} & a_{23} & a_{m1} & a_{m2}\\ a_{j1} & a_{j2} & a_{j3} & a_{j1} & a_{j2}\end{vmatrix}=0\ .$$

On expansion along its first three rows, this yields

$$A\begin{pmatrix} j & k \\ 1 & 2 \end{pmatrix} A\begin{pmatrix} i & j & m \\ 1 & 2 & 3 \end{pmatrix} = A\begin{pmatrix} i & j \\ 1 & 2 \end{pmatrix} A\begin{pmatrix} j & k & m \\ 1 & 2 & 3 \end{pmatrix} + A\begin{pmatrix} j & m \\ 1 & 2 \end{pmatrix} A\begin{pmatrix} i & j & k \\ 1 & 2 & 3 \end{pmatrix}. \tag{5.2.17}$$

Thus if, i,j,k,m are consecutive indices and

$$A\begin{pmatrix} j & k & m \\ 1 & 2 & 3 \end{pmatrix}, \quad A\begin{pmatrix} i & j & k \\ 1 & 2 & 3 \end{pmatrix},$$

are positive, then

$$A\begin{pmatrix} i & j & m \\ 1 & 2 & 3 \end{pmatrix}$$

is positive. Using this relationship we may express any minor of order 3 (with increasing indices) as a positive combination of minors with consecutive indices.

In the general case we suppose first that all but one pair of the i_ν are consecutive. Without loss of generality we may assume the indices to be $1,2,\ldots k-1$, $k+1,\ldots p+2$. Consider the zero $(2p+1)\times(2p+1)$ determinant with terms a_{ij} where $i=1,2,\ldots\ p+1,1,2,\ldots p$; $j=1,2,\ldots p+1,2,3,\ldots\ k-1,k+1,\ldots p+2$.

The expansion of this determinant using the first $(p+1)$ rows has just three non-zero sets of terms, so that

$$A\begin{pmatrix} 2,3, & \cdots & p+1 \\ 1,2, & \cdots & p \end{pmatrix} A\begin{pmatrix} 1,2, & \cdots & k-1,k+1, & \cdots & p+2 \\ 1,2, & \cdots & & & p+1 \end{pmatrix} =$$

$$A\begin{pmatrix} 2,3, & \cdots & k-1,k+1, & \cdots & p+2 \\ 1,2, & \cdots & & & p \end{pmatrix} A\begin{pmatrix} 1,2, & \cdots & p+1 \\ 1,2, & \cdots & p+1 \end{pmatrix} +$$

$$A\begin{pmatrix} 1,2, & \cdots & k-1,k+1, & \cdots & p+1 \\ 1,2, & \cdots & & & p \end{pmatrix} A\begin{pmatrix} 2,3, & \cdots & p+2 \\ 1,2, & \cdots & p+1 \end{pmatrix}. \tag{5.2.18}$$

This expresses the minor of order $p+1$ with one pair of non-consecutive indices as a combination of minors with consecutive indices, the multipliers being minors of order p, assumed positive. Using this device we may now express any minor of order $p+1$ as a positive combination of minors with $p+1$ consecutive indices. ■

Corollary

A necessary and sufficient condition for the matrix **A** to be completely positive is that all minors taken from consecutive rows and consecutive columns be positive.

Examples 5.2

1. If **A** is non-singular then (1.3.20) shows that its inverse $\mathbf{R}=\mathbf{A}^{-1}$ has elements

$$r_{ij}=A_{ji}/\,|\mathbf{A}|=(-)^{i+j}\hat{a}_{ji}/\,|\mathbf{A}|\ .$$

Use (5.2.4) to show that

$$|\mathbf{A}|.R\begin{pmatrix} i_1,i_2,\ldots,i_p \\ k_1,k_2,\ldots,k_p \end{pmatrix} = (-)^s A\begin{pmatrix} \ell_1,\ell_2,\ldots,\ell_{N-p} \\ j_1,j_2,\ldots,j_{N-p} \end{pmatrix}$$

where

$$s=\sum_{m=1}^{p}(i_m+k_m)$$

and j_m, and ℓ_m are as in Theorem 5.2.1.

2. Equations (1.3.10), (1.3.11) are a particular case of *Laplace's expansion* of a determinant, namely

$$|\mathbf{A}| = A\begin{pmatrix} 1,2,3,\ldots,N \\ 1,2,3,\ldots,N \end{pmatrix} = \Sigma(-)^{\sigma} A\begin{pmatrix} i_1,i_2,\ldots,i_r \\ k_1,k_2,\ldots k_r \end{pmatrix} A\begin{pmatrix} j_1,j_2,\ldots,j_{N-r} \\ \ell_1,\ell_2,\ldots,\ell_{N-r} \end{pmatrix}.$$

Here $(i_p)_1^r$ is a fixed set of indices such that $1 \le i_1 < i_2 < \cdots < i_r \le N$, $(j_p)_1^{N-r}$ is the set of indices complementary to $(i_p)_1^r$, the sum is taken over all $(k_p)_1^r$ such that $1 \le k_1 < k_2 < \cdots k_r \le N$, $(\ell_p)_1^{N-r}$ is complementary to $(k_p)_1^r$, and

$$\sigma=\sum_{p=1}^{r}(i_p+k_p)\ .$$

Establish this result and show that there is a similar expansion with $(k_p)_1^r$ fixed and $(i_p)_1^r$ varying.

3. Note that Theorem 5.2.4 is a generalization of the simple result:

$$\text{If } a_i > 0 \ i=1,2,..N\ , \quad \begin{vmatrix} a_i & a_{i+1} \\ b_i & b_{i+1} \end{vmatrix} > 0$$

for $i=1,2,..,N-1$, then

$$\begin{vmatrix} a_i & a_j \\ b_i & b_j \end{vmatrix} > 0 \text{ for } i<j\ .$$

Verify this simple result from first principles.

5.3 Further properties of symmetric matrices

Let $\mathbf{A}$ be real, square, symmetric matrix of order N, i.e., $\mathbf{A}=(a_{ij})_1^N$. We begin this section with two theorems.

Theorem 5.3.1. The matrix $\mathbf{A}$ has N real eigenvectors forming an orthonormal system.

Theorem 5.3.2. To each m-fold eigenvalue λ_0 of $\mathbf{A}$ there correspond m linearly independent eigenvectors.

Before proving these theorems we note that if **A** has *distinct* eigenvalues, then Theorem 1.4.5, applied to the matrix pair (**A**,**I**) (**I** is positive definite!) shows that there are N eigenvectors $(\mathbf{x}^{(i)})_1^N$ which form an orthogonal system. It therefore suffices to prove Theorem 5.3.2.

Proof:

Suppose that λ_0 is an m-fold eigenvalue, i.e., $\Delta(\lambda)=|\mathbf{A}-\lambda\mathbf{I}|$ has λ_0 as an m-fold root, and that the matrix $\mathbf{B}=\mathbf{A}-\lambda_0\mathbf{I}$ has rank r, so that the equation

$$\mathbf{Bx}\equiv(\mathbf{A}-\lambda_0\mathbf{I})x=0 \tag{5.3.1}$$

has $p\equiv N-r$ linearly independent solutions. It is required to prove that $p=m$. Now

$$\Delta(\lambda)=|\mathbf{A}-\lambda\mathbf{I}|=|\mathbf{B}-(\lambda-\lambda_0)\mathbf{I}|=\sum_{i=o}^{N}(-)^i T_j(\lambda-\lambda_0)^i\ ,\ j=N-1, \tag{5.3.2}$$

where T_j is the sum of the jth-order principal minors of **B**, and $T_0=1$. But **B** has rank r so that $T_N=0=T_{N-1}=...T_{r+1}$ and therefore

$$\Delta(\lambda)=\pm(\lambda-\lambda_o)^p\{T_r-(\lambda-\lambda_0)T_{r-1}+\cdots\}\ ,\ r+p=N\ . \tag{5.3.3}$$

Thus it is sufficient to prove that $T_r\neq 0$, for then $\Delta(\lambda)$ will have a p-fold root λ_0, so that $m=p$. Without loss of generality we may assume that the *first* r rows of **B** are linearly independent, so that any row of **B** may be expressed as a linear combination of the first r rows, i.e.,

$$b_{ik}=\sum_{j=1}^{r}c_{ij}b_{jk}\ ,\quad (i,k=1,2,...,N)\ . \tag{5.3.4}$$

Since **B** is symmetric, we have

$$b_{ik}=b_{ki}=\sum_{j=1}^{r}c_{kj}b_{ij}\ ,\quad (i,k=1,2,...,N)\ . \tag{5.3.5}$$

Equation (5.3.4) may be written $\mathbf{B}=\mathbf{CB}^o$, where **C** is $\mathbf{C}(N\times r)$, and $\mathbf{B}^o(r\times N)$ is formed from the first r rows of **B**. Similarly equation (5.3.5) states $\mathbf{B}^T=\mathbf{B}=\mathbf{B}^{oT}\mathbf{C}^T$. Now the Binet-Cauchy theorem gives the following single-term expansions:

$$B\begin{pmatrix} i_1, & i_2, & ..., & i_r \\ k_2, & k_2, & ..., & k_r \end{pmatrix}=C\begin{pmatrix} i_1, & i_2, & ..., & i_r \\ 1, & 2, & ..., & r \end{pmatrix}B\begin{pmatrix} 1, & 2, & ..., & r \\ k_1, & k_2, & ..., & k_r \end{pmatrix} \tag{5.3.6}$$

$$B\begin{pmatrix} 1, & 2, & ..., & r \\ k_1, & k_2, & ..., & r_k \end{pmatrix}=C\begin{pmatrix} k_1, & k_2, & ..., & k_r \\ 1, & 2, & ..., & r \end{pmatrix}B\begin{pmatrix} 1, & 2, & ..., & r \\ 1, & 2, & ..., & r \end{pmatrix} \tag{5.3.7}$$

so that

$$B\begin{pmatrix} i_1, & i_2, & ..., & i_r \\ k_1, & k_2, & ..., & k_r \end{pmatrix} = C\begin{pmatrix} i_1, & i_2, & ..., & i_r \\ 1, & 2, & ..., & r \end{pmatrix} C\begin{pmatrix} k_1, & k_2, & ..., & k_r \\ 1, & 2, & ..., & r \end{pmatrix} B\begin{pmatrix} 1, & 2, & ..., & r \\ 1, & 2, & ..., & r \end{pmatrix} . \quad (5.3.8)$$

Since all the minors of the left cannot all vanish simultaneously (for then **B** would have a rank less than r), we must have

$$B\begin{pmatrix} 1, & 2, & ..., & r \\ 1, & 2, & ..., & r \end{pmatrix} \neq 0 . \quad (5.3.9)$$

But then equation (5.3.8) gives

$$B\begin{pmatrix} i_1, & i_2, & ..., & i_r \\ i_1, & i_2, & ..., & i_r \end{pmatrix} = C\begin{pmatrix} i_1, & i_2, & ..., & i_r \\ 1, & 2, & ..., & r \end{pmatrix}^2 B\begin{pmatrix} 1, & 2, & ..., & r \\ 1, & 2, & ..., & r \end{pmatrix} . \quad (5.3.10)$$

This means that all the rth-order principal minors of **B** have the same sign, and some must be non-zero, since the C-minor must be non-zero for some choice of i_1, i_2,...,i_r. Thus T_r, the sum of the rth-order principal minors of **B** must be non-zero. ■

Theorem 5.3.3. The real symmetric matrix **A** may be expressed in the form

$$\mathbf{A} = \mathbf{U}\mathbf{\Lambda}\mathbf{U}^T \quad (5.3.11)$$

where the columns of **U** are the eigenvectors of **A** and **Λ** is the diagonal matrix

$$\mathbf{\Lambda} = diag\,(\mathbf{\Lambda}_i) . \quad (5.3.12)$$

Proof:

The N mutually orthogonal vectors $(\mathbf{u}^{(i)})_1^N$ satisfy

$$\mathbf{A}\mathbf{u}^{(i)} = \lambda_i \mathbf{u}^{(i)} , \quad \mathbf{u}^{(i)T}\mathbf{u}^{(j)} = \delta_{ij} . \quad (5.3.12)$$

Thus

$$\mathbf{A}\mathbf{U} = \mathbf{U}\mathbf{\Lambda} , \quad (5.3.14)$$

where

$$\mathbf{U}\mathbf{U}^T = \mathbf{U}^T\mathbf{U} = \mathbf{I} . \quad (5.3.15)$$

Thus

$$\mathbf{A} = \mathbf{A}\mathbf{U}\mathbf{U}^T = \mathbf{U}\mathbf{\Lambda}\mathbf{U}^T \;\blacksquare \quad (5.3.16)$$

Suppose $\mathbf{A} = (a_{ij})_1^N$ is a symmetric matrix and let

$$A(\mathbf{x},\mathbf{x}) = \mathbf{x}^T\mathbf{A}\mathbf{x} = a_{11}x_1^2 + 2a_{12}x_1x_2 + \cdots + 2a_{N,N-1}x_Nx_{N-1} + a_{NN}x_N^2 \quad (5.3.17)$$

be the quadratic form associated with **A**. We consider a number of different ways of expressing $A(\mathbf{x},\mathbf{x})$.

Let

$$A_i(\mathbf{x}) = \sum_{j=1}^{N} a_{ij} x_j \;, \quad (i=1,2,\ldots,N) \tag{5.3.18}$$

then

$$A(\mathbf{x},\mathbf{x}) = \sum_{i=1}^{N} x_i A_i(\mathbf{x}) \; . \tag{5.3.19}$$

This yields

$$\begin{vmatrix} a_{11} & a_{12} & \cdots & a_{1N} & A_1(\mathbf{x}) \\ a_{21} & a_{22} & \cdots & a_{2N} & A_2(\mathbf{x}) \\ \cdot & \cdot & \cdot & \cdot & \cdot \\ a_{N1} & a_{N2} & \cdots & a_{NN} & A_N(\mathbf{x}) \\ A_1(\mathbf{x}) & A_2(\mathbf{x}) & \cdots & A_N(\mathbf{x}) & A(\mathbf{x},\mathbf{x}) \end{vmatrix} = 0 \tag{5.3.20}$$

since the last column is a linear combination of the first N columns, and

Theorem 5.3.4. If $|\mathbf{A}| \neq 0$, then

$$A(\mathbf{x},\mathbf{x}) = -\frac{1}{|\mathbf{A}|} \begin{vmatrix} a_{11} & a_{12} & \cdots & a_{1N} & A_1(\mathbf{x}) \\ a_{21} & a_{22} & \cdots & a_{2N} & A_2(\mathbf{x}) \\ \cdot & \cdot & \cdot & \cdot & \cdot \\ a_{N1} & a_{N2} & \cdots & a_{NN} & A_N(\mathbf{x}) \\ A_1(\mathbf{x}) & A_2(\mathbf{x}) & \cdots & A_N(\mathbf{x}) & 0 \end{vmatrix} \tag{5.3.21}$$

Proof:

Expand the zero determinant (5.3.20) along its last row. ■

We introduce the quantities

$$X_1(\mathbf{x}) = A_1(\mathbf{x}), \quad X_i(\mathbf{x}) = \begin{vmatrix} a_{11} & a_{12} & \cdots & a_{1,i-1} & A_1(\mathbf{x}) \\ a_{21} & a_{22} & \cdots & a_{2,i-1} & A_2(\mathbf{x}) \\ \cdot & \cdot & \cdot & \cdot & \cdot \\ a_{i1} & a_{i2} & \cdots & a_{i,i-1} & A_i(\mathbf{x}) \end{vmatrix} \tag{5.3.22}$$

and note

Theorem 5.3.5. If

$$D_p = A\begin{pmatrix} 1,2,\ldots,p \\ 1,2,\ldots,p \end{pmatrix} \neq 0 \quad \text{for } p=1,2,\ldots,N$$

then the $(X_i(\mathbf{x}))_1^N$ are linearly independent, i.e., they can all be zero simultaneously only if the $(x_i)_1^N$ are all zero.

Proof:

By using the expressions (5.3.18) for the $(A_j(x))_1^i$ and using Lemma 1.3.5, we note that $X_i(\mathbf{x})$ is a linear combination of $(x_j)_i^N$ with leading term $D_i \neq 0$, i.e.,

$$X_i(\mathbf{x}) = D_i x_i + \text{ terms in } x_{i+1}, \ldots, x_N . \tag{5.3.23}$$

Thus we see that, in the sequence $X_N(\mathbf{x}), X_{N-1}(\mathbf{x}),\ldots,X_1(\mathbf{x})$, each term involves one more x_j than the previous one. ■

Theorem 5.3.6. (Jacobi)

If $D_o = 1$ and

$$D_p = A\begin{pmatrix}1,2,\ldots,p\\1,2,\ldots,p\end{pmatrix} \neq 0 , \quad (p = 1,2,\ldots,N) ,$$

then

$$A(\mathbf{x},\mathbf{x}) = \sum_{i=1}^{N} \frac{[X_i(\mathbf{x})]^2}{D_i D_{i-1}} . \tag{5.3.24}$$

Note that, on account of Theorem 5.3.5, this equation expresses $A(\mathbf{x},\mathbf{x})$ as a sum of squares of linearly independent combinations of the $(x_j)_1^N$.

Proof:

Put $P_o = 0$ and

$$P_i(\mathbf{x},\mathbf{x}) = \begin{vmatrix} a_{11} & a_{12} & \cdots & a_{1i} & A_1(\mathbf{x}) \\ a_{21} & a_{22} & \cdots & a_{2i} & A_2(\mathbf{x}) \\ \cdot & \cdot & \cdot & \cdot & \cdot \\ a_{i1} & a_{i2} & \cdots & a_{ii} & A_i(\mathbf{x}) \\ A_1(\mathbf{x}) & A_2(\mathbf{x}) & \cdots & A_i(\mathbf{x}) & 0 \end{vmatrix} \tag{5.3.25}$$

and find the recurrence relation linking P_i and P_{i-1}. $P_i(\mathbf{x},\mathbf{x})$ is a determinant of a symmetric matrix of order $i+1$, i.e.,

$$P_i(\mathbf{x},\mathbf{x}) = C\begin{pmatrix}1,2,\ldots,i+1\\1,2,\ldots,i+1\end{pmatrix} \tag{5.3.26}$$

Apply Theorem 5.2.2 to the matrix C, with $p = i-1$, $N = i+1$, then

$$C\begin{pmatrix}1,2,\ldots,i\\1,2,\ldots,i\end{pmatrix} C\begin{pmatrix}1,2,\ldots,i-1,i+1\\1,2,\ldots,i-1,i+1\end{pmatrix} - C\begin{pmatrix}1,2,\ldots,i-1,i\\1,2,\ldots,i-1,i+1\end{pmatrix}^2$$

$$=C\begin{pmatrix}1,2,...,i-1\\1,2,...,i-1\end{pmatrix}C\begin{pmatrix}1,2,...,i+1\\1,2,...,i+1\end{pmatrix}. \tag{5.3.27}$$

But

$$C\begin{pmatrix}1,2,...i\\1,2,...i\end{pmatrix}=D_i\ ,\ C\begin{pmatrix}1,2,...,i-1,i+1\\1,2,...,i-1,i+1\end{pmatrix}=P_{i-1}(\mathbf{x},\mathbf{x}) \tag{5.3.28}$$

$$C\begin{pmatrix}1,2,...,i-1,i\\1,2,...,i-1,i+1\end{pmatrix}=X_i \tag{5.3.29}$$

so that equation (5.3.27) states

$$D_iP_{i-1}-X_i^2=D_{i-1}P_i \tag{5.3.30}$$

or, since the D_p are non-zero,

$$-\frac{P_i}{D_i}=\frac{X_1^2}{D_iD_{i-1}}-\frac{P_{i-1}}{D_{i-1}}. \tag{5.3.31}$$

Now, on summing from $i=1$ to N, using $P_o=0$ and (5.3.21), we deduce (5.3.24). ■

We may now prove a necessary and sufficient condition for the quadratic form $A(\mathbf{x},\mathbf{x})$ to be positive-definite. (See Section 1.4).

Theorem 5.3.7. The necessary and sufficient condition for the symmetric matrix **A** to be *positive definite* (i.e., $A(\mathbf{x},\mathbf{x})>0$ for all $\mathbf{x}\neq 0$) is that $D_i>0$, $i=1,2,...,N$.

Proof:

Suppose **A** is positive definite. The analysis above shows that it has N positive eigenvalues λ_i and corresponding eigenvectors $\mathbf{u}^{(i)}$, satisfying

$$(\mathbf{A}-\lambda_i\mathbf{I})\mathbf{u}^{(i)}=0\ ,\ \lambda_i=\mathbf{u}^{(i)T}\mathbf{A}u^{(\mathrm{i})}/\mathbf{u}^{(i)T}\mathbf{u}^{(i)}. \tag{5.3.32}$$

The product of the λ_i is therefore positive; but

$$\prod_{i=1}^{N}\lambda_i=|\mathbf{A}|$$

so that $|\mathbf{A}|=D_N>0$. But if **A** is positive definite then the matrix obtained by deleting the last row and column of A is also, thus its determinant, D_{N-1}, is positive. Thus all the D_i are positive. Conversely, if all the D_i are positive, then equation (5.3.24) shows that $A(\mathbf{x},\mathbf{x})>0$, since, by Theorem 5.3.5, all the $X_i(\mathbf{x})$ cannot be simultaneously zero. ■

Corollary 1.

If $A(\mathbf{x},\mathbf{x})$ is positive definite then all the principal minors

$$A\begin{pmatrix} i_1,i_2,\ldots,i_p \\ i_1,i_2,\ldots,i_p \end{pmatrix} \quad 1 \le i_1 < i_2 \cdots < i_p \le N$$

are positive.

Corollary 2.

The quadratic form $A(\mathbf{x},\mathbf{x})$ will be positive semi-definite (i.e., non-negative) if and only if all the principal minors of $\mathbf{A}$ are non-negative.

We conclude this section by restating the results described in Section 2.7 in connection with Rayleigh's Principle.

Theorem 5.3.8. If $\mathbf{A}=(a_{ij})_1^N$ is symmetric, then the quadratic forms $\mathbf{x}^T\mathbf{A}\mathbf{x}$ and $\mathbf{x}^T\mathbf{x}$ may be written

$$\mathbf{x}^T\mathbf{A}x = \sum_{i=1}^N \lambda_i y_i^2\,, \quad \mathbf{x}^T\mathbf{x} = \sum_{i=1}^N y_i^2\,. \tag{5.3.33}$$

Proof:

Recall the matrix $\mathbf{U}$ of Theorem 5.3.3: its columns are the eigenvectors $\mathbf{u}^{(i)} = \{u_1^{(i)},u_2^{(i)},\ldots,u_N^{(i)}\}$. Introduce the transformations

$$\mathbf{x}=\mathbf{U}\mathbf{y}=y_1\mathbf{u}^{(1)}+\cdots y_N\mathbf{u}^{(N)}\,, \tag{5.3.34}$$

then

$$\mathbf{x}^T\mathbf{A}\mathbf{x}=\mathbf{y}^T\mathbf{U}^T\mathbf{A}\mathbf{U}y=\mathbf{y}^T\mathbf{\Lambda}\mathbf{y}=\sum_{i=1}^N \lambda_i y_i^2\,, \tag{5.3.35}$$

and

$$\mathbf{x}^T\mathbf{x}=\mathbf{y}^T\mathbf{U}^T\mathbf{U}\mathbf{y}=\mathbf{y}^T\mathbf{y}=\sum_{i=1}^N y_i^2 \quad \blacksquare \tag{5.3.36}$$

Now Rayleigh's Principle and the results of Section 2.7 may be viewed in terms of the stationary values of the quotient $\mathbf{x}^T\mathbf{A}\mathbf{x}/\mathbf{x}^T\mathbf{x}$. Thus, for instance

Theorem 5.3.9. The eigenvalues of the real symmetric matrix $\mathbf{A}$ are the stationary values of the quotient $\mathbf{x}^T\mathbf{A}\mathbf{x}/\mathbf{x}^T\mathbf{x}$.

5.4 Perron's theorem and associated matrices

It may safely be said that *most* matrices appearing in vibration problems are symmetric. It is therefore known that they have real eigenvalues and a complete set of mutually orthogonal eigenvectors. Often the matrices are positive definite, so that their eigenvalues, in addition to being real, are *positive.* However, non-symmetric matrices do arise on occasion. In this section we shall therefore present some results pertaining to a special class of perhaps non-symmetric matrices, following Bellman (1970).

If a vector **x** has all its elements $(x_i)_1^n$ positive (non-negative) we shall say $\mathbf{x}>\mathbf{0}$ $(\geq\mathbf{0})$ and shall say that **x** is *positive (non-negative)*. If **x**,**y** are two vectors then $\mathbf{x}\geq\mathbf{y}$ is equivalent to $\mathbf{x}-\mathbf{y}\geq\mathbf{0}$. The matrix **A** is said to be positive (non-negative) if $a_{ij}>0$ (≥ 0) for $i,j=1,2,\ldots,N$.

The L_1 *norm* of an arbitrary vector **x** is defined to be

$$||\mathbf{x}||_1=\sum_{i=1}^{N}|x_i| \,, \tag{5.4.1}$$

while that of matrix $\mathbf{A}=(a_{ij})$ is

$$||\mathbf{A}||_1=\sum_{i=1}^{N}\sum_{j=1}^{N}|a_{ij}| \,. \tag{5.4.2}$$

Two of the properties of these norms are listed in Ex. 5.4.1.

Theorem 5.4.1. (Perron) If the square matrix **A** is positive, then **A** has a unique eigenvalue ρ which has greatest absolute value. This eigenvalue is positive and simple and its associated eigenvector can be taken to be positive.

Proof:

Let $S(\lambda)$ be the set of all non-negative λ for which there exist non-negative **x** such that

$$\mathbf{A}\mathbf{x}\geq\lambda\mathbf{x} \,. \tag{5.4.3}$$

We shall consider only *normalized* vectors **x**, i.e., such that $||\mathbf{x}||=1$; this automatically excludes the zero vector. If **x** satisfies equation (5.4.3), then Ex. 5.4.1 shows that

$$\lambda||\mathbf{x}||=||\mathbf{A}\mathbf{x}||\leq||\mathbf{A}||\cdot||\mathbf{x}|| \,, \tag{5.4.4}$$

so that $0\leq\lambda\leq||\mathbf{A}||$. This shows that the set $S(\lambda)$ is bounded, and it is clearly not empty since **A** is positive. Let λ_o be the least upper bound of $\lambda\subset S(\lambda)$. Let $\lambda_1,\lambda_2,\ldots,$ be a sequence of λ's in $S(\lambda)$ converging to λ_o, and $\mathbf{x}^{(i)}$ a corresponding sequence of non-negative **x**'s satisfying $\mathbf{A}\mathbf{x}^{(i)}\geq\lambda_i\mathbf{x}^{(i)}$. Since $||\mathbf{x}^{(i)}||=1$, we can choose a subsequence $\{\mathbf{x}^{(\nu_i)}\}$ which converges to a non-negative, non-trivial vector $\mathbf{x}^{(o)}$. This vector will satisfy

$$\mathbf{A}\mathbf{x}^{(o)}\geq\lambda_o\mathbf{x}^{(o)} \,, \tag{5.4.5}$$

so that $\lambda_o\subset S(\lambda)$.

We now show that equality holds in equation (5.4.5) and we do so by reduction to a contradiction.

Let

$$\mathbf{d}=\mathbf{A}\mathbf{x}^{(o)}-\lambda_o\mathbf{x}^{(o)}\geq\mathbf{0}\ , \tag{5.4.6}$$

and suppose that one of the d_i, say d_j, is positive. Put

$$y_i=x_i^{(o)}+[d_j/(2\lambda_o)]\delta_{ij}\ , \tag{5.4.7}$$

then the i-th row of $\mathbf{Ay}-\lambda_o\mathbf{y}$ is

$$e_i=d_i+a_{ij}d_j/(2\lambda_o)-d_j\delta_{ij}/2>0\ . \tag{5.4.8}$$

Now let $\lambda=\lambda_o+\min_i(e_i/y_i)$, then $\lambda>\lambda_o$ and

$$\mathbf{Ay}-\lambda\mathbf{y}=\mathbf{e}-\min(e_i/y_i)\mathbf{y}\geq 0\ , \tag{5.4.9}$$

which contradicts the maximal property of λ_o. Hence all d_i are zero and

$$\mathbf{A}\mathbf{x}^{(o)}=\lambda_o\mathbf{x}^{(o)}\ . \tag{5.4.10}$$

Thus λ_o is an eigenvalue and $\mathbf{x}^{(o)}$ an eigenvector, which is necessarily positive. We will show that $\lambda_o=\rho$.

Suppose that there is another eigenvalue λ such that $|\lambda|\geq\lambda_o$, with $\mathbf{z}$ an associated eigenvector. Then from $\mathbf{Az}=\lambda\mathbf{z}$ we have $|\lambda|\cdot|\mathbf{z}|\leq\mathbf{A}|\mathbf{z}|$, where $|\mathbf{z}|$ denotes the vector with elements $|z_i|$. But then the maximum property of λ_o shows that $|\lambda|\leq\lambda_o$, and therefore $|\lambda|=\lambda_o$ and the argument used in equations (5.4.6)-(5.4.9) shows that $\mathbf{A}|\mathbf{z}|=\lambda_o|\mathbf{z}|=|\mathbf{Az}|$. But the equation $|\mathbf{Az}|=\mathbf{A}|\mathbf{z}|$ can hold only if $\mathbf{z}=c\mathbf{w}$, where c is complex and $\mathbf{w}$ is positive, so that λ also is positive, i.e., $\lambda=\lambda_o$. We now show that $\mathbf{x}^{(o)}$ and $\mathbf{w}$, both positive and both eigenvectors corresponding to λ_o, must be equivalent. Put $\mathbf{z}=\mathbf{x}^{(o)}-\varepsilon\mathbf{w}$ and take $\varepsilon=\min(x_i^{(o)}/w_i)=x_j^{(o)}/w_j$, then $\mathbf{z}$ is a non-negative eigenvector of $\mathbf{A}$ corresponding to λ_o for which $z_j=0$. Hence $\mathbf{z}\equiv 0$ and $\mathbf{x}^{(o)}=\varepsilon\mathbf{w}$. We have shown that λ_o has all the stated properties of ρ. ■

In Section 7.6 we shall need the concept of an *associated* matrix. Let $\mathbf{A}=(a_{ij})_1^N$ be a matrix of order N. We shall define the pth-order associated matrix $\underline{\mathbf{A}}_p$ of the matrix $\mathbf{A}$.

Consider all possible combinations $i_1,i_2,...,i_p$ of the N indices $1,2,...,N$, taken p at a time. There are

$$\underline{N}=\binom{N}{p}=\frac{N!}{p!(N-p)!} \tag{5.4.11}$$

of such combinations. For given N, p the $\underline{N}$ combinations may be arranged in ascending order $1,2,...,\underline{N}$ by associating with the combination $i_1,i_2,...,i_p$ the number with digits $i_1,i_2,...,i_p$ in the scale of $d\equiv N+1$. This procedure associates a specific number $s=s(i_1,i_2,...,i_p)$ with each combination. Thus when $N=5$, $p=3$ then $\underline{N}=10$ and the combinations are 123, 124, 125, 134, 135, 145, 234, 235, 245, 345. Thus $s(1,2,4)=2$, while $s(2,4,5)=9$. The element $\underline{a}_{st}$ of the matrix $\underline{\mathbf{A}}_p$ is then given by

$$\underline{a}_{st} = A\begin{pmatrix} i_1, i_2, \ldots, i_p \\ k_1, k_2, \ldots, k_p \end{pmatrix}, \tag{5.4.12}$$

where $i_1 < i_2 \cdots < i_p$, $k_1 < k_2 < \ldots < k_p$ and $s(i_1, i_2, \ldots, i_p) = s$ $s(k_1, k_2, \ldots, k_p) = t$. The Binet-Cauchy theorem, Theorem 5.2.3, then asserts

Theorem 5.4.2. If **A**, **B**, **C** are square matrices of order n, and $\mathbf{C} = \mathbf{AB}$, then the associated matrices $\underline{\mathbf{A}}$, $\underline{\mathbf{B}}$, $\underline{\mathbf{C}}$, satisfy

$$\underline{\mathbf{C}} = \underline{\mathbf{A}}\underline{\mathbf{B}}. \tag{5.4.13}$$

Proof:

$$C\begin{pmatrix} i_1, i_2, \ldots, i_p \\ k_1, k_2, \ldots, k_p \end{pmatrix} = \Sigma A\begin{pmatrix} i_1, i_2, \ldots, i_p \\ j_1, j_2, \ldots, j_p \end{pmatrix} B\begin{pmatrix} j_1, j_2, \ldots, j_p \\ k_1, k_2, \ldots, k_p \end{pmatrix} \tag{5.4.14}$$

may be written

$$\underline{c}_{st} = \sum_{r=1}^{\underline{N}} \underline{a}_{sr} \underline{b}_{rt} \tag{5.4.15}$$

where

$$s(j_1, j_2, \ldots, j_p) = r. \quad ■$$

Corollary 1

If **A** is a non-singular matrix then the pth-order associated matrix of $\mathbf{A}^{-1}$ is the inverse of $\underline{\mathbf{A}}_p$. For, if $\mathbf{B} = \mathbf{A}^{-1}$, then

$$\mathbf{AB} = \mathbf{I}, \text{ so that } \underline{\mathbf{A}}\underline{\mathbf{B}} = \underline{\mathbf{I}}. \tag{5.4.16}$$

Corollary 2

If **A** is a real symmetric matrix with eigenvalues $(\lambda_i)_1^N$ and eigenmatrix **U**, then the eigenvalues of $\underline{\mathbf{A}}$ are the $\underline{N}$ products $\lambda_{i_1}\lambda_{i_2} \ldots \lambda_{i_p}$ of the N eigenvalues $\lambda_1, \lambda_2, \ldots, \lambda_N$ taken p at a time.

For since

$$\mathbf{A} = \mathbf{U}\mathbf{\Lambda}\mathbf{U}^T, \quad \mathbf{U}\mathbf{U}^T = \mathbf{I} \tag{5.4.17}$$

we have

$$\underline{\mathbf{A}} = \underline{\mathbf{U}}\underline{\mathbf{\Lambda}}\underline{\mathbf{U}}^T, \quad \underline{\mathbf{U}}\underline{\mathbf{U}}^T = \underline{\mathbf{I}} \tag{5.4.18}$$

where $\underline{\mathbf{\Lambda}}$ is the diagonal matrix with elements $\lambda_{i_1}\lambda_{i_2} \cdots \lambda_{i_p}$.

Corollary 3

For a fixed combination $i_1,i_2,...i_p$ and variable combinations $j_1,j_2,...,j_p$, the minors

$$U\begin{pmatrix} j_1,j_2,...,j_p \\ i_1,i_2,...,i_p \end{pmatrix}$$

give all the components of the eigenvector of the matrix $\underline{\mathbf{A}}_p$ corresponding to the eigenvalues $\lambda_{i_1}\lambda_{i_2}\cdots\lambda_{i_p}$.

Examples 5.4

1. Show that the norms defined by equations (5.4.1), (5.4.2) satisfy

$$||\mathbf{x}+\mathbf{y}|| \leq ||\mathbf{x}||+||\mathbf{y}|| \ , \quad ||\mathbf{Ax}|| \leq ||\mathbf{A}||\cdot||\mathbf{x}|| \ .$$

2. When $\mathbf{x}=\mathbf{e}^{(s)}$, i.e., the vector with elements $e_j^{(s)}=\delta_{sj}$, the quadratic form $\mathbf{x}^T\mathbf{A}\mathbf{x}=a_{ss}$. Use this result with equation (5.3.33) and apply it to the associated matrix $\mathbf{A}$ to show that a principal minor of $\mathbf{A}$ may be expressed in the form

$$A\begin{pmatrix} i_1,i_2,...,i_p \\ i_1,i_2,...,i_p \end{pmatrix} = \sum_{k=1}^{N} \lambda_k [u_s^{(k)}]^2$$

where λ_k, $\mathbf{u}^{(k)}$ are the eigenvalues and corresponding eigenvectors of $\underline{A}_p$. Hence deduce that if all the eigenvalues of $\mathbf{A}$ are positive then all the principal minors of $\mathbf{A}$ are positive.

5.5 Oscillatory matrices

In the analysis of vibrating systems up to this point we have encountered basically two kinds of matrices – positive definite (or semi-definite) and Jacobian. While we proved in Chapter 2 that positive definite matrices have real, positive, eigenvalues, we found that the eigenvalues of positive definite *Jacobian* matrices were also *distinct*. In addition, we found that the zeros of the eigenvectors of such matrices had certain interlacing properties. It will be shown in this chapter that the properties of Jacobian matrices are shared with a wider class, the so-called *oscillatory* (or *oscillation*) matrices. First we introduce some definitions.

Completely Positive. The matrix $\mathbf{A}=(a_{ij})_1^N$ is said to be completely positive (completely non-negative) if all its minors are positive (non-negative):

$$A\begin{pmatrix} i_1,i_2,...,i_p \\ k_1,k_2,...,k_p \end{pmatrix} > 0 \ (\geq 0) \ , \quad \begin{matrix} 1\leq i_1<i_2<...<i_p\leq N \\ 1\leq k_1<k_2<...<k_p\leq N \end{matrix}$$

Theorem 5.5.1. If $\mathbf{A}$ is completely non-negative matrix $(a_{ij})_1^n$ and has a zero principal minor D_q, i.e.,

$$D_q = A\begin{pmatrix}1,2,\ldots,q\\1,2,\ldots,q\end{pmatrix} = 0 \quad , \quad < N \ ,$$

then every principal minor bordering D_q is zero, i.e.,

$$D_p = 0 \qquad q \le p \le N \ .$$

Proof:

There are two cases:

(1) $D_1 = a_{11} = 0$.

Then,

$$\begin{vmatrix} a_{11}a_{1k} \\ a_{i1}a_{ik} \end{vmatrix} = -a_{i1}a_{1k} \ge 0 \ , \ a_{i1} \ge 0 \ , \ a_{1k} \ge 0$$

implies that either $(a_{i1})_2^N = 0$ or $(a_{1k})_2^N = 0$. In either case $(D_p)_1^N = 0$.

(2) $a_{11} \ne 0$

Then, for some $p(1 < p \le q)$

$$D_{p-1} \ne 0 \qquad D_p = 0 \ .$$

We introduce bordered determinants

$$b_{ik} = A\begin{pmatrix}1,2,\ldots,p-1,i\\1,2,\ldots,p-1,k\end{pmatrix} \ , i,k = p,p+1,\ldots,N \tag{5.5.1}$$

and form the matrix $\mathbf{B} = (b_{ik})_p^N$. By Sylvester's identity

$$B\begin{pmatrix}i_1,i_2,\ldots,i_g\\k_1,k_2,\ldots,k_g\end{pmatrix} = \left[A\begin{pmatrix}1,2,\ldots,p-1\\1,2,\ldots,p-1\end{pmatrix}\right]^{g-1} A\begin{pmatrix}1,2,\ldots,p-1,i_1,\ldots,i_g\\1,2,\ldots,p-1,k_1,\ldots,k_g\end{pmatrix} \ge 0, \tag{5.5.2}$$

so that $\mathbf{B}$ is completely non-negative. Since $b_{pp} = D_p = 0$ the matrix falls under case 1, and if $r > p$

$$B\begin{pmatrix}p,p+1,\ldots,r\\p,p+1,\ldots,r\end{pmatrix} = \left[A\begin{pmatrix}1,2,\ldots,p-1\\1,2,\ldots,p-1\end{pmatrix}\right]^{r-p} A\begin{pmatrix}1,2,\ldots,r\\1,2,\ldots,r\end{pmatrix} = 0 \ . \tag{5.5.3}$$

but since $D_{p-1} \ne 0$, we have $D_r = 0$. ■

Oscillatory matrices. The matrix $\mathbf{B} = (a_{ij})_1^N$ is said to be oscillatory if it is non-singular, completely non-negative and has positive quasi-principal elements, i.e., $a_{i,i+1} > 0$, $a_{i+1,i} > 0$ for $i = 1,2,\ldots,N-1$.

A completely positive matrix has all its minors positive. Thus a symmetric completely positive (completely non-negative) matrix is necessarily positive definite (positive semi-definite). An oscillatory matrix has all its minors non-negative, i.e., some are positive while others are zero. We now prove:

Theorem 5.5.2. The principal minors of non-singular completely non-negative matrix are positive.

Proof:

If any principal minor, $D_p (p<N)$, were zero then, by Theorem 5.5.1, D_N would be zero. But $D_N>0$ since $\mathbf{A}$ is non-singular. Therefore all $D_p>0$. ■

Corollary 1. A symmetric oscillatory matrix is positive definite.

Corollary 2. If $\mathbf{A}=(a_{ij})_1^N$ is oscillatory then any truncated matrix $(a_{ij})_p^q$ is also oscillatory. (It is clearly completely non-negative, is non-singular by Theorem 5.5.2, and has positive quasi-principal elements.)

Theorem 5.5.3. If $\mathbf{A}=(a_{ij})_1^N$ is completely non-negative and $p<N$ then

$$A\begin{pmatrix}1,2,\ldots,N\\1,2,\ldots N\end{pmatrix}\leq A\begin{pmatrix}1,2,\ldots,p\\1,2,\ldots,p\end{pmatrix}A\begin{pmatrix}p+1,p+2,\ldots,N\\p+1,p+2,\ldots,N\end{pmatrix}. \tag{5.5.4}$$

Proof:

On account of Theorem 5.5.2, we may assume without loss of generality that all the principal minors are positive, for if any were zero, then the inequality would be satisfied trivially. The theorem is true for $n=2$, since

$$A\begin{pmatrix}1 & 2\\1 & 2\end{pmatrix}=a_{11}a_{22}-a_{12}a_{21}\leq a_{11}a_{22}.$$

If $N>2$, then one of the numbers p, $N-p$ is larger than 1; without loss of generality we can assume $p>1$. We prove the theorem by induction and assume it holds for matrices of order $N-1$ or less. We introduce the matrix $\mathbf{B}$ of Theorem 5.5.1, given by (5.5.1). This is a completely non-negative matrix, the order of which is less than N, so that the theorem yields, in particular,

$$B\begin{pmatrix}p,p+1,\ldots,n\\p,p+1,\ldots,n\end{pmatrix}\leq b_{pp}B\begin{pmatrix}p+1,\ldots,N\\p+1,\ldots,N\end{pmatrix}. \tag{5.5.5}$$

But Sylvester's identity yields

$$A\begin{pmatrix}1,2,\ldots,N\\1,2,\ldots,N\end{pmatrix}=B\begin{pmatrix}p,p+1,\ldots,N\\p,p+1,\ldots,N\end{pmatrix}\Big/\left[A\begin{pmatrix}1,2,\ldots,p-1\\1,2,\ldots,p-1\end{pmatrix}\right]^{N-p}$$

$$\leq b_{pp} B\begin{pmatrix} p+1,...,N \\ p+1,...,N \end{pmatrix} \Big/ \left[A\begin{pmatrix} 1,2,...,p-1 \\ 1,2,...,p-1 \end{pmatrix} \right]^{N-p} \tag{5.5.6}$$

Applying Sylvester's identify once more we find

$$A\begin{pmatrix} 1,2,...,N \\ 1,2,...,N \end{pmatrix} \leq A\begin{pmatrix} 1,2,...,p \\ 1,2,...,p \end{pmatrix} A\begin{pmatrix} 1,2,...,p-1, & p+1,...,N \\ 1,2,...,p-1, & p+1,...,N \end{pmatrix} \Big/ \Big\{$$

$$A\begin{pmatrix} 1,2,...,p-1 \\ 1,2,...,p-1 \end{pmatrix} \Big\} \leq A\begin{pmatrix} 1,2,...,p \\ 1,2,...,p \end{pmatrix} A\begin{pmatrix} p+1,...,N \\ p+1,...,N \end{pmatrix} \tag{5.5.7}$$

where, in the last step, we use the fact that the result holds, by hypothesis, for matrices of order $N-1$. ■

Corollary. If $\mathbf{A}=(a_{ij})_1^N$ is completely non-negative then

$$A\begin{pmatrix} 1,2,...,p \\ 1,2,..,p \end{pmatrix} \leq a_{11}a_{22}...a_{pp} \ , \quad p \leq N \tag{5.5.8}$$

Quasi-principal minor. The minor

$$A\begin{pmatrix} i_1,i_2, \ . \ . \ . \ ,i_p \\ k_1,k_2, \ . \ . \ . \ ,k_p \end{pmatrix}$$

is said to be quasi-principal if

$$1 \leq i_1 < i_2 < \cdots < i_p \leq N \qquad 1 \leq k_1 < k_2 < \cdots < k_p \leq N$$

and

$$| \, i_j - k_j \, | \leq 1 \quad , \quad j=1,2,...,p \ .$$

We prove that the quasi-principal minors of an oscillatory matrix are positive; before doing this we need

Theorem 5.5.4. Suppose $\mathbf{A}=(a_{ij})_1^N$ is a completely non-negative matrix. If $\mathbf{A}$ has p linearly dependent rows, of which the first $p-1$ and last $p-1$ are linearly independent, then $\mathbf{A}$ has rank $p-1$.

Proof:

Let the linearly dependent rows be $i_1,i_2,...,i_p$. Without loss of generality we may assume that $1=i_1<i_2< \cdots <i_p=N$. If $p=N$, then $\mathbf{A}$ clearly has rank $p-1=N-1$. If $p<N$ then

$$a_{i_p,k} \equiv a_{N,k} = \sum_{j=1}^{p-1} c_j a_{i_j,k} \ , \quad k=1,2,...,N \tag{5.5.9}$$

and, since rows i_2, $i_3,...,i_{p-1}$, N are linearly independent, $c_1 \neq 0$. Suppose q is any other row, then, for some ℓ, such that $1 \leq \ell < p-1$, we have $i_\ell < q < i_{\ell+1}$. If $k_1,k_2,...,k_p$ are arbitrary integers satisfying $k_1<k_2< \cdots <k_p$, then

$$A\begin{pmatrix} i_2,\ldots,i_\ell, & q, & i_{\ell+1},\ldots,i_p \\ k_1,\ldots, & & k_p \end{pmatrix} = c_1 A\begin{pmatrix} i_2,\ldots,i_\ell, & q, & i_{\ell+1},\ldots,i_{p-1}, & i_1 \\ k_1,\ldots, & & & k_p \end{pmatrix}$$

$$=(-)^{p-1}c_1 A\begin{pmatrix} i_1,\ldots,i_\ell, & q, & i_{\ell+1},\ldots, & i_{p-1} \\ k_1,\ldots, & & & k_p \end{pmatrix} \tag{5.5.10}$$

Since rows $i_2,\ldots,i_p$ are linearly independent, there are integers $k_1^o,\ldots,k_{p-1}^o$, such that $k_1^o<k_2^o\cdots<k_{p-1}^o$, and

$$0<A\begin{pmatrix} i_2,\ldots,i_p \\ k_1^o,\ldots,k_{p-1}^o \end{pmatrix} = c_1(-)^p A\begin{pmatrix} i_1,\ldots,i_{p-1} \\ k_1^o,\ldots,k_{p-1}^o \end{pmatrix} \tag{5.5.11}$$

Thus $(-)^p c_1>0$. But now since both minors in (5.5.10) are ≥ 0, each must be zero. Thus since $k_1,k_2,\ldots,k_p$ are arbitrary, any row, q, must be a linear combination of $i_2,\ldots,i_p$, or equally of $i_1,\ldots,i_{p-1}$. Thus **A** has rank $p-1$. ■

Corollary. If $\mathbf{A}=(a_{ij})_1^N$ is completely non-negative, $1\leq i_1<i_2<\cdots<i_p\leq N$, $1\leq k_1<k_2<\cdots<k_p\leq N$ and

$$A\begin{pmatrix} i_1,i_2,\ldots,i_p \\ k_1,k_2,\ldots,k_p \end{pmatrix}=0$$

while

$$A\begin{pmatrix} i_1,i_2,\ldots,i_{p-1} \\ k_1,k_2,\ldots,k_{p-1} \end{pmatrix}>0\,,\quad A\begin{pmatrix} i_2,i_3,\ldots,i_p \\ k_2,k_3,\ldots,k_p \end{pmatrix}>0$$

then **A** has rank $p-1$.

We now prove

Theorem 5.5.5. The quasi-principal minors of an oscillatory matrix are positive.

Proof:

The quasi-principal minors of order 1 are $a_{i,i}$, $a_{i,i+1}$, $a_{i+1,i}$, all of which are positive. We prove the theorem by induction. Suppose all quasi-principal minors of order $p-1$ were positive while one of order p was zero. The corollary to Theorem 5.5.4 shows that **A** would have rank $p-1$; but **A** is non-singular. This contradiction implies that all quasi-principal minors of order p are positive. ■

The matrix $\mathbf{A}^*$ with elements a_{ij}^* given by

$$a_{ij}^*=(-)^{i+j}a_{ij} \tag{5.5.12}$$

is called the *sign-reverse* of **A**. The matrix **A** is said to be sign-definite (strictly sign-definite) if $\mathbf{A}^*$ is completely non-negative (completely positive). A consideration of (1.3.9) and (1.3.20) yields

Theorem 5.5.6. If the non-singular matrix **A** is sign-definite (strictly sign-definite), then $\mathbf{A}^{-1}$ is completely non-negative (completely positive).

Corollary. If the non-singular matrix **A** is completely non-negative (positive), then so is $(\mathbf{A}^*)^{-1}$.

The matrix **A** is said to be *sign-oscillatory if its sign-reverse is oscillatory.*

Theorem 5.5.7. If $\mathbf{A}^*$ *is oscillatory, then so is* $(\mathbf{A}^*)^{-1}$.

Proof:

$\mathbf{B}\equiv(\mathbf{A}^*)^{-1}$ is certainly non-singular and completely non-negative; it suffices to prove that $b_{i,i+1}$ and $b_{i+l,i}$ are positive. Theorem 5.5.2 shows that $|\mathbf{A}|>0$; $|\mathbf{A}|\cdot b_{i,i+1}$ and $|A|\cdot b_{i+l,i}$ are given by quasi-principal minors of **A** and, by Theorem 5.5.5, these are positive.

Theorem 5.5.8. If the Jacobian matrix **A** has negative off-diagonal elements and is positive definite then it is sign-oscillatory.

Proof:

We prove that $\mathbf{A}^*$ is oscillatory. Consider a minor

$$A^*\begin{pmatrix} i_1,i_2,...,i_p \\ k_1,k_2,...,k_p \end{pmatrix} \quad \begin{matrix} 1\le i_1<i_2<\cdots<i_p\le N \\ 1\le k_1<k_2<\cdots<k_p\le N \end{matrix}$$

There are three cases:

(1) $i_\nu=k_\nu$ for $\nu=1,2,...,p$. In this case the minor is principal, and so positive.

Otherwise there is at least one pair i_q, k_q such that $i_q\neq k_q$, but then

(2) if there is at least one pair (i_q,k_q) such that $|i_q-k_q|>1$, then the minor is zero.

(3) if $|i_\nu-k_\nu|\le1$, so that the minor is quasi-principal, then it may be expressed as a product of principal minors and b's $(b^*_{i+1,i}\equiv b_i)$. ■

Examples 5.5

1. Show that the product $\mathbf{C}=\mathbf{AB}$ of two completely non-negative matrices is completely non-negative, while the product of a completely positive matrix and a completely non-negative matrix is completely positive. Hint: Use the Binet-Cauchy theorem.
2. Use Theorem 5.5.2 to show that if **A** and **B** are oscillatory, then so is $\mathbf{C}=\mathbf{AB}$.
3. Use the Binet-Cauchy expansion of $\mathbf{A}^{N-1}$ to show that if **A** is oscillatory, then $\mathbf{A}^{N-1}$ is completely positive. The *smallest* value of m such that $\mathbf{A}^m$ is completely positive is called the *exponent* of **A**.
4. Show that $\mathbf{A}^*=\mathbf{SAS}$ where $\mathbf{S}=$ *diag*$(1,-1,1,-1,...,(-)^{N-1})$. Hence show that if **A** has eigenvalues $(\lambda_i)_1^N$ and eigenvectors $\mathbf{u}^{(i)}$, then $\mathbf{A}^*$ has eigenvalues $(\lambda_i)_1^N$ and eigenvectors $\mathbf{Su}^{(i)}$.

5.6 Oscillatory systems of vectors

Let u_1, $u_2,...,u_N$ be a sequence of real numbers. If some of the terms are zero we may assign them arbitrarily chosen signs. We can then calculate the number of sign interchanges in the sequence. This number may change depending on the choice of signs for the zero terms. The greatest and the least values of this number are denoted by S_u^+ and S_u^- respectively.

If $S_u^+=S_u^-$, we speak of an exact number of sign interchanges in the sequence, and denote this by S_u. Clearly this case can occur only when:

(1) $u_1u_N\neq 0$

(2) when $u_i=0$ $(i=2,3...,N-1)$, then $u_{i-1}u_{i+1}<0$. In this case, S_u is the number of sign interchanges when the zero terms are removed.

We say that a system of vectors $\mathbf{u}^{(k)}=\{u_{1k},u_{2k},...,u_{Nk}\}$ is an *oscillatory* system, if for any $(c_k)_1^m$ with

$$\sum_{k=1}^{m} c_k^2>0$$

the vector

$$\mathbf{u}=\sum_{k=1}^{m} c_k\mathbf{u}^{(k)} \tag{5.6.1}$$

satisfies $S_u^+\leq m-1$.

Theorem 5.6.1. The necessary and sufficient condition for the system $(\mathbf{u}^{(i)})_1^m$ to be an oscillating system is that all minors

$$U\begin{pmatrix} i_1, & i_2 & ,\ldots, & i_m \\ 1, & 2 & ,\ldots, & m\end{pmatrix}, \ (1\leq i_1<i_2<...<i_m\leq m) \tag{5.6.2}$$

be different from zero and have the same sign.

Proof:

We first prove the necessity. If the minor (5.6.2) were to vanish then we could find numbers $(c_k)_1^m$, not all zero such that

$$\sum_{k=1}^{m} c_k u_{i_j,k}=0\ , \ \ (j=1,2,...,m) \tag{5.6.3}$$

but then the vector (5.6.1) would have m zero terms

$$u_{i_1}=0=u_{i_2}= \cdots =u_{i_m} \tag{5.6.4}$$

so that, by Ex. 5.6.1, $S_u^+\geq m$. In order to show that the minors all have the same sign it is enough to show that all the minors

$$U_j = U\begin{pmatrix}1,2,\ldots, & j-1,j+1,\ldots, & m+1\\ 1,2,\ldots, & & m\end{pmatrix} \tag{5.6.5}$$

have the same sign, which is the sign of

$$U_{m+1} = U\begin{pmatrix}1,2,\ldots,m\\ 1,2,\ldots,m\end{pmatrix}. \tag{5.6.6}$$

Introduce a vector $\mathbf{u}^{(m+1)}$ such that

$$u_{i,m+1} = \begin{cases}(-)^{j-1}U_{m+1}, & i=j\\ (-)^{m+1}U_j, & i=m+1\\ 0 & \text{otherwise}.\end{cases} \tag{5.6.7}$$

Then

$$U\begin{pmatrix}1,2,\ldots,m+1\\ 1,2,\ldots,m+1\end{pmatrix} = (-)^{m+1}\{(-)^{m+1}u_{m+1,m+1}U_{m+1} + (-)^j u_{j,m+1}U_j\} = 0 \tag{5.6.8}$$

so that we can find $(c_k)_1^{m+1}$, not all zero such that

$$\sum_{k=1}^{m+1} c_k u_{i,k} = 0\,, \quad (i=1,2,\ldots,m+1)\,. \tag{5.6.9}$$

But then the vector (5.6.1) will have coordinates

$$u_i = -c_{m+1}u_{i,m+1}\,, \quad (i=1,2,\ldots,m+1)\,. \tag{5.6.10}$$

The quantity c_{m+1} cannot be zero, for then $\mathbf{u}$ would have $m+1$ zero terms and hence $S_u^+ \geq m$; choose c_{m+1} such that $c_{m+1}U_{m+1} > 0$. Then, $(u_i)_1^{j-1} = 0$, $u_j = (-)^j c_{m+1}U_{m+1}$, $(u_i)_{j+1}^m = 0$ $u_{m+1} = (-)^m c_{m+1}U_j$. If U_{m+1}, U_j had opposite signs then we may, for all $i=1,2,\ldots m+1$, assign u_i the sign $(-)^i$, so that $S_u^+ = m$. But $S_u^+ \leq m-1$, so that U_j and U_{m+1}, and hence all u_j must have the same sign. This proves the necessity.

Now suppose that all the minors (5.6.2) are non-zero and have the same sign, which we may take to be positive. We prove that the vector $\mathbf{u}$ of (5.6.1) satisfies $S_u^+ \leq m-1$, by assuming the contrary, i.e., $S_u^+ \geq m$. We could then find $m+1$ components $u_{i_1}, u_{i_2}, \ldots, u_{i_{m+1}}$ such that

$$u_{i_j}u_{i_{j+1}} \leq 0 \qquad (j=1,2,\ldots,m)\,. \tag{5.6.11}$$

The $(u_{i_j})_1^m$ cannot be simultaneously zero, for then the $(c_k)_1^m$ with

$$\sum_{s=1}^{m} c_k^2 > 0$$

would satisfy equation (5.6.3), the determinant of which is non-zero.

Now consider the zero determinant

$$\begin{vmatrix} u_{i_1,1} & u_{i_1,2} & \cdots & u_{i_1,m} & u_{i_1} \\ u_{i_2,1} & u_{i_2,2} & \cdots & u_{i_2,m} & u_{i_2} \\ & & & & \\ u_{i_{m+1},1} & u_{i_{m+1},2} & \cdots & u_{i_{m+1},m} & u_{i_{m+1}} \end{vmatrix} = 0 \qquad (5.6.12)$$

and expand it along its last column

$$\sum_{k=1}^{m+1} u_{i_k}(-)^{m+1+k} U\begin{pmatrix} i_1, & i_2 & ,\ldots, & i_{k-1}, i_{k+1}, & \ldots, & i_{m+1} \\ 1, & 2 & ,\ldots, & & & m \end{pmatrix} = 0 \,. \qquad (5.6.13)$$

But this is impossible since all the minors are positive and, by (5.6.11), the quantities $(-)^k u_{i_k}$ all have the same sign. ■

Examples 5.6

1. Consider the real sequence $u_1, u_2,\ldots,u_N$. Show that if $(u_i)_1^N=0$ then $S_u^-=0$, $S_u^+=N-1$. Show also that if $m(0\le m<N)$ of the u_i are zero then

$$S_u^-\le N-m-1 \;, \quad N-1\ge S_u^+\ge m$$

while if $m(1<m<N)$ successive u_i are zero then $S_u^+-S_u^-\ge m$.

5.7 Eigenvalues of oscillatory matrices

We are now in a position to prove some important theorems regarding eigenvalues and eigenvectors of oscillatory matrices.

Theorem 5.7.1. The eigenvalues of an oscillatory matrix **A** are positive and distinct, i.e.,

$$\lambda_1>\lambda_2 \cdots >\lambda_N>0 \,.$$

Proof:

Let the eigenvalues of **A** be $\lambda_1, \lambda_2,\ldots, \lambda_N$. Ex. 5.5.3 shows that there is an exponent $m_o\le N-1$ such that $\mathbf{A}^m\equiv\mathbf{B}$ is completely positive for all $m\ge m_o$. The eigenvalues of **B** are $\mu_1, \mu_2,\ldots,\mu_N$, where $\mu_i=\lambda_i^m$. For any $q\le N$ all the elements of $\underline{\mathbf{B}}_q$, the q-th associated matrix for **B** will be positive. Thus Perron's Theorem (Theorem 5.4.1) applies to $\underline{\mathbf{B}}_q$. If the eigenvalues of **B** are numbered in order of decreasing modulus, i.e., $|\mu_1|\ge|\mu_2|\ge\cdots\ge|\mu_N|$, then the eigenvalue of largest modulus of $\underline{\mathbf{B}}_q$ is $\mu_1\,\mu_2\ldots\,\mu_q$. Thus Perron's theorem yields

$$\mu_1\mu_2\cdots\mu_q>0 \,, \quad (q=1,2,\ldots,N) \qquad (5.7.1)$$

$$\mu_1\mu_2\cdots\mu_q>|\mu_1\mu_2\cdots\mu_{q-1}\mu_{q+1}| \;, (q=1,2,\ldots,N-1) \,. \qquad (5.7.2)$$

From the first inequality it follows that all the μ_i are real and positive, and from the second that

$$\mu_q > \mu_{q+1} , \quad (q=1,2,...,N-1) . \tag{5.7.3}$$

Thus,

$$\mu_1 > \mu_2 > \cdots > \mu_N > 0 . \tag{5.7.4}$$

But this means that

$$\lambda_1^m > \lambda_2^m > \cdots > \lambda_N^m > 0 , \quad m \geq m_o . \quad \blacksquare \tag{5.7.5}$$

Theorem 5.7.2. Let **A** be an oscillating matrix with eigenvalues $\lambda_1, \lambda_2,...,\lambda_N$, ordered so that $\lambda_1 > \lambda_2 > \cdots > \lambda_N > 0$. Let $\mathbf{u}^{(k)} = \{u_{1k}, u_{2k},...,u_{Nk}\}$ be an eigenvector of **A** corresponding to λ_k and let

$$\mathbf{u} = \sum_{k=p}^{q} c_k \mathbf{u}^{(k)} , \quad 1 \leq p \leq q \leq N , \quad \sum_{k=p}^{q} c_k^2 > 0, \tag{5.7.6}$$

then the number of sign interchanges among the components of **u** for differing $(c_k)_p^q$ satisfies

$$p-1 \leq S_u^- \leq S_u^+ \leq q-1 . \tag{5.7.7}$$

Proof:

Since the eigenvectors of **A** are also the eigenvectors of $\mathbf{A}^m$, and since $\mathbf{A}^m$ is completely positive for some $m \geq m_o$, we lose no generality by assuming that **A** itself is completely positive.

The eigenvalues of **A** are distinct, so that each $\mathbf{u}^{(k)}$ is determined except for a multiplicative factor. For given q, $1 \leq q \leq N$, and for all choices $i_1, i_2,...,i_q$ satisfying $1 \leq i_1 < i_2 < \cdots < i_q \leq N$, the minors

$$U\begin{pmatrix} i_1, & i_2, & ..., & i_q \\ 1, & 2, & ..., & q \end{pmatrix}$$

are the coordinates of the eigenvector of the associated matrix $\mathbf{A}_q$ corresponding to its maximum eigenvalue $\lambda_1\lambda_2 \cdots \lambda_q$. By Perron's theorem, all these numbers are non-zero and have the same sign. If the sign of the q-th set of minors is ε_q then, by multiplying the vectors $(\mathbf{u}^{(k)})_1^N$ by $\varepsilon_1, \varepsilon_2/\varepsilon_1,...\varepsilon_N/\varepsilon_{N-1}$ respectively, we may make

$$U\begin{pmatrix} i_1, & i_2, & ..., & i_q \\ 1, & 2, & ..., & q \end{pmatrix} > 0 , \quad (q=1,2,...,N) . \tag{5.7.8}$$

Theorem 5.6.1 now shows that $S_u^+ \leq q-1$.

To prove the second part of the inequality we put $\mathbf{B} = (\mathbf{A}^*)^{-1}$. Theorem 5.5.6 shows that **B** is completely positive, and Ex. 5.5.4 shows that it has eigenvalues μ_k, where $\mu_1 > \mu_2 > \cdots > \mu_N > 0$ and $\mu_k = 1/\lambda_\ell$, and eigenvectors $\mathbf{v}^{(k)} = \mathbf{S}\mathbf{u}^{(\ell)}$, where $\ell = N+1-k$. Thus

$$\mathbf{v}^{(k)}=\{v_{1k},v_{2},...,v_{Nk}\}=\{u_{1\ell},-u_{2\ell},u_{3\ell},...,(-)^{N-1}u_{N\ell}\}\ . \tag{5.7.9}$$

The result already proved shows that the number of sign interchanges in

$$\mathbf{v}=\sum_{k=N+1-q}^{N+1-p} c_{N+k-k}v^{(k)}=\mathbf{S}\sum_{k=p}^{q} c_k u^{(k)}=\mathbf{Su} \tag{5.7.10}$$

satisfies $S_v^+\leq N-p$. But since $v_i=(-)^{i-1}u_i$, we have $S_v^+ + S_u^- = N-1$ so that $S_u^-\geq p-1$. ■

Corollary 1. The vector $\mathbf{u}^{(k)}$ has exactly $k-1$ sign reversals.

Corollary 2. $u_N^{(k)}\neq 0$.

The argument used in this theorem leads directly to

Corollary 3. For each p, q such that $1\leq p$, $q\leq N$ the minors

$$U\begin{pmatrix} i_1, & i_2, & ..., & i_q \\ 1, & 2, & ..., & q\end{pmatrix} \quad 1\leq i_1<i_2< \cdots <i_q\leq N$$

have the same sign for all i_1, i_2,...,i_q, and the minors

$$V\begin{pmatrix} i_1, & i_2, & ..., & i_p \\ 1, & 2, & ..., & p\end{pmatrix} \quad 1\leq i_1<i_2< \cdots <i_p\leq N$$

have the same sign for all i_1, i_2,...,i_p.

The minor in Corollary 3 to the last theorem relates to components i_1, i_2,...,i_q of the first q eigenvectors. We now prove a result in which component and eigenvector indices are reversed; this theorem will play a vital role in the inverse problem of the vibrating beam (Chapter 7). Before stating the theorem we comment on the relation between oscillatory and sign-oscillatory matrices.

If **A** is oscillatory with eigenvalues $(\lambda_i)_1^N$ ordered so that $\lambda_1>\lambda_2>...\lambda_N>0$ then its eigenvectors $(\mathbf{u}^{(k)})_1^N$ satisfy Theorem 5.7.2 so that, in particular, $\mathbf{u}^{(k)}$ has exactly $k-1$ sign interchanges. If **A** is *sign-oscillatory* and we label its eigenvalues $(\lambda_i)_1^N$ in reverse order, i.e., so that $0<\lambda_1<\lambda_2<...<\lambda_N$, then its eigenvectors $(\mathbf{u}^{(k)})_1^N$ again satisfy Theorem 5.7.2 so that, in particular, $\mathbf{u}^{(k)}$ has $(k-1)$ sign interchanges. We will phrase the final theorem for a sign-oscillatory matrix.

Theorem 5.7.3. If **A** is sign-oscillatory, with eigenvalues $(\lambda_i)_1^N$ satisfying $0<\lambda_1<\lambda_2<...<\lambda_N$, then its eigenvectors $(\mathbf{u}^{(i)})_1^N$ may be chosen so that

$$\underline{u}_{\underline{N}}^{(s)}\equiv U\begin{pmatrix} N-p+1, & N-p+2, & ..., & N \\ i_1, & i_2, & ..., & i_p\end{pmatrix}>0 \tag{5.7.11}$$

for $p=1,2,...,N$, and each combination i_1, i_2,...,i_p satisfying $1\leq i_1<i_2<...<i_p\leq N$.

Proof:

Note that the analysis of Section 5.4 shows that $\underline{u}_N^{(s)}$ is the last component of the eigenvector of the associated matrix $\underline{\mathbf{A}}_p$ corresponding to the eigenvalue $\underline{\lambda}_s = \lambda_{i_1}\lambda_{i_2}\cdots\lambda_{i_p}$. The more general statement of the theorem is that all the elements $\underline{u}_N^{(s)}$ have the *same sign*, which is thus the sign chosen for the case $p=1$, i.e., for $u_N^{(i)}$.

The proof is by induction on p. Corollary 2 to Theorem 5.7.2 shows that $u_N^{(i)}\neq 0$. Choose $u_N^{(i)}>0$; the theorem then holds for $p=1$. Suppose the result holds for p. Corollary 1 to Theorem 5.7.2 shows that $(-)^{i-1}u_1^{(i)}>0$. Choose $(c_j)_i^{p+1}$ so that

$$\mathbf{u}=\sum_{j=i}^{p+i} c_j\mathbf{u}^{(j)} \tag{5.7.12}$$

and $u_{N-p+1}=0=u_{N-p+2}=...=u_N$, using the choice

$$c_i=U\begin{pmatrix} N-p+1, & N-p+2, & ..., & N \\ i+1, & i+2, & ..., & i+p\end{pmatrix},$$

$$c_{i+1}=-U\begin{pmatrix} N-p+1, & N-p+2, & ..., & N \\ i, & i+2, & ..., & i+p\end{pmatrix}, \tag{5.7.13}$$

$$c_{i+p}=(-)^p U\begin{pmatrix} N-p+1, & n-p+2, & ..., & N \\ i, & i+1, & ..., & i+p-1\end{pmatrix}. \tag{5.7.14}$$

The vector $\mathbf{u}=(u_1,u_2,...,u_{N-p},0,0,...,0)$ has the first element

$$u_1=c_iu_1^{(i)}+c_{i+1}u_1^{(i+1)}+...+c_{i+p}u_1^{(i+p)}. \tag{5.7.15}$$

Since, by hypothesis, the result is true for p, the coefficients c_j satisfy $(-)^jc_{i+j}>0$; this and the inequality $(-)^{i+j-1}u_1^{(i+j)}>0$ yield $(-)^{i-1}c_{i+j}u_1^{(i+j)}>0$, so that $(-)^{i-1}u_1>0$. By Theorem 5.7.2,

$$i-1\leq S_u^-\leq S_u^+\leq p+i-1 \tag{5.7.16}$$

and since the last p elements of $\mathbf{u}$ are zero, there must be exactly $i-1$ sign interchanges in the first $N-p$ elements of $\mathbf{u}$; but $(-)^{i-1}u_1>0$, so that the last non-zero element, u_{N-p}, must be positive, i.e.,

$$u_{N-p}=U\begin{pmatrix} N-p, & N-p+1, & ..., & N \\ i, & i+1, & ..., & i+p\end{pmatrix}>0\,,\quad i+p\leq N\,. \tag{5.7.17}$$

This shows that all $(p+1)$-order minors with *consecutive* indices $i,i+1,...,i+p$ are positive, and Theorem 5.2.4 shows that *all* $(p+1)$-order minors are positive. (bx

Examples 5.7

1. Rephrase Theorem 5.7.2 for the case in which $\mathbf{A}$ is sign-oscillatory. Suppose that $\mathbf{A}$ has eigenvalues $(\lambda_i)_1^N$ ordered so that $\lambda_1<\lambda_2<...<\lambda_N$, and corresponding eigenvalues $\mathbf{u}^{(i)}$. Show that the number of sign interchanges in $\mathbf{u}$ still

satisfies (5.7.7).

2. Show that if $\mathbf{u}^{(i)}$, $\mathbf{u}^{(i+1)}$ are eigenvectors of a (sign -) oscillatory matrix then $u_{n-1}^{(i)}\ u_n^{(i+1)} - u_{n-1}^{(i+1)}\ u_n^{(i)}$ is non-zero, and has the same sign for $n=2,3,...,N$.
3. Find two vectors $\mathbf{u}$, $\mathbf{v}$ such that $S_u=1$, $S_v=2$, but $u_{N-1}\ v_N - v_{N-1}\ u_N=0$ for some N.

5.8 u-Line analysis

We recall the concept of the u-line corresponding to the vector $\mathbf{u}=\{u_1,u_2,...,u_N\}$; it is the broken line made up of the links joining the points with coordinates $(x,y)=(i,u_i)$, so that

$$u(x)=(i+1-x)u_i+(x-i)u_{i+1}\ , \quad i\leq x\leq i+1\ . \tag{5.8.1}$$

Theorem 5.8.1. Let $\mathbf{u}^{(k)}$ be an eigenvector corresponding to an oscillatory matrix $\mathbf{A}$. Then the corresponding $u^{(k)}$-line has no links on the x-axis, and has just $k-1$ nodes, i.e., simple zeros at which $u(x)$ changes sign.

Proof:

A link of a u-line can lie along the x-axis only if two successive u_i are zero, but then $S_u^+ - S_u^- \geq 2$ which cannot happen when $\mathbf{u}=\mathbf{u}^{(k)}$, since $S_u^+ = S_u^-$. Since $S_u=k$, the $u^{(k)}$-line has just $k-1$ nodes. ■

Corollary. If α,β are successive zeros of a $u^{(k)}$-line then $|\alpha-\beta|>1$.

Theorem 5.8.2. Two u-lines corresponding to two successive eigenvectors of an oscillatory matrix cannot have a common node.

Proof:

Suppose, if possible, that $u^{(k)}(\alpha)=0=u^{(k+1)}(\alpha)$ and put

$$u(x)=cu^{(k)}(x)-u^{(k+1)}(x) \tag{5.8.2}$$

then Theorem 5.7.2 shows that

$$k-1\leq S_u^- \leq S_u^+ \leq k\ . \tag{5.8.3}$$

The Corollary to Theorem 5.8.1 shows that $u^{(k)}(x)$ and $u^{(k+1)}(x)$ will both be non-zero in $(\alpha,\alpha+1]$. Choose γ so that $\alpha<\gamma\leq\alpha+1$ and put $c=u^{(k+1)}(\gamma)/u^{(k)}(\gamma)$. Then $u(x)$ will have two zeros α, γ such that $|\gamma-\alpha|<1$, it must therefore have a link along the x-axis, and two successive u_i must vanish, and therefore $S_u^+ - S_u^- \geq 2$, contradicting (5.8.3). ■

Theorem 5.8.3. The nodes of the u-lines corresponding to two successive eigenvectors $\mathbf{u}^{(k)}$, $\mathbf{u}^{(k+1)}$ of an oscillatory matrix interlace.

Proof:

Suppose that α, β are two successive nodes of the $\mathbf{u}^{(k+1)}$-line, then $u^{(k+1)}(\alpha)=0=u^{(k+1)}(\beta)$ and $\beta-\alpha>1$. Suppose if possible that the $u^{(k)}$-line has no node in (α,β). Without loss of generality we may assume that

$$u^{(k)}(x)>0 \text{ in } [\alpha,\beta]\ ,\ \ u^{(k+1)}(x)>0 \text{ in } (\alpha,\beta). \tag{5.8.4}$$

Put

$$u(x)=cu^{(k)}(x)-u^{(k+1)}(x) \tag{5.8.5}$$

then

$$k-1\leq S_u^-\leq S_u^+\leq k\ . \tag{5.8.6}$$

For sufficiently large c, $u(x)>0$ in $[\alpha,\beta]$. Decrease c to a certain value c_o at which $u(x)$ first vanishes at least once, at a point γ in $[\alpha,\beta]$. Clearly $c_o>0$ and

$$u_o(x)=c_o u^{(k)}(x)-u^{(k+1)}(x) \tag{5.8.7}$$

does not vanish at α or β, so that $\alpha<\gamma<\beta$. Thus $u_o(x)\geq 0$ in $[\alpha,\beta]$ and $u_o(\gamma)=0$. The broken line $u_o(x)$ cannot have a complete link on the x-axis – for then it would be zero at two successive $u_o(i)$ and $S_{u_o}^+-S_{u_o}^-\geq 2$. Thus $u_o(x)$ has a simple zero (not a node) at γ, which must therefore be a junction, i, between links. In this case there must be three points $i-1$, i, $i+1$, such that $u_o(i-1)>0$, $u_o(i)=0$, $u_o(i+1)>0$ and again $S_{u_o}^+-S_{u_o}^-\geq 2$ in contradiction to (5.7.16). We conclude that between any two nodes of $u^{(k+1)}(x)$ there must be *at least one* node of $u^{(k)}x$. But the $u^{(k)}$-line has only $k-1$ nodes, while the $u^{(k+1)}$-line has k nodes. Thus $u^{(k)}(x)$ has *no more than* one node between two nodes of $u^{(k+1)}(x)$. ■

CHAPTER 6

SOME APPLICATIONS OF THE THEORY OF OSCILLATORY MATRICES

Memory is necessary for all the operations of reason.
Pascal's Pensées

6.1 The inverse mode problem for a Jacobian matrix

In this section we consider the problem of constructing a Jacobian matrix which has one or more specified eigenvectors. Following Vijay (1972), Gladwell (1985b), we prove:

Theorem 6.1.1. The necessary and sufficient condition for **u** to be an eigenvector of a positive definite Jacobian matrix with negative off-diagonal elements is that $S_u^+=S_u^-$.

Proof:

The necessity follows from Theorem 3.3.1. To prove sufficiency we need to show first that if $S_u^+=S_u^-$ then we can find $(a_r)_1^N>0$ and $(b_r)_1^{N-1}>0$ such that

$$(a_1-\lambda)u_1-b_1u_2=0 \tag{6.1.1}$$

$$-b_{r-1}u_{r-1}+(a_r-\lambda)u_r-b_ru_{r+1}=0\ ,\quad (r=1,2,...,N-1) \tag{6.1.2}$$

$$-b_{N-1}u_{N-1}+(a_N-\lambda)u_N=0\ . \tag{6.1.3}$$

If all $(u_r)_1^N\neq 0$ then we may take $(b_r)_1^{N-1}=1$ and

$$a_r=\lambda+c_r\ ,\quad c_r=(u_{r-1}+u_{r+1})/u_r\ ,\quad (r=1,2,...,N)\ , \tag{6.1.4}$$

where $u_o \equiv 0 \equiv u_{N+1}$. Thus

$$\mathbf{A}=\lambda\mathbf{I}+\mathbf{C}\ , \tag{6.1.5}$$

where

$$\mathbf{C}=\begin{bmatrix} c_1 & -1 & & \\ -1 & c_2 & -1 & \\ & & & \\ & & -1 & c_N \end{bmatrix}. \tag{6.1.6}$$

The matrix **C**, having non-zero off-diagonal elements, will have distinct eigenvalues $(\nu_i)_1^N$, one of which will be zero, so that **A** will have eigenvalues $(\lambda+\nu_i)_1^N$. Choose λ so that

$$\lambda > \max_{1\le i\le N}(-\nu_i) \tag{6.1.7}$$

then **A**, having positive eigenvalues will be positive definite.

The condition $S_u^+=S_u^-$ implies $u_1\neq 0$, $u_N\neq 0$. Suppose $u_m=0$ for some m satisfying $1<m<N$, then $u_{m-1}u_{m+1}<0$ and

$$b_{m-1}u_{m-1}+b_m u_{m+1}=0 \tag{6.1.8}$$

so that a_m, b_{m-1}, b_m may be taken so that

$$a_m=\lambda\ ,\quad b_{m-1}=1\ ,\quad b_m=-u_{m-1}/u_{m+1}\ . \tag{6.1.9}$$

The remaining b_r may be chosen so that $(b_r)_1^{m-1}=1$, $(b_r)_m^N=b_m$, and then λ may be chosen as before. ■

This argument may easily be generalized to the case when two or more (non-consecutive) u_r are zero.

Theorem 6.1.2. Let

$$s_n=\sum_{r=1}^{n}u_r v_r\ ,\quad t_n=v_n u_{n+1}-u_n v_{n+1}$$

then the necessary and sufficient conditions for **u**, **v** to be, for some i, j, unspecified except that $i<j$, the i-th and j-th eigenvectors of a positive definite Jacobian matrix with negative off-diagonal elements are:

(i) $S_u^+=S_u^-$, $S_v^+=S_v^-$

(ii) $s_N=0$

(iii) either $s_n=0=t_n$; or $s_n t_n>0$, $n=1,2,...N-1$. $n=1,2,...,N-1$.

Proof:

The conditions are necessary. For Theorems 5.5.7 and Corollary 1 to Theorem 5.7.2 yields (i). Condition (ii) states that $\mathbf{u}^T\mathbf{v}=0$, while condition (iii) follows from equation (3.3.13), namely

$$b_n t_n=(\mu-\lambda)s_n \ . \tag{6.1.10}$$

Note that (i) implies that u_1, v_1, u_N, v_N are all non-zero, so that $s_1t_1>0$ and $s_{N-1}t_{N-1}>0$.

Condition (i) also shows that if $s_{m-1}=0=s_m$ then $u_m=0=v_m$. For if $s_{m-1}=0=s_m$ then $u_mv_m=0$ and $v_mu_{m+1}-u_mv_{m+1}=0$. If $v_m\neq 0$ then $u_{m+1}=0$, contradicting (i). We deduce that three consecutive s_r $(r=m-1,m,m+1)$ cannot be simultaneously zero.

The conditions are sufficient. For equations (3.3.10) show that if $s_nt_n>0$ for $n=1,2,...,N-1$, then

$$a_n=\lambda+(\mu-\lambda)\left\{\frac{v_nu_{n+1}}{t_n}+\frac{s_{n-1}(v_{n-1}u_{n+1}-u_{n-1}v_{n+1})}{t_{n-1}t_n}\right\} \ . \tag{6.1.11}$$

Equations (6.1.11), (6.1.10) show that

$$a_n=\lambda+(\mu-\lambda)c_n \ , \ \ b_n=(\mu-\lambda)d_n \tag{6.1.12}$$

so that

$$\mathbf{A}=\lambda I+(\mu-\lambda)\mathbf{C} \tag{6.1.13}$$

where

$$\mathbf{C}=\begin{bmatrix} c_1 & -d_1 & 0 & \cdots & 0 & 0 \\ -d_1 & c_2 & -d_2 & \cdots & 0 & 0 \\ . & . & . & . & . & . \\ 0 & 0 & 0 & \cdots & -d_{N-1} & c_N \end{bmatrix} \tag{6.1.14}$$

Thus $\mathbf{C}$, having non-zero off-diagonal elements, will have distinct eigenvalues $(\nu_i)_1^N$. Two of the eigenvalues will be 0, 1, with corresponding eigenvectors $\mathbf{u}$, $\mathbf{v}$. Thus $\mathbf{A}$ will have eigenvalues $\lambda+(\mu-\lambda)\nu_i$, and $\mathbf{A}$ will be positive definite if $\lambda+(\mu-\lambda)\min\nu_i>0$.

If $s_m=0$, then $t_m=0$, and b_m may be chosen arbitrarily, e.g., $b_m=\mu-\lambda$. Then $s_{m-1}=-u_mv_m$ and (6.1.11) must be replaced by

$$a_m=\lambda-(\mu-\lambda)u_{m-1}v_m/t_{m-1} \ , \tag{6.1.15}$$

provided that $s_{m-1}\neq 0$ (i.e., $t_{m-1}\neq 0$). If $s_m=0=s_{m-1}$ then a_m may be chosen arbitrarily, e.g., $a_m=\lambda$. In any case the previous argument is not essentially affected. ■

We note that if $j=i+1$ then the t_n are all non-zero, and have the same sign, for $n=1,2,...,N-1$. (Ex. 5.7.2).

6.2 The inverse problem for a single mode of a spring-mass system

The eigenmodes $\mathbf{u}^{(i)}$ of the system of Figure 2.2.1, i.e, the fixed-free rod, are the eigenvectors of the equation

$$\omega^2\mathbf{Mu}=\mathbf{EKE}^T\mathbf{u} \tag{6.2.1}$$

which may be reduced to the standard form

$$\mathbf{Av}=\lambda\mathbf{v} \tag{6.2.2}$$

by the substitutions

$$\mathbf{v}=\mathbf{M}^{1/2}\mathbf{u}\ ,\quad \mathbf{A}=\mathbf{M}^{-1/2}\mathbf{EKE}^T\mathbf{M}^{-1/2}\ ,\quad \lambda=\omega^2\ . \tag{6.2.3}$$

Theorem 5.5.7 shows that the matrix $\mathbf{A}$ is sign-oscillatory, so that the analysis of Section 5.7 applies to the eigenvectors $\mathbf{v}^{(i)}$, and hence to $\mathbf{u}^{(i)}$.

Write $w_n=u_n-u_{n-1}$ so that

$$\mathbf{w}=\mathbf{E}^T\mathbf{u}\ ,\quad \text{i.e., } \mathbf{u}=\mathbf{E}^{-T}\mathbf{w}\ , \tag{6.2.4}$$

then equation (6.1.1) may be written

$$\omega^2\mathbf{ME}^{-T}\mathbf{w}=\mathbf{EKw}\ , \tag{6.2.5}$$

i.e.,

$$\omega^2\mathbf{w}=(\mathbf{E}^T\mathbf{M}^{-1}\mathbf{E})\mathbf{Kw}\ , \tag{6.2.6}$$

which may be written

$$\mathbf{Bz}=\lambda\mathbf{z}\ , \tag{6.2.7}$$

where

$$\mathbf{z}=\mathbf{K}^{1/2}\mathbf{w}\ ,\quad \mathbf{B}=\mathbf{K}^{1/2}\mathbf{E}^T\mathbf{M}^{-1}\mathbf{EK}^{1/2}\ ,\quad \lambda=\omega^2\ . \tag{6.2.8}$$

The matrix $\mathbf{B}$ is sign-oscillatory so that the eigenvectors $\mathbf{z}^{(i)}$, and hence $\mathbf{w}^{(i)}$, have the properties listed in Section 5.7 for the eigenvectors of a sign-oscillatory matrix.

We first prove two theorems regarding the shape of the vector $\mathbf{u}^{(i)}$.

Theorem 6.2.1. The mode $\mathbf{u}^{(i)}$ can assume a relative maximum (minimum) at $\mathbf{u}_n^{(i)}$ $(n=1,2,...,N-1)$ only if $u_n^{(i)}>0(<0)$.

Proof:

We drop the suffix i. The recurrence relations are

$$m_n\omega_i^2u_n=\theta_n-\theta_{n+1}\ ,\quad \theta_n=k_n(u_n-u_{n-1})\ .$$

Suppose u has a relative maximum at u_n. Then $u_n\geq u_{n-1}$ and $u_n\geq u_{n+1}$, so that $\theta_n\geq 0$ and $\theta_{n+1}\leq 0$, and hence $u_n\geq 0$. In fact, since θ_n, θ_{n+1} cannot be simultaneously zero, $u_n > 0$. ■

Theorem 6.2.2. Two neighbouring u_n can be equal only at a relative maximum or minimum.

Proof:

Suppose $u_n = u_{n-1}$, then $\theta_n = 0$, so that θ_{n-1} and θ_{n+1} are non-zero and have opposite signs, i.e.,

$$(u_n - u_{n-2})(u_n - u_{n+1}) > 0 \ . \ ■$$

These theorems show that $\mathbf{u}^{(i)}$ will consist of $i-1$ portions which are concave towards the axis, and a final portion that bends away from the axis, as shown in Figure 6.2.1

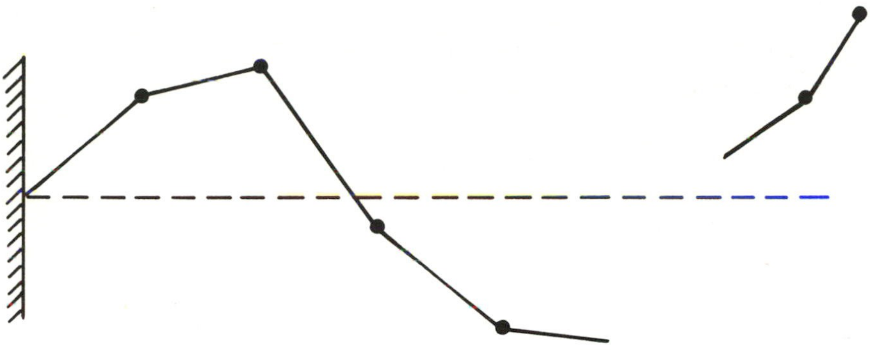

Figure 6.2.1 – The ith mode of the spring-mass system

Theorem 6.2.3. The necessary and sufficient conditions for $\mathbf{u}$ to be the ith mode of a spring-mass system in the fixed-free configuration are that

(i) $S_u^+ = S_u^- = S_w^+ = S_w^- = i-1$,

(ii) $u_1 w_1 > 0$,

where $w_n = u_n - u_{n-1}$.

Proof:

The necessity of these conditions has already been established. To prove sufficiency we first note that conditions (i), (ii) imply $u_n w_n > 0$ and that no two of u_n, u_{n-1} and w_n can be simultaneously zero; now we construct a system.

The mode will have a shape like that shown in Fig. 6.2.1. Thus u_n will start positive ($u_n > 0$, $w_n > 0$) until an index r, the first for which

$$u_r = 0 \ , \quad w_r \geq 0 \ , \quad w_{r+1} < 0 \ .$$

Then u_n will decrease ($u_n > 0$, $w_n < 0$) until an index s, the first for which

$$u_s \geq 0 \ , \ \ u_{s+1} < 0 \ , \ \ w_s < 0 \ .$$

Now u_n will continue to decrease ($u_n < 0$, $w_n < 0$) until an index t, the first for which

$$u_t < 0 \ , \ \ w_t \leq 0 \ , \ \ w_{t+1} > 0 \ ,$$

and then proceed to increase again.

The governing equation (6.2.1) may be written

$$\mathbf{Kw} = \lambda \mathbf{E}^{-1}\mathbf{Mu} \tag{6.2.9}$$

and since $\mathbf{E}^{-1}$ is given by equation (2.2.14), we have

$$k_n w_n = \lambda \sum_{i=n}^{N} m_i u_i = \lambda \sigma_n \ . \tag{6.2.10}$$

This shows that σ_n has the same sign as w_n.

Now refer to Fig. 6.2.2. Choose the $m_n > 0$ so that

(i) $\sigma_r \geq 0$, with $\sigma_r = 0$ if and only if $w_r = 0$. Then

$$\sigma_n = \sigma_r + \sum_{i=n}^{r-1} m_i u_i > 0 \ , \quad \text{for } n = 1,2,...,r-1 \ , \tag{6.2.11}$$

(ii) $\sigma_{r+1} < 0$, then $\sigma_n < 0$ for $n = r+1,...,s$,

(iii) $\sigma_t \leq 0$, with $\sigma_t = 0$ if and only if $w_t = 0$,

(iv) $\sigma_{t+1} > 0$,

and so on. This procedure yields a set of σ_n having the same sign as w_n. If $w_n \neq 0$ then k_n is given by equation (6.2.10), while if $w_n = 0$, then k_n may be given an arbitrary positive value. ■

The question of reconstructing a spring mass system from model data was considered by Porter (1970, 1971), but he did not discuss the necessary or sufficient conditions on the modes for the masses and stiffnesses to be positive.

6.3 The reconstruction of a spring-mass system from two modes

The reconstruction descibed in Section 6 was far from unique. In this section we shall show that, provided certain conditions are satisfied, there is a unique system for which two given modes are eigenmodes.

We first prove that two modes $\mathbf{u}^{(i)}$, $\mathbf{u}^{(j)}$ may be given which satisfy the conditions of Theorem 6.2.3, but for which there is *no* system for which they are both eigenmodes. We produce a counter example. Suppose

$$u^{(1)} = \{1,2,3\} \ , \ \ w^{(1)} = \{1,1,1\} \ , \ \ u^{(3)} = \{1,-1,4\} \ , \ \ w^{(3)} = \{1,-2,5\} \tag{6.3.1}$$

then the conditions of Theorem 6.2.3 are satisfied. The recurrence relations are

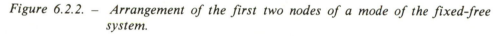

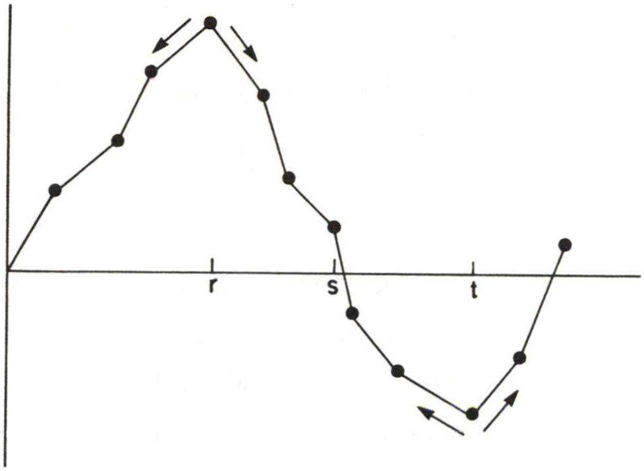

Figure 6.2.2. – Arrangement of the first two nodes of a mode of the fixed-free system.

$$\lambda_1 m_1 = k_1 - k_2 \ , \quad 2\lambda_1 m_2 = k_2 - k_3 \ , \quad 3\lambda_1 m_3 = k_3 \ , \tag{6.3.2}$$

$$\lambda_3 m_1 = k_1 + 2k_2 \ , \quad \lambda_3 m_2 = 2k_2 + 5k_3 \ , \quad 4\lambda_3 m_3 = 5k_3 \ . \tag{6.3.3}$$

Thus, $\lambda_3/\lambda_1 = 15/4$ so that

$$\frac{2k_2+5k_3}{k_2-k_3} = \frac{\lambda_3}{2\lambda_1} = \frac{15}{8} \tag{6.3.4}$$

i.e., $k_2 = -55k_3$, which is unrealizable.

In order to derive the conditions on the modes we formalize the elimination procedure used in the example. For convenience we set

$$\mathbf{u}^{(i)} = u \ , \quad \mathbf{u}^{(j)} = \mathbf{v} \ , \quad \mathbf{w}^{(i)} = \mathbf{w} \ , \quad \mathbf{w}^{(j)} = \mathbf{z} \ , \tag{6.3.5}$$

then the recurrence relations are

$$\lambda_i m_n u_n = k_n w_n - k_{n+1} w_{n+1} \ , \quad n = 1,2,\ldots,N-1 \ , \tag{6.3.6}$$

$$\lambda_j m_n v_n = k_n z_n - k_{n+1} z_{n+1} \ , \quad n = 1,2,\ldots,N-1 \ , \tag{6.3.7}$$

and

$$\lambda_i m_N u_N = k_N w_N \ , \quad \lambda_j m_N v_n = k_N z_N \ . \tag{6.3.8}$$

Thus

$$\frac{\lambda_i}{\lambda_j}=\frac{w_N}{u_N}\cdot\frac{v_N}{z_N}\ , \tag{6.3.9}$$

and the condition $\lambda_i<\lambda_j$ means that

$$s_{N-1}\equiv u_N z_N - v_N w_N>0\ . \tag{6.3.10}$$

Since $w_N=u_N-u_{N-1}$, $z_N=v_N-v_{N-1}$, this inequality may be written

$$u_N(v_N-v_{N-1})-v_N(u_N-u_{N-1})>0\ ,$$

i.e.,

$$s_{N-1}\equiv u_{N-1}v_N-u_Nv_{N-1}>0\ . \tag{6.3.11}$$

Eliminating k_n, k_{n+1} in turn from equations (6.3.6), (6.3.7), we find

$$m_n[\lambda_i u_n z_n-\lambda_j v_n w_n]=k_{n+1}[w_n z_{n+1}-w_{n+1}z_n]\ ,$$

$$m_n[\lambda_i u_n z_{n+1}-\lambda_j v_n w_{n+1}]=k_n[w_n z_{n+1}-w_{n+1}z_n]\ ,$$

so that on substituting for λ_i/λ_j from equation (6.3.9) we find

$$\lambda_i m_n p_n=k_{n+1}w_N v_N r_n\ , \tag{6.3.12}$$

$$\lambda_i m_n q_n=k_n w_N v_N r_n\ , \tag{6.3.13}$$

$$p_n=u_n v_N w_N z_n-u_N v_n w_n z_N\ , \tag{6.3.14}$$

$$q_n=u_n v_N w_N z_{n+1}-u_N v_n w_{n+1} z_N\ , \tag{6.3.15}$$

$$r_n=w_n z_{n+1}-w_{n+1}z_n\ . \tag{6.3.16}$$

Thus we may state

Theorem 6.3.1. The necessary and sufficient conditions for $\mathbf{u}^{(i)}=\mathbf{u}$ and $\mathbf{u}^{(j)}\equiv\mathbf{v}$ to be the ith and jth modes of the spring mass system for some $i<j$ are

(i) $S_u=S_w<S_v=S_z$,

(ii) $v_N w_N>0$,

(iii) $s_{N-1}\equiv u_{N-1}v_N-u_Nv_{N-1}>0$,

(iv) for each value of n the three quantities p_n, q_n, r_n have the same sign or are identically zero; this sign need not be the same for all $n=1,2,\ldots,N-1$.

Proof:

The necessity of (i) has already been proved. The necessity of (ii) follows from (6.3.8), and that of (iii) from (6.3.11). Equations (6.3.12), (6.3.13) show the necessity of (iv). On the other hand, equations (6.3.12), (6.3.13) show that if p_n, q_n, r_n are all *non-zero* and of the same sign and (ii), (iii) hold then there is a system for which $\mathbf{u}$, $\mathbf{v}$ are eigenvectors corresponding to two eigenvalues whose ratio is given by (6.3.9). These eigenvectors will then satisfy (i). If some of the triplets p_n, q_n, r_n are identically zero then (i) has to be retained. When the conditions of the theorem are satisfied, equations (6.3.8), (6.3.12), (6.3.13) yield unique values of λ_i m_n, k_n, $n=1,2,...,N$, apart from a single scale factor. ■

A careful analysis shows that $r_{N-1}>0$ so that one must have $p_{N-1}>0$, $q_{N-1}>0$. On the other hand $p_1= -u_1v_1s_{N-1}\neq 0$ so that $q_1p_1>0$, $s_1p_1>0$.

In the particular case $j=i+1$ the conditions may be made sharper, to give

Theorem 6.3.2. The necessary and sufficient conditions for $\mathbf{u}^{(i)}\equiv\mathbf{u}$ and $\mathbf{u}^{(j)}\equiv\mathbf{v}$ to be the ith and $(i+1)$th eigenmodes of the spring mass system are that

(i) $v_Nz_N>0$,

(ii) $s_{N-1}>0$,

(iii) $(p_n,q_n,r_n)_1^{N-1}>0$.

Proof:

The necessity of (i) and (ii) follow from equations (6.3.8), (6.3.9). The necessity of $(r_n)_1^{N-1}>0$ follows from Ex. 5.7.2. Then equations (6.3.12), (6.3.13) show that $(p_n,q_n)_1^{N-1}>0$. The sufficiency of (i) - (iii) follows as before. ■

6.4 A note on the matrices appearing in a finite element model of a rod

The analysis of oscillatory matrices in Chapter 5 allows us to obtain some qualitative properties of the eigenvalues and eigenvectors obtained by using a finite element model of a rod.

In Section 2.4, it was shown that, for such a model, the inertia and stiffness matrices, **A**, **C** were both tridiagonal. Moreover equation (2.4.9) shows that the matrix **A**, in addition to being positive definite, has positive off-diagonal elements $a_{r,r+1}$. Equation (2.4.10) shows that **C** will be positive definite if the rod is anchored (fixed) at at least one end, and has negative off-diagonal elements.

$$\mathbf{A}=\begin{bmatrix} a_1 & b_1 & & & \\ b_1 & a_2 & b_2 & & \\ \cdot & \cdot & \cdot & \cdot & \cdot \\ & & & b_{N-1} & a_N \end{bmatrix}, \quad \mathbf{C}=\begin{bmatrix} c_1 & -d_1 & & & \\ -d_1 & c_2 & -d_2 & & \\ \cdot & \cdot & \cdot & \cdot & \cdot \\ & & & -d_{N-1} & c_N \end{bmatrix} \tag{6.4.1}$$

and the eigenvalue equation is

$$(\mathbf{C}-\lambda\mathbf{A})\mathbf{y}=0\,. \tag{6.4.2}$$

In order to reduce this equation to one for a single matrix **B** we factorize **C**, writing

$$\mathbf{C}=\mathbf{L}\mathbf{L}^T\,, \tag{6.4.3}$$

where **L** is lower triangular. Clearly

$$\mathbf{L}=\begin{bmatrix} \ell_{11} & & & \\ -\ell_{21} & \ell_{22} & & \\ \cdot & \cdot & \cdot & \cdot \\ & & -\ell_{N,N-1} & \ell_{NN} \end{bmatrix}. \tag{6.4.4}$$

Since $\mathbf{L}$ is sign-definite, its inverse $\mathbf{L}^{-1}$ is completely non-negative; it is also non-singular where all $\ell_{ij}>0$. Put

$$\mathbf{L}^T\mathbf{y}=\mathbf{v}\ , \tag{6.4.5}$$

and then

$$\mathbf{L}^{-1}(\mathbf{C}-\lambda\mathbf{A})\mathbf{y}=(\mathbf{I}-\lambda\mathbf{L}^{-1}\mathbf{A}\mathbf{L}^{-T})\mathbf{v}=0\ . \tag{6.4.6}$$

The matrix $\mathbf{L}^{-1}\mathbf{A}$, being the product of non-singular, completely non-negative matrices, is completely non-negative (Ex. 5.5.1), and so is $\mathbf{B}=(\mathbf{L}^{-1}\mathbf{A})\mathbf{L}^{-T}$. Moreover, $\mathbf{B}$ has non-zero off-diagonal elements, and so is oscillatory. Thus the eigenvalue problem reduces to

$$(\mathbf{I}-\lambda\mathbf{B})\mathbf{v}=0\ , \tag{6.4.7}$$

where $\mathbf{B}$ is oscillatory. It can also be written

$$(\mathbf{B}^{-1}-\lambda\mathbf{I})\mathbf{v}=0\ , \tag{6.4.8}$$

where $\mathbf{B}^{-1}$ is sign-oscillatory. In either case its eigenproperties follow from the analysis of Section 5.7. Thus we may conclude that the eigenvalues obtained from a finite element model of the rod are all simple, while the eigenvectors $\mathbf{v}^{(i)}$ have the oscillatory properties described in Theorem 5.7.2.

The procedure described above is the one that is used in numerical work; it yields a symmetric matrix $\mathbf{B}$. Unfortunately, Theorem 5.7.2 yields results concerning the vectors $\mathbf{v}^{(i)}$ and not the physically meaningful vectors $\mathbf{y}^{(i)}$. To obtain such results we multiply equation (6.4.2) by $\mathbf{C}^{-1}$ to obtain

$$(\mathbf{I}-\lambda\mathbf{C}^{-1}\mathbf{A})\mathbf{y}=0\ .$$

Since $\mathbf{C}$ is sign-oscillatory, its inverse $\mathbf{C}^{-1}$ is oscillatory and hence, by Ex. 5.5.2, $\mathbf{C}^{-1}\mathbf{A}$ is oscillatory (and $\mathbf{A}^{-1}\mathbf{C}$ is sign-oscillatory). Hence the eigenvectors $\mathbf{y}^{(i)}$ will have the oscillatory properties described in Theorem 5.7.2. Note that all the theory of oscillatory matrices applies to non-symmetric as well as symmetric matrices because it stems from Perron's theorem.

CHAPTER 7

THE INVERSE PROBLEM FOR THE DISCRETE VIBRATING BEAM

There is enough light for those who only desire to see, and enough obscurity for those who have a contrary disposition.

Pascal's Pensées

7.1 Introduction

In this chapter we shall present in detail the solution of the inverse problem for the discrete spring-mass-rod model of a vibrating beam discussed in Section 2.3. This model is important because it is the simplest model – it is in effect a finite-difference approximation – for a beam with continuously distributed mass. The inverse problem for a continuous beam will be considered in Chapter 10. The inverse problem for a discrete beam was first considered by Barcilon (1979, 1982). He established that the reconstruction of such a system would require three spectra, corresponding to three different end conditions. The necessary and sufficient conditions for this spectral data to correspond to a realizable system, one with positive masses, lengths and stiffnesses, were derived by Gladwell (1984).

Two papers by Sweet (1969, 1971) consider the discrete model of a beam obtained by using the so-called 'method of straight lines'; he shows that the coefficient matrix obtained in this procedure is (similar to) an oscillatory matrix.

The plan of the chapter is as follows. In Section 7.2 we show that natural frequencies of the system are the eigenvalues of an oscillatory matrix. This means that the eigenvalues are distinct and the eigenvectors $\mathbf{u}^{(i)}$ have all the properties derived in Section 5.7. It is found, moreover, that not only $\mathbf{u}^{(i)}$, but also $\boldsymbol{\theta}^{(i)}$, $\tau^{(i)}$,

$\phi^{(i)}$, the slopes, moments and shearing forces, have these same properties (Theorem 7.2.2 and Ex. 7.2.1). Theorem 7.2.3 derives an additional result, that the beam always bends away from the axis at a free end. In Section 7.4 the oscillatory properties of the eigenvectors are used in the ordering (7.4.19) of the natural frequencies of the system corresponding to different end conditions. In Section 7.5.5 it is shown that while it is possible to take three spectra as the data for the reconstruction, it is better to take one spectrum, that corresponding to a free end, and the end values $u_N^{(i)}$, $\theta_N^{(i)}$ of the normalized eigenvectors, as the basic data. In this way, the conditions on the data may be written as determinantal inequalities. In Section 7.6 a procedure for inversion is presented and it is shown that the conditions (Theorem 7.5.1) which were put forward earlier are in fact sufficient to ensure that all the physical parameters, masses, lengths and stiffnesses, will be positive. Finally, in Sections 7.7-7.8 a numerical procedure, based on the block Lanczos algorithm, is described for the actual computation of the physical parameters.

7.2 The eigenanalysis of the clamped-free beam

The equations governing the response of the beam were derived in Section 2.3. Equation (2.3.11) shows that vibration with frequency ω is governed by the equation

$$\lambda \mathbf{M}\mathbf{u}=\mathbf{C}\mathbf{u}-\phi_N\mathbf{e}_N-\ell_N^{-1}\tau_N\mathbf{E}\mathbf{e}_N\ ,\quad \lambda=\omega^2\ . \tag{7.2.1}$$

This shows that the free vibrations satisfy

$$\lambda \mathbf{M}\mathbf{u}=\mathbf{C}\mathbf{U} \tag{7.2.2}$$

which may be reduced to standard form

$$\mathbf{A}\mathbf{v}=\lambda\mathbf{v} \tag{7.2.3}$$

by the substitution

$$\mathbf{v}=\mathbf{M}^{1/2}\mathbf{u}\ ,\quad \mathbf{A}=\mathbf{M}^{-1/2}\mathbf{C}\mathbf{M}^{-1/2}\ . \tag{7.2.4}$$

Theorem 7.2.1. The matrix **A** is sign-oscillatory.

Proof:

Equation (2.3.12) shows that $\mathbf{C}=\mathbf{E}\mathbf{L}^{-1}\mathbf{E}\mathbf{K}\mathbf{E}^T\mathbf{L}^{-1}\mathbf{E}^T$; **E** is sign-definite so that $\mathbf{E}^*$ is completely non-negative. Ex. 5.5.1 shows that $\mathbf{B}^*=\mathbf{E}^*\mathbf{L}^{-1}\mathbf{E}^*$ is completely non-negative as is its transpose, and hence also $\mathbf{C}^*=\mathbf{B}^*\mathbf{K}\mathbf{B}^{*T}$. But $\mathbf{C}^*$ is non-singular and a simple calculation shows that its quasi-principal elements are positive. Hence $\mathbf{C}^*$, and $\mathbf{A}^*$, are oscillatory. ■

Theorem 7.2.1 has important consequences. It means that the eigenvalues $(\lambda_i)_1^N$ are distinct (Theorem 5.7.1); that the last element $u_N^{(i)}$ in each eigenvector $\mathbf{u}^{(i)}$ of equation (7.2.2) may be chosen to be positive; (Corollary 2 to Theorem 5.7.2; note that $v_N^{(i)}= m_N^{1/2}u_N^{(i)}$, so that $v_N^{(i)}>0$ implies $u_N^{(i)}>0$); the $u_r^{(i)}$ will satisfy the inequalities (5.7.11). We now prove

Theorem 7.2.2. The vectors $(\theta^{(i)})_1^N$ are the eigenvectors of a sign-oscillatory matrix.

Proof:

Since $\boldsymbol{\theta}=\mathbf{L}^{-1}\mathbf{E}^T\mathbf{u}$ and $\mathbf{u}=\mathbf{E}^{-T}\mathbf{L}\boldsymbol{\theta}$ we have

$$\lambda\mathbf{Mu}=\omega^2\mathbf{ME}^{-T}\mathbf{L}\boldsymbol{\theta}=\mathbf{Cu}=\mathbf{EL}^{-1}\mathbf{EKE}^T\boldsymbol{\theta}\ , \tag{7.2.5}$$

so that

$$\lambda\mathbf{G}\boldsymbol{\theta}=\mathbf{H}\boldsymbol{\theta}\ , \tag{7.2.6}$$

where

$$\mathbf{G}=\mathbf{LE}^{-1}\mathbf{ME}^{-T}\mathbf{L}\ ,\quad \mathbf{H}=\mathbf{EKE}^T\ . \tag{7.2.7}$$

The matrix $\mathbf{G}$ is oscillatory so that $(\mathbf{G}^{-1})^*$ is oscillatory (Theorem 5.5.6). $\mathbf{H}$ is sign-oscillatory so that $\mathbf{H}^*$ is oscillatory. Therefore, by Ex. 5.5.2, $(\mathbf{G}^{-1}\mathbf{H})^*$ is oscillatory and thus $\mathbf{G}^{-1}\mathbf{H}$ is sign-oscillatory. ■

Theorem 7.2.2 means that the $\boldsymbol{\theta}^{(i)}$ must satisfy all the requirements for the eigenvectors of a sign-oscillatory matrix, e.g., $\theta_N^{(i)}\neq 0$. We now show that, for the particular sign-oscillatory matrix system governing the beam, if the $u_N^{(i)}$ are chosen so that $u_N^{(i)}>0$, so that also all the minors $\underline{u}_{\underline{N}}^{(s)}$ of Theorem 5.7.3 are positive, then $\theta_N^{(i)}$, and hence all the corresponding minors

$$\underline{\theta}_{\underline{N}}^{(s)}=\theta\begin{pmatrix}N-p+1, & N-p+2, & \ldots, & N\\ i_1, & i_2, & \ldots, & i_p\end{pmatrix} \tag{7.2.8}$$

will be positive. It is sufficient to prove

Theorem 7.2.3. Each eigenvector of the clamped-free beam satisfies the inequality

$$u_N^{(i)}\theta_N^{(i)}>0\ .$$

Proof:

Choose $\mathbf{u}^{(i)}$ so that $u_N^{(i)}>0$. There is an index r $(1\leq r\leq N-1)$ such that

(i) $u_s^{(i)}>0,\ s=r,r+1,\ldots,N,$

(ii) $u_{r-1}^{(i)}\leq 0.$

(Note that when $i=1$, then $r=1$; note that equation (2.3.6) gives $u_o\equiv 0$).

Then

$$\theta_r^{(i)}=(u_r^{(i)}-u_{r-1}^{(i)})/\ell_r>0\ . \tag{7.2.9}$$

Since

$$\phi^{(i)}=\lambda_i \mathbf{E}^{-1}\mathbf{M}\mathbf{u}^{(i)} \ , \tag{7.2.10}$$

then, because of the form of $\mathbf{E}^{-1}$ given in equation (2.2.14) we have

$$\phi_s^{(i)}>0 \ , \quad s=r-1,r,...,N-1 \ . \tag{7.2.11}$$

But

$$\tau^{(i)}=\mathbf{E}^{-1}\mathbf{L}\phi^{(i)} \ , \tag{7.2.12}$$

so that

$$\tau_s^{(i)}>0 \ , \quad s=r-1,r,...,N-1 \ . \tag{7.2.13}$$

Now consider equation (2.3.3), namely

$$\theta_{s+1}-\theta_s=k_{s+1}^{-1}\tau_s \tag{7.2.14}$$

and sum from $s=r$ to $N-1$ to obtain

$$\theta_N^{(i)}-\theta_r^{(i)}=\sum_{s=r}^{N-1} k_{s+1}^{-1}\tau_s^{(i)} \ > 0 \ , \tag{7.2.15}$$

so that $\theta_N^{(i)}>0$. ■

Theorem 7.2.2, while showing that the $\boldsymbol{\theta}^{(i)}$ are eigenvectors of a sign-oscillatory matrix, shows that $\mathbf{u}^{(i)}$ and $\boldsymbol{\theta}^{(i)}$ must both have exactly $i-1$ sign-interchanges. This means that the first mode, $\mathbf{u}^{(1)}$ will steadily increase, i.e.,

$$0<u_1^{(1)}<u_2^{(1)}< \cdots <u_N^{(1)} \ , \tag{7.2.16}$$

as shown in Figure 7.2.1, while the i-th mode ($i>1$) will have ($i-1$) portions that are convex towards the axis, and one final portion that bends away from the axis, as shown in Figure 7.2.2.

Examples 7.2

1. Show that $\tau^{(i)}$ and $\phi^{(i)}$ are eigenvectors of the equations

$$\lambda\mathbf{K}^{-1}\tau=\mathbf{E}^T\mathbf{L}^{-1}\mathbf{E}^T\mathbf{M}^{-1}\mathbf{E}\mathbf{L}^{-1}\mathbf{E}\tau$$

$$\lambda\mathbf{L}\mathbf{E}^{-T}\mathbf{K}^{-1}\mathbf{E}^{-1}\mathbf{L}\phi=\mathbf{E}^T\mathbf{M}^{-1}\mathbf{E}\phi$$

   and that each is the eigenvector of a sign-oscillatory matrix.

7.3 The forced response of the beam

The equation governing the response to an end shearing force and bending moment is equation (2.3.11), which for vibration of frequency ω becomes

$$\lambda\mathbf{M}\mathbf{u}=\mathbf{C}\mathbf{u}-\phi_N\mathbf{e}_N-\ell_N^{-1}\tau_N\mathbf{E}\mathbf{e}_N \ . \tag{7.3.1}$$

Since the eigenvalues $\mathbf{u}^{(i)}$ of the clamped-free beam span the space of N-vectors and are orthogonal w.r.t. $\mathbf{M}$ and $\mathbf{C}$ we may write

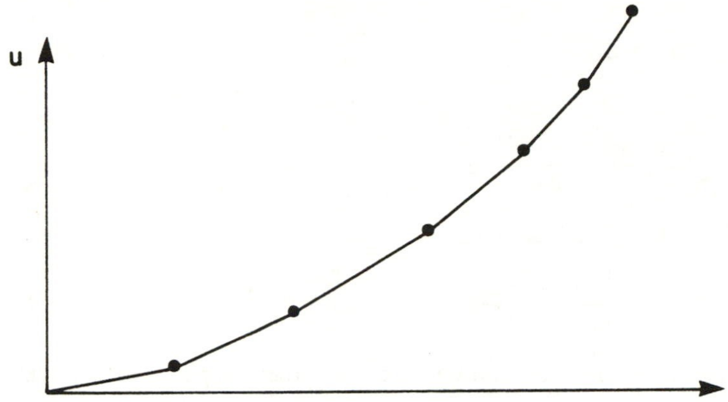

Figure 7.2.1 – A possible shape for $\mathbf{u}^{(1)}$

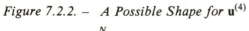

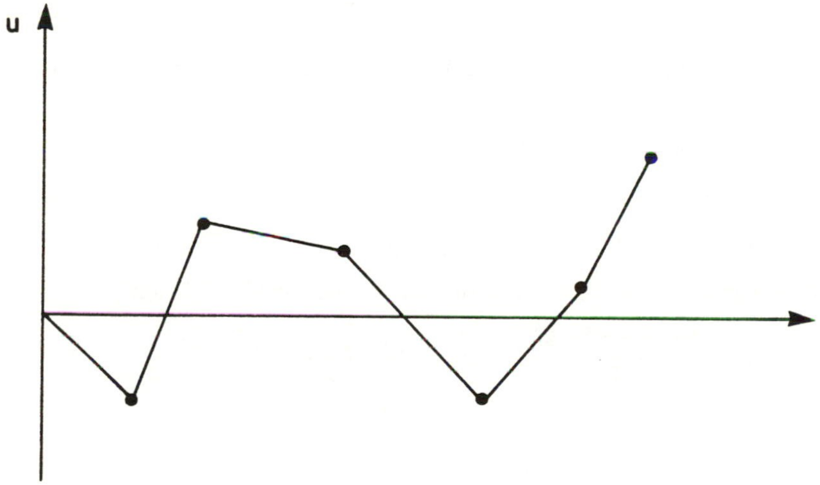

Figure 7.2.2. – A Possible Shape for $\mathbf{u}^{(4)}$

$$u = \sum_{i=1}^{N} \alpha_i \mathbf{u}^{(i)} \tag{7.3.2}$$

and find

$$\alpha_i = [\phi_N u_N^{(i)} + \tau_N \theta_N^{(i)}]/(\lambda_i - \lambda) , \tag{7.3.3}$$

where

$$\mathbf{u}^{(i)T}\mathbf{M}\mathbf{u}^{(j)} = \delta_{ij} . \tag{7.3.4}$$

Thus,

$$\mathbf{u} = \sum_{i=1}^{N} \frac{[\phi_N u_N^{(i)} + \tau_N \theta_N^{(i)}]}{\lambda_i - \lambda} \mathbf{u}^{(i)} \tag{7.3.5}$$

and on multiplying by $\mathbf{L}^{-1}\mathbf{E}^T$ we find

$$\boldsymbol{\theta} = \sum_{i=1}^{N} \frac{[\phi_N u_N^{(i)} + \tau_N \theta_N^{(i)}]}{\lambda_i - \lambda} \boldsymbol{\theta}^{(i)} . \tag{7.3.6}$$

These two equations completely characterize the forced response of the beam.

There is another, completely independent, but equivalent way of describing the forced response. The governing equations for the beam are fourth-order difference equations so that there will be, for any frequency ω, two independent solutions $u^I(\lambda)$ and $u^{II}(\lambda)$ satisfying the clamped-conditions at the left hand end. The solutions may be chosen so that

$$\text{I} \ : u_{-1} = 0 = \theta_o = \phi_{-2} , \quad \tau_{-1} = 1 \tag{7.3.7}$$

$$\text{II} \ : u_{-1} = 0 = \theta_0 = \tau_{-1} , \quad \phi_{-2} = 1 , \tag{7.3.8}$$

Any mode of vibration corresponding to the left hand end clamped will be a combination of I and II, i.e.,

$$u_n = \alpha u_n^I(\lambda) + \beta u_n^{II}(\lambda) . \tag{7.3.9}$$

If $\mathbf{u}$ is the forced response due to ϕ_N and τ_N, then α, β must be chosen so that

$$\phi_N = \alpha\phi_N^I + \beta\phi_N^{II} , \quad \tau_N = \alpha\tau_N^I + \beta\tau_N^{II} , \tag{7.3.10}$$

so that, on solving for α, β, we find the end displacement and slope

$$u_N = \frac{\phi_N \Pi_N + \tau_N B_N}{\Phi_N} , \quad \theta_N = \frac{\phi_N B_N^* + \tau_N \Sigma_N}{\Phi_N} , \tag{7.3.11}$$

where for general n

$$\Pi_n(\omega^2) = \begin{vmatrix} \tau_n^I & u_n^I \\ \tau_n^{II} & u_n^{II} \end{vmatrix} = \{\tau_n, u_n\} , \tag{7.3.12}$$

$$B_n(\lambda) = \{u_n, \phi_n\} , \quad \Phi_n(\lambda) = \{\tau_n, \phi_n\} , \tag{7.3.13}$$

$$B_n^*(\lambda) = \{\tau_n, \theta_n\} , \quad \Sigma_n(\lambda) = \{\theta_n, \phi_n\} . \tag{7.3.14}$$

On comparing equation (7.3.11) with equations (7.3.5), (7.3.6), we see that

$$\frac{\Pi_N}{\Phi_N}=\sum_{i=1}^{N}\frac{[u_N^{(i)}]^2}{\lambda_1-\lambda}\ ,\quad \frac{\Sigma_N}{\Phi_N}=\sum_{i=1}^{N}\frac{[\theta_N^{(i)}]^2}{\lambda_i-\lambda}\ , \tag{7.3.15}$$

$$\frac{B_N}{\Phi_N}=\frac{B_N^*}{\Phi_N}=\sum_{i=1}^{N}\frac{u_N^{(i)}\theta_N^{(i)}}{\lambda_i-\lambda}\ . \tag{7.3.16}$$

In order to find the 'brackets' Φ_n, Π_n, Σ_n and B_n we develop the solutions I and II starting from the conditions (7.3.7), (7.3.8). They are:

I.

$$\begin{aligned}
&u_{-1}=0=u_o\ ,\ u_1=\ell_1K_1\ ,\ u_2=\ell_1K_1+\ell_2(K_1+K_2)\\
&\theta_o=0\ ,\ \theta_1=K_1\ ,\ \theta_2=K_1+K_2\\
&\tau_{-1}=1=\tau_o=\tau_1\ ,\ \tau_2=1+m_1\ell_1\ell_2K_1\lambda\\
&\phi_{-2}=0=\phi_{-1}=\phi_o\ ,\ \phi_1=-m_1\ell_1K_1\lambda\ ,
\end{aligned} \tag{7.3.17}$$

II.

$$\begin{aligned}
&u_{-1}=0=u_o\ ,\ u_1=-K_1\ell_o\ell_1\ ,\\
&u_2=-K_1\ell_o(\ell_1+\ell_2)-K_2\ell_2(\ell_o+\ell_1)\ ,\\
&\theta_o=0\ ,\ \theta_1=-K_1\ell_o\ ,\ \theta_2=-K_1\ell_o-K_2(\ell_o+\ell_1)\\
&\tau_{-1}=0\ ,\ \tau_o=-\ell_o\ ,\ \tau_1=-\ell_o-\ell_1\\
&\phi_{-2}=1=\phi_{-1}=\phi_o\ ,\ \phi_1=1+m_1\ell_o\ell_1K_1\lambda\ .
\end{aligned} \tag{7.3.18}$$

It is convenient to introduce two new brackets, namely

$$\Gamma_n=\{\theta_n,u_n\}\ ,\ H_n=\{\theta_{n+1},\phi_n\}\ . \tag{7.3.19}$$

Then

$$\begin{aligned}
&B_o=0=B_o^*\ ,\ \Pi_o=0\ ,\ \Sigma_o=0\ ,\ \Gamma_o=0\ ,\ H_o=K_1\ ,\ \Phi_o=1\ ,\\
&B_1=\ell_1K_1=B_1^*\ ,\ \Pi_1=\ell_1^2K_1\ ,\ \Sigma_1=K_1\ ,\ \Gamma_1=0\ ,\\
&H_1=K_1+K_2+m_1\ell_o\ell_1K_1K_2\lambda\ ,\ \Phi_1=1-m_1\ell_1^2K_1\lambda\ ,\quad \text{etc.,}
\end{aligned} \tag{7.3.20}$$

where $K_n=1/k_n$. The brackets Φ_n, H_n are polynomials in λ of degree n; Π_n, $B_n\equiv B_n^*$, Σ_n of degree $n-1$ and Γ_n of degree $n-2$. The polynomials are related by recurrence equations; for instance

$$\begin{aligned}
\Phi_n &= \{\tau_n,\phi_n\}=\{\tau_n,\phi_{n-1}-m_n\lambda u_n\}\\
&= \{\tau_{n-1}-\ell_n\phi_{n-1},\phi_{n-1}\}-m_n\lambda\{\tau_n,u_n\}
\end{aligned} \tag{7.3.21}$$

The full set of relations is

$$\Gamma_n = \Gamma_{n-1} + K_n \Pi_{n-1} \, , \quad \Pi_n = \Pi_{n-1} + \ell_n B_{n-1} + \ell_n B_n \, , \tag{7.3.22}$$

$$\Phi_n = \Phi_{n-1} - m_n \lambda \Pi_n \, , \quad \Sigma_n = H_{n-1} - m_n \lambda \Gamma_n \, , \tag{7.2.23}$$

$$H_n = \Sigma_n + K_{n+1} \Phi_n \, , \quad B_n = B_{n+1} + \ell_n H_{n-1} \, . \tag{7.3.24}$$

The brackets are linked by the quadratic identity

$$\Gamma_n \Phi_n + B_n^2 - \Pi_n \Sigma_n = 0 \, . \tag{7.3.25}$$

Examples 7.3

1. Verfiy that the quantities Π_n, B_n, Σ_n, Γ_n and Φ_n are independent of ℓ_o.
2. Establish the quadratic identity (7.3.25) on the basis of the recurrence relations (7.3.22) - (7.3.24).
3. Use the quadratic identity and the equations (7.3.15) - (7.3.16) to obtain an expression for Γ_N / Φ_N.

7.4 The spectra of the beam

Now suppose that the left hand end of the beam remains clamped while the conditions at the right hand end are varied. The possible end conditions and eigenvalues i.e., (eigenfrequencies)^2 are:

free: $\phi_N = 0 = \tau_N \quad (\lambda_i)_1^N$

pinned: $u_N = 0 = \tau_N \quad (\mu_i)_1^{N-1}$

sliding: $\theta_N = 0 = \phi_N \quad (\sigma_i)_1^{N-1}$

anti-resonant: $\left\{ \begin{array}{l} u_N = 0 = \phi_N \\ \text{or:} \\ \theta_N = 0 = \tau_N \end{array} \right. \quad (\nu_i)_1^{N-1}$

clamped: $u_N = 0 = \theta_N \quad (\gamma_i)_1^{N-2}$

Note that the anti-resonant frequencies are the frequencies at which the application of an end bending moment produces no end displacement; we will show that there are $N-1$ such (real) frequencies and that they are also the frequencies at which the application of an end shearing force produces no end rotation.

The various sets of eigenvalues are the zeros of the brackets introduced in the previous section. In fact, by carrying on the process of equations (7.3.20), we find

$$\Phi_N = \prod_{i=1}^{N} (1 - \lambda/\lambda_i) \, , \quad \Pi_N = F_2 \prod_{i=1}^{N-1} (1 - \lambda/\mu_i) \, , \tag{7.4.1}$$

$$\Sigma_N=\prod_{i=1}^{N-1}(1-\lambda/\sigma_i)\ , \quad B_N=F_1\prod_{i=1}^{N-1}(1-\lambda/\nu_i)\ , \tag{7.4.2}$$

$$\Gamma_N=(F_oF_2-F_1^2)\prod_{i=1}^{N-2}(1-\lambda/\gamma_i)\ , \tag{7.4.3}$$

where

$$F_k=\sum_{n=1}^{N}x_n^kK_n \quad (k=0,1,2)\ , \ x_n=\sum_{s=n}^{N}\ell_s\ . \tag{7.4.4}$$

These expansions are similar to those used by Barcilon (1979, 1982). We note that equations (7.4.1), (7.4.2) show that $(\lambda_i)_1^N$ are the poles of the u_N/ϕ_N, u_N/τ_N, θ_N/ϕ_N and θ_N/τ_N response functions while $\{\mu_i,\nu_i,\nu_i,\sigma_i\}_1^{N-1}$, respectively, are the zeros.

Now return to the expansions generated in Section 7.3. Equations (7.3.15), (7.4.1) show that the $(\mu_i)_1^{N-1}$ are the zeros of the equation

$$\sum_{i=1}^{N}\frac{[u_N^{(i)}]^2}{\lambda_i-\lambda}=0\ . \tag{7.4.5}$$

This equation was studied in Section 2.8. We have

Theorem 7.4.1. If $(p_i)_1^N>0$ and $x_1<x_2<\cdots<x_N$, then the equation

$$f(x)\equiv\sum_{i=1}^{N}\frac{p_i}{x_i-x}=0 \tag{7.4.6}$$

has $N-1$ real zeros ξ_i satisfying

$$x_i<\xi_i<x_{i+1}\ , \quad i=1,2,\ldots,N-1\ . \tag{7.4.7}$$

Proof:

In each interval (x_i,x_{i+1}), $f(x)$ is strictly increasing from $-\infty$ to $+\infty$ and will cross the x-axis just once. ■

Since $u_N^{(i)}>0$, Theorem 7.4.1 applied to equation (7.4.5) yields the result that the eigenvalues of the clamped-pinned and the clamped-free beams interlace, i.e.,

$$\lambda_i<\mu_i<\lambda_{i+1}\ , \quad i=1,2,\ldots,N-1\ . \tag{7.4.8}$$

The eigenfrequencies $(\sigma_i)_1^{N-1}$ of the clamped-sliding beam are the zeros of Σ_N, i.e.,

$$\sum_{i=1}^{N}\frac{[\theta_N^{(i)}]^2}{\lambda_i-\lambda}=0\ . \tag{7.4.9}$$

Since $\theta_N^{(i)}>0$, these zeros satisfy

$$\lambda_i < \sigma_i < \lambda_{i+1}\ , \quad i=1,2,\ldots,N-1\ . \tag{7.4.10}$$

In order to find the relative positions of the μ_i and σ_i we need

Theorem 7.4.2. Suppose $(p_i)_1^N > 0$, $(q_i)_1^N > 0$, $x_1 < x_2 < \cdots < x_N$,

$$f(x) \equiv \sum_{i=1}^{N} \frac{p_i}{x_i - x}\ , \quad g(x) = \sum_{i=1}^{N} \frac{q_i}{x_i - x} \tag{7.4.11}$$

and that $(\xi_i)_1^{N-1}$, $(\eta_i)_1^{N-1}$ are the zeros of $f(x)$, $g(x)$ respectively. If $p_i q_j - p_j q_i > 0$ for $i < j$, then $\xi_i > \eta_i$ for $i = 1,2,\ldots,N-1$.

Proof:

$$p_j g(x) - q_j f(x) = \sum_{i=1}^{N} \frac{p_j q_i - p_i q_j}{x_i - x}\ . \tag{7.4.12}$$

Put $x = \xi_j$, so that $x_j < \xi_j < x_{j+1}$, and divide the sum into two parts, thus

$$p_j g(\xi_j) = \sum_{i=1}^{j-1} \frac{p_i q_j - p_j q_i}{\xi_j - x_i} + \sum_{i=j+1}^{N} \frac{p_j q_i - p_i q_j}{x_i - \xi_j}\ . \tag{7.4.13}$$

Under the stated conditions, each of the numerators and denominators on the right will be positive, so that $g(\xi_j) > 0$, i.e., $g(x)$ has already become positive when $f(x)$ is zero, i.e. $\xi_j > \eta_j$. ■

Note that it is sufficient to have $p_i q_{i+1} - p_{i+1} q_i > 0$ for $i = 1,2,\ldots,N-1$, for then $p_i q_j - p_j q_i > 0$ for all $i < j$. The converse of this theorem is *not* true – see Ex. 7.4.1.

In order to apply this theorem to equations (7.4.5), (7.4.9) we note that, since $\theta_N = \ell_N^{-1}(u_N - u_{N-1})$,

$$u_N^{(i)} \theta_N^{(j)} - u_N^{(j)} \theta_N^{(i)} = \ell_N^{-1}(u_{N-1}^{(i)} u_N^{(j)} - u_N^{(i)} u_{N-1}^{(j)}) > 0 \tag{7.4.14}$$

on account of Theorem 5.7.3. Thus

$$\sigma_i < \mu_i\ , \quad i = 1,2,\ldots,N-1\ . \tag{7.4.15}$$

This may be interpreted physically as saying that the sliding end condition is less constraining than the pinned.

The anti-resonant frequencies $(\nu_i)_1^{N-1}$ are the zeros of

$$\sum_{i=1}^{N} \frac{u_N^{(i)} \theta_N^{(i)}}{\lambda_i - \lambda} = 0\ . \tag{7.4.16}$$

Theorem 7.2.3 shows that these zeros interlace with the $(\lambda_i)_1^N$, while Theorem 7.4.2 and the inequality (7.4.14) show that

$$\sigma_i < \nu_i < \mu_i \;, \quad i=1,2,\ldots,N-1 \;. \tag{7.4.17}$$

The final set of frequencies $(\gamma_i)_1^{N-2}$ corresponds to the right end clamped. This is equivalent to applying one more constraint to either the pinned or to the sliding beam. The eigenfrequencies will therefore satisfy

$$\sigma_i < \gamma_i < \sigma_{i+1} \text{ and } \mu_i < \gamma_i < \mu_{i+1} \;. \tag{7.4.18}$$

Combining the inequalities (7.4.8), (7.4.10), (7.4.17), (7.4.18), we deduce that

$$\lambda_1 < \sigma_1 < \nu_1 < \mu_1 < (\lambda_2,\gamma_1) < \sigma_2 < \nu_2 < \cdots < (\lambda_{N-1},\gamma_{N-2}) < \sigma_{N-1} < \nu_{N-1} < \mu_{N-1} < \lambda_N \;.$$

Note that the relative position of γ_i and λ_{i+1} is (so far) indeterminate; in numerical experiments it was always found that $\gamma_i > \lambda_{i+1}$.

Examples 7.4

1. Construct a counter example to show that the converse to Theorem 7.4.2 is not true. Take $N=3$, $(x_1,x_2,x_3) = (1,4,7)$, $(p_1,p_2,p_3) = (4,1,4)$, $(q_1,q_2,q_3) = (5,1,7)$. Find ξ_1, ξ_2, η_1, η_2 and show that $g(\xi_1)>0$, $g(\xi_2)>0$ so that $\xi_1>\eta_1$, $\xi_2>\eta_2$ but $p_1q_2-p_2q_1<0$, $p_2q_3-p_3q_2>0$.
2. Show that if $p_i>0$, $q_i>0$, $p_iq_{i+1}-q_ip_{i+1}>0$ for $i=1,2,\ldots,N-1$, then $p_iq_j-q_ip_j>0$ for all $i,j=1,2,\ldots,N$ such that $i<j$. Compare with Theorem 5.2.4.
3. Show that if the zeros of (7.4.5), (7.4.9) are $(\mu_i)_1^{N-1}$, $(\sigma_i)_1^{N-1}$, respectively, then

$$[u_N^{(j)}]^2 = C_1\lambda_j \frac{\prod_{i=1}^{N-1}(1-\lambda_j/\mu_i)}{\prod_{i=1}^{N}{}'(1-\lambda_j/\lambda_i)}$$

$$[\theta_N^{(j)}]^2 = C_2\lambda_j \frac{\prod_{i=1}^{N-1}(1-\lambda_j/\sigma_i)}{\prod_{i=1}^{N}{}'(1-\lambda_j/\lambda_i)}$$

where C_1, C_2 are constants.

7.5 Conditions on the data

In the inverse problem it is required to construct a beam with given eigenfrequencies. Barcilon showed (for his model) that the beam cannot be uniquely determined from two spectra, and attempted to prove that it could be so determined (apart from a scale factor) from three properly chosen spectra. His procedure (in our notation) was to start from $(\lambda_i,\nu_i,\mu_i)_1^N$ (and note that he had N of the ν_i, μ_i, not $N-1$ as in the model of Figure 2.3.1) satisfying

$$\lambda_1<\nu_1<\mu_1<\lambda_2<\cdots<\lambda_N<\nu_N<\mu_N \tag{7.5.1}$$

and construct the frequencies $(\sigma_i)_1^N$ and $(\gamma_i)_1^{N-1}$ by using the recurrence relations (similar to (7.3.22) - (7.3.24)) and the quadratic identity (7.3.25). For his model it is not possible to prove that the frequencies so constructed satisfy the complete set of inequalities (similar to) (7.4.19). Subsidiary conditions had to be placed on $(\lambda_i,\nu_i,\mu_i)_1^N$ in order for the inequalities to be satisfied. His second step was a stripping procedure for calculating the parameters ℓ_N, k_N, m_N of the last segment and for computing the corresponding frequencies $(\lambda_i^*,\nu_i^*,\mu_i^*)_1^{N-1}$ of the truncated system obtained by deleting the last segment. The ℓ_N, k_N, m_N were all found to be positive but, even with the subsidiary conditions on the $(\lambda_i,\nu_i,\mu_i)_1^N$, it was not possible to prove that the new frequencies satisfied the necessary orderings, which meant that if the stripping procedure were continued, negative masses, stiffnesses or lengths might be encountered at some stage. He concluded that further conditions must be placed on the data, preferably conditions which could be applied *ab initio*, so eliminating the need for checks at each stage of the stripping procedure. We shall now state such conditions and construct a new stripping procedure.

The frequency spectra, from which will be drawn the data for the inverse problem, may be divided into three parts (i) $(\lambda_i)_1^N$; (ii) $(\sigma_i,\nu_i,\mu_i)_1^{N-1}$; (iii) $(\gamma_i)_1^{N-2}$. Suppose that (i) is given. Each spectrum which is given from (ii) then determines, to within an arbitrary multiplier, the set of coefficients $[\theta_N^{(i)}]^2$ $u_N^{(i)}\theta_N^{(i)}$ or $[u_N^{(i)}]^2$ respectively, in the frequency equations (7.4.9), (7.4.16) and (7.4.5). If *any two* of these spectra are given, then the two sets of coefficients yield the third set and hence the third spectra. (Note that since $u_N^{(i)}\theta_N^{(i)}>0$, there is no ambiquity in taking the square root of $[u_N^{(i)}]^2$ $[\theta_N^{(i)}]^2$). However, if two given spectra, say $(\nu_i)_1^{N-1}$, $(\mu_i)_1^{N-1}$ satisfy the appropriate ordering (e.g., $\nu_i<\mu_i$) then the third set $(\sigma_i)_1^{N-1}$ need not satisfy its appropriate ordering ($\sigma_i<\nu_i$). Two counter-examples are provided in Ex. 7.5, and these clearly show that the ordering requirements on the given spectra are insufficient for the existence of a real model – they do not even ensure the ordering of the remaining spectra. We now prove the fundamental

Theorem 7.5.1. A necessary conditions for the existence of a real (i.e. positive) model corresponding to data sets $(\lambda_i,u_N^{(i)},\theta_N^{(i)})_1^N$ is that the $(N+1)\times N$ matrix

$$\mathbf{P}=\begin{bmatrix} u_n^{(1)} & u_N^{(2)} & \cdots & u_N^{(N)} \\ \theta_N^{(1)} & \theta_N^{(2)} & \cdots & \theta_N^{(N)} \\ \lambda_1 u_N^{(1)} & \lambda_2 u_N^{(2)} & \cdots & \lambda_N u_N^{(N)} \\ \lambda_1\theta_N^{(1)} & \lambda_2\theta_N^{(2)} & \cdots & \lambda_N\theta_N^{(N)} \\ \lambda_1^2 u_N^{(1)} & \lambda_2^2 u_N^{(2)} & \cdots & \lambda_N^2 u_N^{(N)} \\ \cdot & \cdot & & \cdot \end{bmatrix} \tag{7.5.2}$$

should be completely positive, i.e., all its minors are positive. Note that the last row of $\mathbf{P}$ is

$$\lambda_1^r u_N^{(1)} \ \lambda_2^r u_N^{(2)} \cdots \lambda_N^r u_N^{(N)} \quad \text{or} \quad \lambda_1^r \theta_N^{(1)} \ \lambda_2^r \theta_N^{(2)} \cdots \lambda_N^r \theta_N^{(N)} ,$$

according to whther N is even or odd respectively, where $r=[N/2]$.

Proof:

Because of the repetitive character of the rows of **P**, the corollary to Theorem 5.2.4 shows that the matrix will be completely positive if and only if

$$P\begin{pmatrix}1, & 2, & \cdots & p \\ i, & i+1, & \cdots & i+p-1\end{pmatrix}>0 \quad P\begin{pmatrix}2, & 3, & \cdots & p+1 \\ i, & i+1, & \cdots & i+p-1\end{pmatrix}>0 \tag{7.5.3}$$

for $p=1,2,...,N$, $i=1,2,...,N-p+1$.

The proof follows directly from Theorem 5.7.3, for

$$U\begin{pmatrix}N-1, & N \\ i, & i+1\end{pmatrix}=\begin{vmatrix}u_{N-1}^{(i)} & u_{N-1}^{(i+1)} \\ u_N^{(i)} & u_N^{(i+1)}\end{vmatrix} > 0 . \tag{7.5.4}$$

But the recurrence $u_{N-1}=u_N-\ell_N\theta_N$ yields

$$U\begin{pmatrix}N-1 & N \\ i, & i+1\end{pmatrix}=\begin{vmatrix}u_N^{(i)}-\ell_N\theta_N^{(i)} & u_N^{(i+1)}-\ell_N\theta_N^{(i+1)} \\ u_N^{(i)} & u_N^{(i+1)}\end{vmatrix} = \ell_N\begin{vmatrix}u_N^{(i)} & u_N^{(i+1)} \\ \theta_N^{(i)} & \theta_N^{(i+1)}\end{vmatrix} , \tag{7.5.5}$$

which we write symbolically as

$$[u_{N-1},u_N]=[u_N-\ell_N\theta_N,u_N]=\ell_N[u_N,\theta_N]>0 . \tag{7.5.6}$$

Similarly the relations (2.3.2)-(2.3.5) yield

$$\begin{aligned} 0 &< [\theta_{N-1},\theta_N]=[\theta_N-K_N\tau_{N-1},\theta_N]=-K_N[\tau_{N-1},\theta_N]=-K_N[\ell_N\phi_{N-1},\theta_N] \\ &= -K_N\ell_N[m_N\lambda u_N,\theta_N]=K_N\ell_N m_N[\theta_N,\lambda u_N] \end{aligned} \tag{7.5.7}$$

i.e.,

$$P\begin{pmatrix}2, & 3 \\ i, & i+1\end{pmatrix}=\begin{vmatrix}\theta_N^{(i)} & \theta_N^{(i+1)} \\ \lambda_i u_N^{(i)} & \lambda_{i+1}u_N^{(i+1)}\end{vmatrix} > 0 . \tag{7.5.8}$$

All the u- and θ- inequalities in Theorem 5.7.3 may be related in this way to the minors (7.5.3) of the matrix **P**. Thus

$$U\begin{pmatrix}N-2, & N-1, & N \\ i, & i+1, & i+2\end{pmatrix}=P_{N-2}[u_N,\theta_N,\lambda u_N]=P_{N-2}P\begin{pmatrix}1, & 2, & 3 \\ i, & i+1, & i+2\end{pmatrix} , \tag{7.5.9}$$

$$\Theta\begin{pmatrix}N-2, & N-1 & N \\ 1, & i+1, & i+2\end{pmatrix}=Q_{N-2}[\theta_N,\lambda u_N,\lambda\theta_N]=Q_{N-2}P\begin{pmatrix}2, & 3, & 4 \\ i, & i+1, & i+2\end{pmatrix} , \tag{7.5.10}$$

where

$$P_{N-2}=K_N\ell_N\ell_{N-1}m_N \ , \quad Q_{N-2}=K_NK_{N-1}\ell_N^2\ell_{N-1}m_Nm_{N-1} \ . \qquad (7.5.11)$$

Note that the quantities P_{N-p+1}, Q_{N-p+1} are products of the m_n, ℓ_n, K_n for the indices $n=N-p+2,...,N$ only. ■

It will be shown below that the conditions (7.5.3) are also *sufficient* for the existence of a real model.

Clearly, if $(\lambda_i)_1^N$, and say $(\mu_i,\nu_i)_1^{N-1}$ are given, the chances that the computed values of $(u_N^{(i)},\theta_N^{(i)})_1^N$ will satisfy all these conditions, are extremely remote. A more systematic procedure is to specify *two* spectra, say $(\lambda_i)_1^N$ and $(\mu_i)_1^{N-1}$, compute $(u_N^{(i)})_1^N$, and then find a set of values of $(\theta_N^{(i)})_1^N$ satisfying all the conditions. The condition (7.5.4) will ensure that the corresponding $(\sigma_i,\nu_i)_1^{N-1}$ satisfy the ordering $\lambda_i<\sigma_i<\nu_i<\mu_i<\lambda_{i+1}$ and the $(\gamma_i)_1^{N-2}$ will automatically satisfy $\mu_i<\gamma_i<\sigma_{i+1}$. This is the procedure that will be followed. Two algorithms will be presented for the computation of the m_n, ℓ_n, k_n and it will be shown that, under the conditions (7.5.3), the m_n, ℓ_n, k_n so computed will all be positive.

Examples 7.5

1. Construct a counterexample to show that $\lambda_i<\nu_i<\mu_i<\lambda_{i+1}$ does not imply $\lambda_i<\sigma_i<\nu_i<\mu_i<\lambda_{i+1}$. Take $N=3$, $(\lambda_1,\lambda_2,\lambda_3)=(1,4,7)$, $(u_3^{(1)},u_3^{(2)},u_3^{(3)})=(2,1,2)$ so that $(\mu_1,\mu_2)=(3,5)$. Take $\theta_3^{(1)}=3/2$, $\theta_3^{(2)}=1$ and find $\theta_3^{(3)}$ so that $\nu_1<\mu_1$, $\nu_2<\mu_2$, $\sigma_1<\mu_1$ but $\sigma_2>\mu_2$.
2. With the same λ_i and $u_3^{(i)}$ data but with $\theta_3^{(1)}=1$, find $\theta_3^{(3)}$ so that $\sigma_1<\mu_1$, $\sigma_2<\mu_2$, but $\nu_2<\mu_2$, but $\nu_1>\mu_1$.

7.6 Inversion by using orthogonality

In this section we show how the system parameters may be found, at least in theory, from the eigenvalue data, and establish necessary and sufficient conditions on the data for the system parameters to be positive. An effective numerical scheme for calculating the parameters is described in Sections 7.7-7.8.

The algorithm is based on the orthogonality properties of the $\mathbf{u}^{(i)}$ (Theorem 1.4.4). The $\mathbf{u}^{(i)}$ satisfy

$$\lambda_i\mathbf{M}u^{(i)}=\mathbf{C}\mathbf{u}^{(i)} \ , \qquad (7.6.1)$$

so that if $\mathbf{U}=[\mathbf{u}^{(2)},\mathbf{u}^{(2)},...,\mathbf{u}^{(N)}]$ and $\mathbf{\Lambda}=diag\,(\lambda_i)$, then

$$\mathbf{MU\Lambda}=\mathbf{CU} \ . \qquad (7.6.2)$$

The orthogonality of the $\mathbf{u}^{(i)}$ w.r.t. $\mathbf{M}$ and $\mathbf{C}$ yields

$$\mathbf{U}^T\mathbf{MU}=\mathbf{I} \ , \quad \mathbf{U}^T\mathbf{CU}=\mathbf{\Lambda} \ . \qquad (7.6.3)$$

The first of equations (7.6.3) yields

$$\mathbf{M}^{-1}=\mathbf{U}\mathbf{U}^T . \tag{7.6.4}$$

The matrix on the left is a diagonal matrix with n-th element m_n^{-1}; that on the right must therefore also be diagonal, and therefore

$$\frac{1}{m_n}=\sum_{i=1}^{N}[u_n^{(i)}]^2 , \quad n=1,2,...,N . \tag{7.6.5}$$

Since the $[u_N^{(i)}]^2$ are known apart from a constant factor – they are the coefficients in equation (7.4.5) – m_N may be computed. The n, $n-1$ off-diagonal term on the right of (7.6.4) must be zero so that

$$\sum_{i=1}^{N} u_n^{(i)}u_{n-1}^{(i)}=0 \tag{7.6.6}$$

which, together with $u_{n-1}=u_n-\ell_n\theta_n$, yields

$$\sum_{i=1}^{N}[u_n^{(i)}]^2-\ell_n\sum_{i=1}^{N}u_n^{(i)}\theta_n^{(i)}=0 . \tag{7.6.7}$$

The first term is m_n^{-1}, so that

$$\frac{1}{m_n\ell_n}=\sum_{i=1}^{N}u_n^{(i)}\theta_n^{(i)} , \quad n=1,2,...,N . \tag{7.6.8}$$

Since the $u_N^{(i)}\theta_N^{(i)}$ are known (apart from a constant factor) and are positive, this equation immediately yields ℓ_N. The crucial point is to prove that the remaining $(\ell_n)_1^{N-1}$ are always positive: this will be proved below.

The next step of the algorithm is the determination of the k_n. For this it is necessary to use the form of $\mathbf{C}$ given in equation (2.3.12). Thus equation (2.3.12) gives

$$\mathbf{C}=\mathbf{E}\mathbf{L}^{-1}\mathbf{E}\mathbf{K}\mathbf{E}^T\mathbf{L}^{-1}\mathbf{E}^T , \quad \mathbf{C}=\mathbf{U}^{-T}\boldsymbol{\Lambda}\mathbf{U}^{-1} , \tag{7.6.9}$$

so that

$$\mathbf{C}^{-1}=\mathbf{E}^{-T}\mathbf{L}\mathbf{E}^{-T}\mathbf{K}^{-1}\mathbf{E}^{-1}\mathbf{L}\mathbf{E}^{-1}=\mathbf{U}\boldsymbol{\Lambda}^{-1}\mathbf{U}^T , \tag{7.6.10}$$

and thus

$$\mathbf{K}^{-1}=\mathbf{E}^T\mathbf{L}^{-1}\mathbf{E}^T\mathbf{U}\boldsymbol{\Lambda}^{-1}\mathbf{U}^T\mathbf{E}\mathbf{L}^{-1}\mathbf{E} . \tag{7.6.11}$$

But equation (2.3.7) shows that

$$\boldsymbol{\Theta}=[\boldsymbol{\theta}^{(1)},\boldsymbol{\theta}^{(2)},\ldots,\boldsymbol{\theta}^{(N)}]=\mathbf{L}^{-1}\mathbf{E}^T\mathbf{U} , \tag{7.6.12}$$

so that

$$\mathbf{K}^{-1}=\mathbf{E}^T\boldsymbol{\Theta}\boldsymbol{\Lambda}^{-1}\boldsymbol{\Theta}^T\mathbf{E} , \tag{7.6.13}$$

which yields

$$\frac{1}{k_n}=\sum_{i=1}^{N}\frac{[\theta_n^{(i)}-\theta_{n-1}^{(i)}]^2}{\lambda_i}\ , \quad n=1,2,...,N \tag{7.6.14}$$

or alternatively,

$$k_n=\sum_{i=1}^{N}\frac{[\tau_{n-1}^{(i)}]^2}{\lambda_i}\ , \quad n=1,2,...,N\ . \tag{7.6.15}$$

Since $\tau_{N-1}^{(i)}=\ell_N\phi_{N-1}^{(i)} = m_N\ell_N\lambda_i u_N^{(i)}$, this yields

$$k_N=m_N^2\ell_N^2\sum_{i=1}^{N}\lambda_i[u_N^{(i)}]^2\ . \tag{7.6.16}$$

The steps in the algorithm are thus

(i) set $n=N$

(ii) $u_n^{(i)}$, $\theta_n^{(i)}$, $\tau_n^{(i)}\equiv 0\equiv\ \phi_n^{(i)}$ are known from data

(iii) compute m_n, ℓ_n from equations (7.6.5), (7.6.8)

(iv) compute $u_{n-1}^{(i)}=u_n^{(i)}-\ell_n\theta_n^{(i)}$, $\phi_{n-1}^{(i)}=\phi_n^{(i)}+m_n\lambda_i u_n^{(i)}$, $\tau_{n-1}^{(i)}=\tau_n^{(i)}+\ell_n\phi_{n-1}^{(i)}$

(v) compute k_n from equation (7.6.15)

(vi) compute $\theta_{n-1}^{(i)}=\theta_n^{(i)}-\tau_{n-1}^{(i)}/k_n$

(vii) set $n=n-1$. If $n>1$ go to step (iii) stop.

We note that the quantities $\{u_N^{(i)},\theta_N^{(i)}\}_1^N$ are known only to within arbitrary multiplying factors. If a second, primed, set is related to the first by

$$[u_n^{(i)}]'=\alpha u_N^{(i)}\ , \quad [\theta_N^{(i)}]'=\beta\theta_N^{(i)}\ , \tag{7.6.17}$$

then the algorithm yields

$$m_n'=m_n/\alpha^2\ , \quad k_n'=k_n/\beta^2\ , \quad \ell_n'=\alpha\ell_n/\beta\ . \tag{7.6.18}$$

This equation therefore defines the equivalence class of systems with the given eigenfrequencies. The validity of the inversion procedure is based on

Theorem 7.6.1. The complete positivity of the matrix **P** of (7.5.2) is necessary and sufficient for the existence of a real model having the given three spectra;, i.e., $(\lambda_i)_1^N$ and two of $(\sigma_i,\nu_i,\mu_i)_1^{N-1}$.

Proof:

The necessity was proved in Section 7.5. We prove the sufficiency. Consider the equations

$$\mathbf{M}^{-1}=\mathbf{U}\mathbf{U}^T\ , \quad \mathbf{\Theta}=\mathbf{L}^{-1}\mathbf{E}^T\mathbf{U} \tag{7.6.19}$$

and construct the matrix

$$\mathbf{B} \equiv \mathbf{L}^{-1}\mathbf{E}^T\mathbf{M}^{-1} = \mathbf{L}^{-1}\mathbf{E}^T\mathbf{U}\mathbf{U}^T = \mathbf{\Theta}\mathbf{U}^T \ . \tag{7.6.20}$$

The corresponding associated matrices satisfy

$$\underline{\mathbf{B}}_p = \underline{\mathbf{L}}_p^{-1}\underline{\mathbf{E}}_p^T\underline{\mathbf{M}}_p^{-1} = \underline{\mathbf{\Theta}}_p\underline{\mathbf{U}}_p^T \tag{7.6.21}$$

on account of Theorem 5.3.4. Since $\underline{\mathbf{L}}_p^{-1}$ and $\underline{\mathbf{M}}_p^{-1}$ are diagonal matrices, and each principal minor of $\underline{\mathbf{E}}_p^T$ is unity, the bottom right-hand element of $\underline{\mathbf{B}}_p$ is

$$\underline{b}_{\underline{N}\underline{N}} = \prod_{k=N-p+1}^{N} (m_k \ell_k)^{-1} = \sum_{s=1}^{N} \underline{\theta}_{\underline{N}}^{(s)} \underline{u}_{\underline{N}}^{(s)} \ . \tag{7.6.22}$$

We now proceed by induction. Suppose that conditions (7.5.3) are satisfied and that $\ell_N, \ell_{N-1}, \ldots, \ell_{N-p+2}$ are all positive. Each $\underline{u}_{\underline{N}}^{(s)}$, $\underline{\theta}_{\underline{N}}^{(s)}$ may be expressed, as in equation (7.5.9), (7.5.10) as a product of terms involving m_n, K_n, which are all positive, and terms involving $\ell_N, \ell_{N-1}, \ldots, \ell_{N-p+2}$ which are positive by hypothesis. Each such $\underline{u}_{\underline{N}}^{(s)}$, $\underline{\theta}_{\underline{N}}^{(s)}$ is thus positive. Therefore equation (7.6.22) shows that $\ell_{N-p+1} > 0$. But $\ell_N > 0$, therefore all the ℓ_n are positive. ■

7.7 The block-Lanczos algorithm

The algorithm described in Section 7.6 has theoretical value: it enables one to find necessary and sufficient conditions for the existence of a system corresponding to given data, but it does *not* yield a numerically stable procedure for calculating the k_n, ℓ_n, m_n. Numerical experiments show that for $N > 5$, the method breaks down due to cancellation of leading digits. Golub and Boley (1977) have developed a block form of the Lanczos algorithm described in Section 4.2. This algorithm is outlined below. In Section 7.8 we show how it may be used to calculate the model parameters k_n, ℓ_n, m_n.

The starting point is the recognition that any band matrix may be displayed as a tridiagonal matrix of blocks. Thus for a pentadiagonal matrix the blocks are 2×2, and there are two arrangements, depending on whether N is even or odd. These are shown diagrammatically for $N = 5$ and 6 below.

$$\begin{bmatrix} x & x & x & & \\ x & x & x & x & \\ & & & & \\ x & x & x & x & x \\ & x & x & x & x \\ & & & & \\ & & x & x & x \end{bmatrix} \qquad \begin{bmatrix} x & x & x & & & \\ x & x & x & x & & \\ & & & & & \\ x & x & x & x & x & \\ & x & x & x & x & x \\ & & \cdot & & & \\ & & x & x & x & x \\ & & & x & x & x \end{bmatrix}$$

For a symmetric band matrix $\mathbf{A}$ in which

$$a_{ij}=0 \text{ when } |i-j|>p \ , \tag{7.7.1}$$

($p=2$ for a pentadiagonal matrix) we display $\mathbf{A}$ as a tridiagonal matrix with $p\times p$ blocks, except for the last set of rows and columns. Thus

$$\mathbf{A}=\begin{bmatrix} \mathbf{P}_1 & \mathbf{R}_1^T & \mathbf{0} & \cdot & \cdot \\ \mathbf{R}_1 & \mathbf{P}_2 & \mathbf{R}_2^T & & \\ \mathbf{0} & \mathbf{R}_2 & \cdot & & \\ \cdot & \cdot & & & \\ \cdot & \cdot & & \mathbf{P}_\ell & \mathbf{R}_\ell^T \\ \cdot & \cdot & & \mathbf{R}_\ell & \mathbf{P}_{\ell+1} \end{bmatrix} \tag{7.7.2}$$

If $N=\ell\times p+m$, $1\leq m\leq p$, then $\mathbf{R}_\ell$ is a $m\times p$ matrix and $\mathbf{P}_{\ell+1}$ is $m\times m$.

We now suppose that we know the eigenvalues $(\lambda_i)_1^N$ of $\mathbf{A}$, and the first p elements $q_n^{(i)}$, $n=1,2,...,p$, of the normalized eigenvectors $(\mathbf{q}^{(i)})_1^N$. In the notation of Section 4.2, we know the vectors $(\mathbf{x}^{(r)})_1^p$; they must satisfy the orthogonality condition (4.2.8).

We partition the matrix $\mathbf{X}$ in the form

$$\mathbf{X}=[\mathbf{x}^{(1)},\mathbf{x}^{(2)},\ldots,\mathbf{x}^{(N)}]=[\mathbf{X}_1,\mathbf{X}_2,\ldots,\mathbf{X}_\ell,\mathbf{X}_{\ell+1}] \ , \tag{7.7.3}$$

where $\mathbf{X}_s$ ($s=1,2,...,\ell$) is $N\times p$, and $\mathbf{X}_{\ell+1}$ is $N\times m$. The data $(\mathbf{x}^{(r)})_1^p$ make up $\mathbf{X}_1$, which must satisfy

$$\mathbf{X}_1^T\mathbf{X}_1=\mathbf{I}_p \ . \tag{7.7.4}$$

Equation (4.2.10) is now

$$[\mathbf{X}_1 \ \ \mathbf{X}_2 \ \ \cdots \ \ \mathbf{X}_{\ell+1}]\begin{bmatrix} \mathbf{P}_1 & \mathbf{R}_1^T & \cdots & 0 & 0 \\ \mathbf{R}_1 & \mathbf{P}_2 & \cdots & 0 & 0 \\ \cdot & \cdot & \cdots & \cdot & \cdot \\ \mathbf{0} & \mathbf{0} & \cdots & \mathbf{P}_\ell & \mathbf{R}_\ell^T \\ \mathbf{0} & \mathbf{0} & \cdots & \mathbf{R}_\ell & \mathbf{P}_{\ell+1} \end{bmatrix}=\boldsymbol{\lambda} \ [\mathbf{X}_1 \ \ \mathbf{X}_2 \ \ \cdots \ \ \mathbf{X}_{\ell+1}] \tag{7.7.5}$$

The matrices $\mathbf{X}_s$ must satisfy

$$\mathbf{X}_r^T\mathbf{X}_s=\begin{cases} \mathbf{I}_p & \text{if } r=s\leq\ell \ , \\ \mathbf{I}_m & \text{if } r=s=\ell+1 \ , \\ 0 & \text{if } r\neq s \ . \end{cases} \tag{7.7.6}$$

The first (block) column of equation (7.7.5) is

$$\mathbf{X}_1\mathbf{P}_1+\mathbf{X}_2\mathbf{R}_1=\boldsymbol{\Lambda}\mathbf{X}_1 \tag{7.7.7}$$

so that on multiplying by $\mathbf{X}_1^T$ and using (7.7.6), we find

$$\mathbf{P}_1 = \mathbf{X}_1^T \mathbf{\Lambda} \mathbf{X}_1 \, . \tag{7.7.8}$$

Having found $\mathbf{P}_1$, we write equation (7.7.7) as

$$\mathbf{X}_2\mathbf{R}_1 = \mathbf{\Lambda}\mathbf{X}_1 - \mathbf{X}_1\mathbf{P}_1 \equiv \mathbf{Z}_1 \, . \tag{7.7.9}$$

We must now compute matrices $\mathbf{X}_2$, $\mathbf{R}_1$ from this equation, knowing that $\mathbf{X}_2$ is an orthogonal matrix, i.e., $\mathbf{X}_2^T\mathbf{X}_2 = \mathbf{I}_p$, and $\mathbf{R}_1$ is upper triangular; this can be accomplished by using the familiar Gram-Schmidt process. Column s of equation (7.7.5) is

$$\mathbf{X}_{s-1}\mathbf{R}_{s-1}^T + \mathbf{X}_s\mathbf{P}_s + \mathbf{X}_{s+1}\mathbf{R}_s = \mathbf{\Lambda}\mathbf{X}_s \, . \tag{7.7.10}$$

At this stage $\mathbf{X}_{s-1}$, $\mathbf{X}_s$, $\mathbf{R}_{s-1}$ have been found, and $\mathbf{X}_{s-1}$, $\mathbf{X}_s$ satisfy equation (7.7.6). Thus on multiplying by $\mathbf{X}_s^T$ we find

$$\mathbf{P}_s = \mathbf{X}_s^T\mathbf{\Lambda}\mathbf{X}_s \tag{7.7.11}$$

and then $\mathbf{X}_{s+1}$ and $\mathbf{R}_s$ must be found, as before, from

$$\mathbf{X}_{s+1}\mathbf{R}_s = \mathbf{\Lambda}\mathbf{X}_s - \mathbf{X}_{s-1}\mathbf{R}_{s-1}^T - \mathbf{X}_s\mathbf{P}_s \, . \tag{7.7.12}$$

This process may be continued until the whole matrix $\mathbf{A}$ and the eigenvector matrix $\mathbf{X}$ is found. It may be verified, as in Section 4.2, that the matrices $\mathbf{X}_s$ so produced do satisfy equation (7.7.6) even though $\mathbf{X}_s$ was explicitly enforced to be orthogonal only to its neighbours $\mathbf{X}_{s-1}$ and $\mathbf{X}_{s+1}$.

Further studies on the block-Lanczos algorithm have been carried out by Underwood (1975) and Golub and Underwood (1977). See also Mattis and Hochstadt (1981). A completely different and highly efficient procedure for the solution of band matrix inverse problems has been developed by Biegler-König (1980), (1981). See also the review papers Friedland (1977), Friedland (1979).

7.8 A numerical procedure for the beam inverse problem

We suppose that the data consists of three spectra, corresponding to the clamped-free, clamped-pinned and clamped-sliding beams; the natural frequencies are $(\lambda_i)_1^N$, $(\mu_i)_1^{N-1}$ and $(\sigma_i)_1^{N-1}$ respectively. It was shown in Section 7.4 (Ex. 7.4.3) that from these quantities we may compute

$$\{u_N^{(i)} \, , \; \theta_N^{(i)}\}_1^N \, , \tag{7.8.1}$$

apart from two constant factors, one for each sequence. Now equations (7.6.5)-(7.6.8) show how we may calculate $\{u_{N-1}^{(i)}\}_1^N$, ℓ_N, m_N and m_{N-1}.

The search for a set of quantities $(\lambda_i, u_N^{(i)}, \theta_N^{(i)})_1^N$ satisfying all the determinantal inequalities described in Section 7.5 may be carried out in the fashion described in Section 10.8 below.

The eigenvalue equation is

$$\mathbf{Cu}=\lambda\mathbf{Mu}\ . \tag{7.8.2}$$

Put

$$\mathbf{q}=\mathbf{M}^{1/2}\mathbf{u}\ , \tag{7.8.3}$$

then we may reduce the equation to

$$\mathbf{Aq}=\lambda\mathbf{q}\ , \tag{7.8.4}$$

where

$$\mathbf{A}=\mathbf{M}^{-1/2}\mathbf{C}\mathbf{M}^{-1/2}\ . \tag{7.8.5}$$

The matrix $\mathbf{A}$ is pentadiagonal and has eigenvalues $(\lambda_i)_1^N$. The last two elements of the eigenvectors $\mathbf{q}^{(i)}$ are

$$q_N^{(i)}=m_N^{1/2}u_N^{(i)}\ ,\quad q_{N-1}^{(i)}=m_{N-1}^{1/2}u_{N-1}^{(i)}\ , \tag{7.8.6}$$

and these are known, so that $\mathbf{A}$ may be computed from the block-Lanczos algorithm.

To compute the remaining masses, lengths and stiffnesses, we use the static behaviour of the system. To find the m_n, we proceed as we did with the second-order system in Section 4.4. By applying external forces only to masses 1 and 2, we may deform the system as shown in Figure 7.8.1. For this configuration $\mathbf{u}=\{1,1,...,1\}$ so that

$$\mathbf{q}=\mathbf{q}^o=\mathbf{M}^{1/2}\mathbf{u}=\{m_1^{1/2},m_2^{1/2},...,m_{N-1}^{1/2},m_N^{1/2}\}\ . \tag{7.8.7}$$

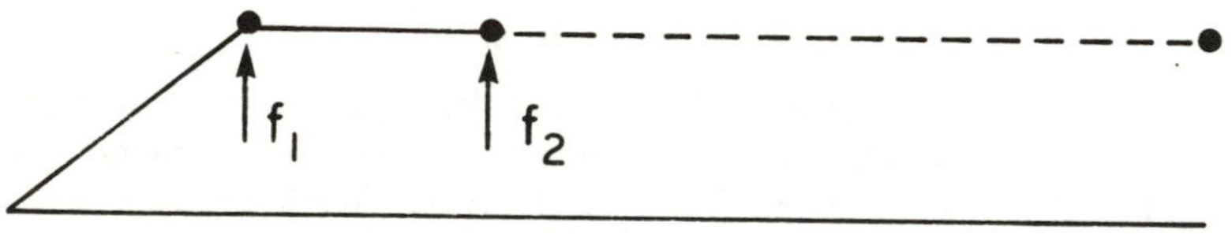

Figure 7.8.1 – This configuration may be obtained by applying forces to masses 1 and 2.

But with $\omega=0$, the static behaviour of the beam due to external forces f_1, f_2 is governed by

$$\mathbf{Cu}=\{f_1,f_2,0,0,...,0\} \tag{7.8.8}$$

i.e.,

$$\mathbf{Aq} \equiv \mathbf{M}^{-1/2}\mathbf{Cu} = \mathbf{M}^{-1/2}\{f_1, f_2, 0, 0, ..., 0\} . \tag{7.8.9}$$

Now **A** is pentadiagonal, so that using this equation, and knowing the final components $m_{N-1}^{1/2}$, $m_N^{1/2}$ of **q**, we may calculate the remaining terms $(m_s^{1/2})_1^{N-2}$; and find $f_1 = f_1^o$, $f_2 = f_s^o$. To calculate the $(\ell_s)_1^N$ we note that the configuration of Figure 7.8.2 may be obtained by applying a single force to m_1. In fact if

$$\mathbf{f} = \{k_1/\ell_1, 0, 0, ..., 0\} , \tag{7.8.10}$$

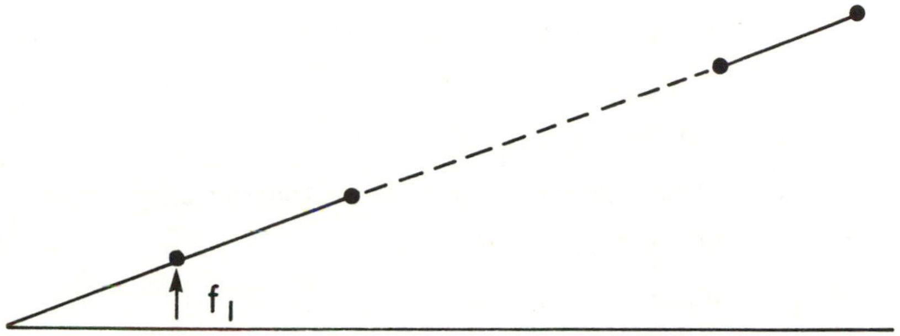

Figure 7.8.2. – This configuration may be obtained by applying a single force to mass 1.

then

$$\mathbf{u} = \{\ell_1, \ell_1 + \ell_2, \ldots, \ell - \ell_N, \ell\} , \tag{7.8.11}$$

where

$$\ell = \sum_{s=1}^{N} \ell_s .$$

For this **u**, the corresponding **q** is

$$\mathbf{q} = \mathbf{M}^{1/2}\mathbf{u} = \ell\mathbf{q}^0 - \mathbf{q}^1 , \tag{7.8.12}$$

where

$$\mathbf{q}^1 = \{m_1^{1/2}(\ell_2 + ... + \ell_N), m_2^{1/2}(\ell_3 + ... \ell_N), \ldots, m_{N-1}^{1/2}\ell_N, 0\} \tag{7.8.13}$$

satisfies an equation of the form

$$\mathbf{Aq}^1 = \mathbf{M}^{-1/2}\{f_1, f_2, 0, ..., 0\} . \tag{7.8.14}$$

Again, knowing the last two components of $\mathbf{q}^1$ (the last is zero), we may find the others, and find $f_1 = f_1^1$, $f_2 = f_2^1$. Now we may find **q**, because

$$\mathbf{Aq} = \ell \mathbf{Aq}^0 - \mathbf{Aq}^1 = \{\ell f_1^0 - f_1^1,\ \ell f_2^0 - f_2^1,\ 0, \ldots, 0\}\ , \tag{7.8.15}$$

and ℓ is determined by the condition

$$\ell f_2^0 - f_2^1 = 0\ . \tag{7.8.16}$$

At this stage the $(m_n)_1^N$, $(\ell_n)_1^N$ are known, and these values make

$$\mathbf{AM}^{1/2}\{\ell_1, \ell_1 + \ell_2, \ldots, \ell_1 + \ell_2 + \ldots \ell_N\} = \{m_1^{-1/2} k_1/\ell_1,\ 0,0,\ldots,0\}\ , \tag{7.8.17}$$

so that k_1 is known. Using the form of $\mathbf{E}^{-1}$ in equation (2.2.14) we may write equation (7.8.17) as

$$\mathbf{AM}^{1/2}\mathbf{E}^{-T}\mathbf{L}\{1,1,\ldots 1\} = \{m_1^{-1/2} k_1/\ell_1,\ 0,0, \ldots, 0\}\ , \tag{7.8.18}$$

i.e.,

$$\mathbf{LE}^{-1}\mathbf{M}^{1/2}\mathbf{AM}^{1/2}\mathbf{E}^{-T}\mathbf{L}\{1,1,\ldots 1\} = \{k_1,\ 0,0, \ldots, 0\}, \tag{7.8.19}$$

It may easily be verified that the matrix on the left is tridiagonal, and hence

$$\mathbf{E}^{-1}(\mathbf{LE}^{-1}\mathbf{M}^{1/2}\mathbf{AM}^{1/2}\mathbf{E}^{-T}\mathbf{L})\mathbf{E}^{-T} = \mathbf{K} \tag{7.8.20}$$

is the required diagonal matrix of stiffnesses.

CHAPTER 8

GREEN'S FUNCTIONS AND INTEGRAL EQUATIONS

Mathematicians who are only mathematicians have exact minds, provided all things are explained to them by means of definitions and axioms; otherwise they are inaccurate and unsufferable, for they are only right when the principles are quite clear.

Pascal's Pensées

8.1 Introduction

In this and subsequent chapters we shall be concerned with the vibration of, and in particular the inverse problems for, three systems with continuously distributed mass: the taut vibrating string, and the rod in longitudinal or torsional vibration. In the first three sections of this chapter we shall state the governing differential equations, discuss some transformations linking them, derive some elementary properties of their solutions, and formulate them as integral equations, using the concept of a Green's function. Section 8.4 lists some classical properties of integral equations with symmetric kernels while Section 8.5 introduces the concept of an oscillatory kernel. The physical meaning of this term is given in Theorem 8.5.7. Section 8.6 is concerned with classical results on completeness while Sections 8.7, 8.8 lay the groundwork for a description of the oscillatory properties of eigenfunctions. Section 8.9 introduces Perron's theorem and the concept of an associted kernel. The reader may note that the presentation mirrors that given earlier for oscillatory *matrices*. In Section 8.10 we discuss the interlacing of eigenvalues, i.e, how the eigenvalues corresponding to one set of boundary conditions lie in between those for another.

Section 8.11 is concerned with asymptotic properties, while Section 8.12 is virtually separate from the earlier part of the chapter and is concerned with impulse responses, i.e., the behaviour of the system in the time, rather than frequency, domain. This description is taken up again in Section 9.8.

The equation governing the free (infinitesimal, undamped) vibration of a taut string having tension T, mass per unit length $\rho(x)$, vibrating with frequency ω is

$$u''(x)+\lambda\rho(x)u(x)=0\ , \tag{8.1.1}$$

where $\lambda=\omega^2/T$ and $'\equiv d/dx$. The end conditions will be assumed to be

$$u'(0)-hu(0)=0=u'(\ell)+Hu(\ell)\ , \tag{8.1.2}$$

where h, $H\geq 0$ and h, H are not both zero; This means that the ends $x=0$, $x=\ell$ are attached to fixed supports by the use of springs having stiffnesses Th and TH respectively.

The free longitudinal vibrations of a thin straight rod of cross-sectional area $A(x)$, density ρ and Young's modulus E are governed by the equation

$$(A(x)v'(x))'+\lambda A(x)v(x)=0\ , \tag{8.1.3}$$

where $\lambda=\rho\omega^2/E$. The end conditions are

$$v'(0)-hv(0)=0=v'(\ell)+Hv(\ell)\ , \tag{8.1.4}$$

where again h, $H\geq 0$ and h, H are not both zero.

The free torsional vibrations of a thin straight rod of second moment of area $J(x)$, density ρ and shear modulus G are governed by the equation

$$(J(x)\theta'(x))'+\lambda J(x)\theta(x)=0\ , \tag{8.1.5}$$

where $\lambda=\rho\omega^2/G$. The end conditions are

$$\theta'(0)-h\theta(0)=0=\theta'(\ell)+H\theta(\ell)\ . \tag{8.1.6}$$

There is clearly a direct correspondence $(E,A,\rho,v)\longleftrightarrow(G,J,\rho,\theta)$ between the longitudinal and torsional systems, but we now show that, by means of a transformation of variables, all these equations may be reduced to the same basic equation.

In equation (8.1.3) introduce a new variable ξ, where

$$\xi'(x)=B(x)\equiv 1/A(x)\ ,\ \ v(x)=u(\xi)\ . \tag{8.1.7}$$

Then $A(x)v'(x)=A(x)\dot{u}(\xi)\xi'(x)=\dot{u}(\xi)$ where $\dot{}\equiv d/d\xi$ and hence $(Av')'=B\ddot{u}$ so that equation (8.1.3) becomes

$$\ddot{u}(\xi)+\lambda\rho(\xi)u(\xi)=0\ ,\ \ \rho(\xi)=A^2(x)\ . \tag{8.1.8}$$

This has exactly the same form as (8.1.1). If

$$\xi(x)=\int_o^x B(t)dt \ , \ \ L=\int_o^\ell B(t)dt \ , \tag{8.1.9}$$

then the end conditions (8.1.4) become

$$B(0)\dot{u}(0)-hu(0)=0=B(L)\dot{u}(L)+Hu(L) \ . \tag{8.1.10}$$

Since $A(x)$, $B(x)$ are positive, these equations have the same form as (8.1.2).

There is another important transformation of equation (8.1.3). Put

$$y(x)=f(x)v(x) \ , \tag{8.1.11}$$

then

$$\begin{aligned}(Av')' &= [A(f^{-1}y'-f'f^{-2}y)]' \\ &= (Af^{-1})y''+\{(Af^{-1})'-Af'f^{-2}\}y'-(Af'f^{-2})'y \ .\end{aligned} \tag{8.1.12}$$

The function f may be chosen to make the terms in y' vanish. We take $(Af^{-1})'-Af'f^{-2}\equiv A'f^{-1}-2Af'f^{-2}=0$, i.e., $(fA^{-1/2})'=0$ or $f=A^{1/2}$. Then

$$(Av')'+\lambda Av=fy''-f''y+\lambda fy=0 \tag{8.1.13}$$

or

$$y''(x)+\{\lambda-q(x)\}y(x)=0 \ , \ \ q(x)=f''(x)/f(x) \ . \tag{8.1.14}$$

We note that since (8.1.3) may be transformed into (8.1.1), the latter may be transformed into (8.1.14). In fact if

$$\begin{aligned}&u(x)=v(\xi)=y(\xi)/f(\xi) \ , \ \ f(\xi)=A^{1/2}(x)=\rho^{1/4}(x) \ , \\ &\xi'(x)=f^2(\xi) \ ,\end{aligned} \tag{8.1.15}$$

then

$$u'=\dot{v}f^2=f\dot{y}-\dot{f}y \ , \ \ u''=f^2(f\ddot{y}+\ddot{f}y) \ , \tag{8.1.16}$$

and

$$u''+\lambda\rho u=f^2(f\ddot{y}-\ddot{f}y)+\lambda f^4f^{-1}y=0 \ , \tag{8.1.17}$$

so that

$$\ddot{y}+\{\lambda-q(\xi)\}y=0 \ , \ \ q(\xi)=\ddot{f}(\xi)/f(\xi) \ . \tag{8.1.18}$$

8.2 Sturm-Liouville systems

The Sturm-Liouville operator L is defined by

$$L\equiv-(p(x)y'(x))'+q(x)y(x) \ . \tag{8.2.1}$$

We define a Sturm-Liouville system as one governed by the equation

$$L\,y=\lambda\rho y \tag{8.2.2}$$

where p, p', q, ρ are continuous functions in$[0,\ell]$, $p>0$, $\rho>0$ and y satisfies the end conditions

$$y'(0)-hy(0)=0=y'(\ell)+Hy(\ell)\,. \tag{8.2.3}$$

The system will be said to be *positive* if $q\geq 0$, h, $H\geq 0$ and $h+H>0$. The equations governing the free vibrations of a string and of a rod in longitudinal or torsional vibration are positive Sturm-Liouville systems, provided that at least one of the end spring stiffnesses is non-zero, i.e., provided that the system is anchored.

We first state *Green's identity.* If $u(x)$, $v(x)$ are in C^2 and $0\leq x_1<x_2\leq \ell$, then

$$\int_{x_1}^{x_2}\{vL\,u-uL\,v\}dx=\int_{x_1}^{x_2}\{u(pv')'-v(pu')'\}dx=[p(uv'-vu')]_{x_1}^{x_2} \tag{8.2.4}$$

Introduce the *inner product*

$$(u,v)=\int_o^{\ell}u(x)v(x)dx\,. \tag{8.2.5}$$

Theorem 8.2.1. The Sturm-Liouville operator is *self-adjoint*, i.e., $(u,L\,v)=(v,L\,u)$ under the end-conditions (8.2.3).

Proof:

Green's identity yields

$$(v,L\,u)-(u,L\,v)=[p(uv'-vu')]_o^{\ell}=0 \tag{8.2.6}$$

since u, v satisfy (8.2.3). ■

The equation (8.2.2) will have a non-trivial solution satisfying (8.2.3) only for certain values of λ, called *eigenvalues*; the corresponding solutions are called *eigenfunctions.*

Theorem 8.2.2. Eigenvalues of a Sturm-Liouville system are simple.

Proof:

Note that we have not yet proved that the equation has any eigenvalues; we merely show that if it has any, then they are simple. Suppose that $y_1(x)$, $y_2(x)$ are two linearly independent solutions of equation (8.2.2) corresponding to the same λ and the same end conditions (8.2.3). Then Green's identity yields

$$p(y_1y'_2-y_2y'_1)=const.\,. \tag{8.2.7}$$

But under the end conditions (8.2.3), the left hand side is zero at each end, so that $y_1y'_2-y_2y'_1\equiv 0$, which contradicts the assumption of linear independence. ■

Theorem 8.2.3. Eigenvalues of a positive Sturm-Liouville system are positive.

Proof:

Suppose λ is a (possibly complex) eigenvalue and $y(x)$ is the corresponding eigenfunction. Then,

$$(Ly,\bar{y})=\lambda(\rho y,\bar{y})\ . \tag{8.2.8}$$

But, by Theorem 8.2.1, $(Ly,\bar{y})=(L\bar{y},y)=\overline{(Ly,\bar{y})}$, so that $(Ly,\bar{y})$, like $(\rho y,\bar{y})$ is real. Thus all eigenvalues λ of a Sturm-Liouville system are *real*. Now

$$\begin{aligned}(Ly,y) &= -\int_o^\ell (py')'y\,dx+\int_o^\ell qy^2dx\\ &= [-py'y]_o^\ell+\int_o^\ell (py'^2+qy^2)dx\ .\end{aligned} \tag{8.2.9}$$

For a positive system, the integrated term is non-negative so that λ, like (Ly,y) and $(\rho y,y)$, is positive. ■

Theorem 8.2.4. Eigenfunctions of a Sturm-Liouville system corresponding to different eigenvalues λ_1, λ_2, are orthogonal w.r.t. ρ, i.e., $(\rho y_1,y_2)=0$.

Proof:

$Ly_1=\lambda_1\rho y_1$ and $Ly_2=\lambda_2\rho y_2$ give

$$(\lambda_1-\lambda_2)(\rho y_1,y_2)=(Ly_1,y_2)-(Ly_2,y_1)=0\ .$$ ■

Examples 8.2

1. Show that the eigenvalues and eigenfunctions of $y''+\lambda y=0$ corresponding to the end conditions (8.2.3) are given by $\lambda_n=\omega_n^2$, $\omega_n\ell=\alpha_n+\beta_n+n\pi$ where $\alpha_n = \arctan(h/\omega_n)$, $\beta_n = \arctan(H/\omega_n)$, and $y_n=\cos(\omega_n x-\alpha_n)$, where $n=0,1,2,\ldots$. Hence show that ω_n is an increasing function of h and H and that, when h, H are positive, there is just one eigenvalue in each interval $(n\pi/\ell,(n+1)\pi/\ell)$, $n=0,1,2,\ldots$,.
2. Consider various special cases of Ex. 8.2.1. Thus

   (i) $h=0=H$, then $\omega_n\ell=n\pi$, $n=0,1,2,\ldots$. Note: In this case, which was explicitly excluded earlier, there is a zero eigenvalue.

   (ii) $h=0$, $H=\infty$, then $\omega_n\ell = (n+1/2)\pi$, $y_n=\cos\lambda_n x$, $n=0,1,2,\ldots$.

   (iii) $h=\infty$, $H=\infty$, then $\omega_n\ell = (n+1)\pi$, $y_n=\sin\lambda_n x$, $n=0,1,2,\ldots$.

   (iv) h,H finite, then for large n

   $$\omega_n\ell=n\pi+\frac{(h+H)\ell}{n\pi}+0(\frac{1}{n^3}).$$

8.3 Green's functions

The idea of a Green's function is perhaps most easily introduced by considering the static deflection of a string with fixed ends due to a distributed load $f(x)$. The governing equation is

$$-Tu''(x)=f(x) \tag{8.3.1}$$

and the end conditions are $u(0)=0=u(\ell)$. If instead of a distributed load we consider a single unit concentrated load at $x=s$, then the string will be straight on each side of $x=s$, and have a discontinuity in its derivative at $x=s$ as shown in Figure 8.3.1. Thus

$$u(x)=\begin{cases} Ax & 0\leq x\leq s \\ B(\ell-x) & s<x\leq\ell \end{cases} . \tag{8.3.2}$$

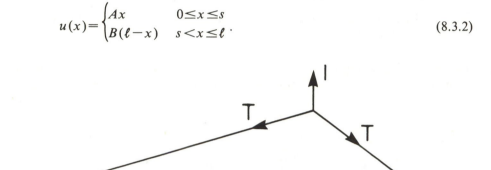

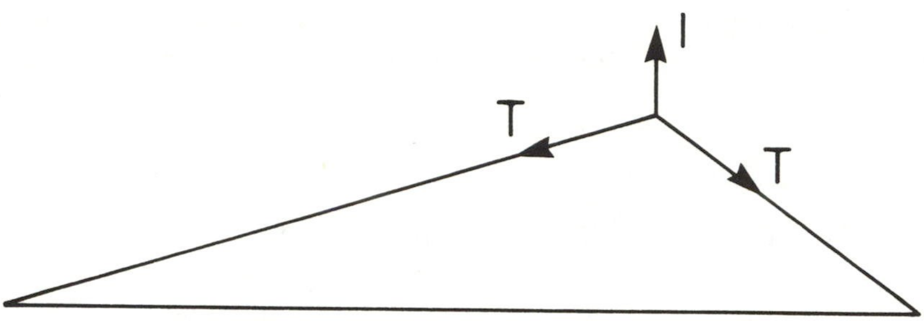

Figure 8.3.1 – The displacement of a string due to a point load

Equilibrium of the two portions gives

$$T\frac{\partial u}{\partial x}\Big|_{x=s+}-T\frac{\partial u}{\partial x}\Big|_{x=s-}=-1 , \tag{8.3.3}$$

while continuity yields $As=B(\ell-s)$. Thus

$$u(x)=G(x,s)=\begin{cases} x(\ell-s)/\ell T & 0\leq x<s \\ s(\ell-x)/\ell T & s<x\leq\ell \end{cases} . \tag{8.3.4}$$

The deflection under the action of a distributed load $f(x)$ will therefore be

$$u(x)=\int_0^\ell G(x,s)f(s)ds . \tag{8.3.5}$$

The form of the Green's function $G(x,s)$ for the more general end conditions (8.2.3) is given in Ex. 8.3.1.

If zero is not an eigenvalue, then the Sturm-Liouville operator (8.2.1) has a Green's function $G(x,s)$ with the following properties:

(1) $G(x,s)$ is, for fixed s, a continuous function of x and satisfies the end conditions (8.2.3).

(2) Except at s, the first and second derivatives of G with respect to x are continuous in $[0,\ell]$. At $x=s$, the first derivative has a jump discontinuity given by

$$\left.\frac{\partial G(x,s)}{\partial x}\right|_{x=s-}^{x=s+} = -\frac{1}{p(s)} . \tag{8.3.6}$$

(3) $L_x G=0$ for $0\le x<s$, $L_s G=0$ for $s<x\le\ell$. (Here $L_x G$ means that differentiation is w.r.t. x while s is kept constant).

In order to construct the Green's function we may proceed as follows. Let $\phi(x)$, $\psi(x)$ satisfy

$$L\phi=0 \text{ and } \phi'(0)-h\phi(0)=0 , \tag{8.3.7}$$

$$L\psi=0 \text{ and } \psi'(\ell)+H\psi(\ell)=0 . \tag{8.3.8}$$

We note that the argument used in Theorem 8.2.2 yields

$$p(\psi'\phi-\phi'\psi)= \textit{constant} . \tag{8.3.9}$$

The constant is not zero, for otherwise $\phi(x)$, $\psi(x)$ would be proportional so that $\phi(x)$ would be a solution of $L\phi=0$ satisfyng both end-conditions, i.e., zero would be an eigenvalue of $L\phi=\lambda\phi$, contrary to hypothesis. Choose $\phi(x)$, $\psi(x)$ so that

$$p(s)(\psi'(s)\phi(s)-\phi'(s)\psi(s))=-1, \tag{8.3.10}$$

then

$$G(x,s)=\begin{cases}\phi(x)\psi(s) & 0\le x<s\\ \phi(s)\psi(x) & s\le x\le\ell\end{cases} \tag{8.3.11}$$

has properties (1)-(3).

Theorem 8.3.1. If $f(x)$ is piecewise continuous, then

$$u(x)=\int_o^\ell G(x,s)f(s)ds \tag{8.3.12}$$

is a solution of

$$L(u)=f(x) \tag{8.3.13}$$

and satisfies the end conditions (8.2.3). Conversely, if $u(x)$ satisfies (8.3.13) and the end conditions (8.2.3) then it can be represented by (8.3.12). The proof may be found in Courant and Hilbert (1953), p. 353. The form (8.3.11) shows that the Green's function is symmetric. An alternative proof of this fact is as follows.

Theorem 8.3.2. The Green's function of the Sturm-Liouville equation under the end conditions (8.2.3) is *symmetric*, i.e.,

$$G(s,t)=G(t,s)\ . \tag{8.3.14}$$

Proof:

Put $u=G(x,s)$, $v=G(x,t)$ and suppose $s<t$. Since $Lu=0=Lv$, Green's identity gives

$$(Lu,v)-(Lv,u)=[p(v'u-u'v)]_o^s+[p(v'u-u'v)]_s^t+[p(v'u-u'v)]_t^\ell=0\ . \tag{8.3.15}$$

Now u' has a discontinuity at s and v' at t, so that (8.3.15) gives

$$p(s)[u']_{s-}^{s+}v(s)-p(t)[v']_{t-}^{t+}u(t)=0 \tag{8.3.16}$$

so that, on account of (8.3.6),

$$v(s)=G(s,t)=u(t)=G(t,s)\ . \ \blacksquare \tag{8.3.17}$$

For our purposes the most important use of the Green's function is that it reduces the free vibration problem

$$Lu=\lambda\rho u\ , \quad (\rho(x)>0) \tag{8.3.18}$$

to the integral equation

$$u(x)=\lambda\int_o^\ell G(x,s)\rho(s)u(s)ds\ . \tag{8.3.19}$$

We note that by the change of variables

$$v(x)=\rho^{1/2}(x)u(x) \quad H(x,s)=\rho^{1/2}(x)\rho^{1/2}(s)G(x,s) \tag{8.3.20}$$

the equation may be written in the symmetrical form

$$v(x)=\lambda\int_o^\ell H(x,s)v(s)ds\ . \tag{8.3.21}$$

Examples 8.3

1. Show that the Green's function for equation (8.1.3) corresponding to the end conditions $v(0)=0=v(\ell))$ is

$$G(x,s)=k\int_o^{x^-}B(t)dt\int_{x^+}^{o}B(t)dt$$

where $B(t)=1/A(t)$, $k=\{\int_o^\ell B(t)dt\}^{-1}$, $x^-=min(x,s)$, $x^+=max(x,s)$. When $A(t)\equiv T$, this equation reduces to (8.3.4).

2. Show that the Green's function of equation (8.3.1) corresponding to the end conditions (8.2.3) is

$$G(x,s)=\begin{cases}(1+hx)\{1+H(\ell-s)\}/[(h+H+hH\ell)T]\ , & 0\le x<s\\ (1+hs)\{1+H(\ell-x)\}/[h+H+hH\ell)T]\ , & s<x\le\ell\end{cases}$$

provided that not both of h, H are zero.

3. Show that if $u(x)$ satisfies

$$u''+\lambda\rho u=0\ ,\quad u(0)=0=u'(\ell)\ ,\quad \rho(x)>0\ ,$$

then $v=u'$ satisfies $(\rho^{-1}v')'+\lambda v=0$, $v'(0)=0=v(\ell)$. Hence show that $v(x)$ is an eigenfunction of the integral equation

$$v(x)=\lambda\int_o^\ell K(x,s)v(s)ds\ ,$$

where

$$K(x,s)=\int_{x^+}^\ell \rho(t)dt\ ,\quad x^+=max(x,s)\ .$$

4. Show that if $u(x)$ satisfies

$$(Au')'+\lambda Au=0\ ;\quad u(0)=0=u'(\ell)\ ,\quad A>0$$

then $v=Au'$ satisfies $(Bv')'+\lambda Bv=0$ where $B=1/A$ and $v'(0)=0=v(\ell)$. Hence show that v satisfies

$$v(x)=\lambda\int_o^\ell L(x,s)B(s)v(s)ds\ ,\quad \text{where } L(x,s)=\int_{x^+}^\ell A(t)dt\ .$$

8.4 Symmetric kernels and their eigenvalues

A comprehensive study of integral equations and of their eigenvalues may be found in Courant and Hilbert (1953). Here we state some basic results which will be needed in the following analysis.

First consider the integral equation

$$\phi(s)=\lambda\int_o^\ell K(s,t)\phi(t)dt\ ,\quad 0\le s\le\ell\ , \tag{8.4.1}$$

where $K(s,t)$ is defined and continuous in the square $0\le s\le\ell$, $0\le t\le\ell$. A value of λ for which the equation has a non-trivial solution is called an *eigenvalue*, and the corresponding solution an *eigenfunction*.

Unless explicitly stated otherwise, we shall assume that the kernel K is *symmetric*, i.e.,

$$K(s,t)=K(t,s)\ . \tag{8.4.2}$$

For an existence proof for the eigenvalues of a symmetric kernel, see e.g., Courant and Hilbert (1953), p. 122.

Theorem 8.4.1. The eigenvalues of (8.4.1) are real.

Proof:

$$\int_o^\ell \phi(s)\phi(\bar{s})ds=\lambda\int_o^\ell\int_o^\ell K(s,t)\phi\overline{(s)}\phi(t)dsdt\ .$$

Both integrals are real, so that λ is real. ■

Theorem 8.4.2. Eigenfunctions corresponding to different eigenvalues λ_i, λ_j are *orthogonal*, i.e.,

$$\int_o^\ell \phi_i(s)\phi_j(s)ds = 0 \; .$$

Proof:

$$(\lambda_i - \lambda_j)\int_o^\ell \phi_i(s)\phi_j(s)ds = \lambda_i\lambda_j \int_o^\ell \int_o^\ell K(s,t)[\phi_i(s)\phi_j(t) - \phi_j(s)\phi_i(t)]dsdt = 0 \; . \blacksquare$$

If there is more than one eigenfunction corresponding to a given eigenvalue then, by using the Gram-Schmidt procedure, we may form linear combinations which are *orthonormal*. From now on we therefore assume that the $\phi_i(s)$ have been chosen so that

$$\int_o^\ell \phi_i(s)\phi_j(s)ds = \delta_{ij} \; . \tag{8.4.3}$$

Theorem 8.4.3. The equation (8.4.1) has a finite number of eigenvalues $\lambda_0, \lambda_1, \ldots, \lambda_n$ if and only if it is *degenerate*, i.e., it may be expressed in the form

$$K(s,t) = \sum_{i=o}^{p} \alpha_i(s)\beta_i(t) \; , \tag{8.4.4}$$

where $\{\alpha_i(s)\}_1^p$ and $\{\beta_i(t)\}_1^p$ are sets of linearly independent functions. A symmetric degenerate kernel may be written

$$K(s,t) = \sum_{i=o}^{n} \frac{\phi_i(s)\phi_i(t)}{\lambda_i} \; . \tag{8.4.5}$$

Proof:

The theory of integral equations with degenerate kernels parallels that of algebraic eigenvalue problems. A formal proof of the theorem may be found in Courant and Hilbert (1953). We merely note that

$$K_m^*(s,t) \equiv K(s,t) - \sum_{i=o}^{m} \frac{\phi_i(s)\phi_i(t)}{\lambda_i} \tag{8.4.6}$$

has the eigenvalues $(\lambda_i)_{m+1}^n$ and eigenfunctions $(\phi_i)_{m+1}^n$; for

$$\begin{aligned} \lambda_j \int_o^\ell K_m^*(s,t)\phi_j(t)dt &= \phi_j(s) - \lambda_j \sum_{i=o}^{m} \frac{\phi_i(s)}{\lambda_i} \int_o^\ell \phi_i(t)\phi_j(t)dt \\ &= \phi_j(s) - \delta_{ij}\phi_i(s) \; . \; \blacksquare \end{aligned} \tag{8.4.7}$$

Theorem 8.4.4. A non-degenerate symmetric kernel has an infinity of eigenvalues. They have no finite point of accumulation, so that their absolute values increase beyond all bounds and only a finite number can be equal to each other.

Proof:

Consider the integral inequality

$$I = \int_o^\ell [K(s,t) - \sum_{i=o}^{n} c_i\phi_i(t)]^2 dt = \int_o^\ell K^2(s,t)dt$$

$$- 2\sum_{i=o}^{n} c_i \int_o^\ell K(s,t)\phi_i(t)dt + \sum_{i=o}^{n} c_i^2 \geq 0. \tag{8.4.8}$$

Put $c_i = \int_o^\ell K(s,t)\phi_i(t)dt = \phi_i(s)/\lambda_i$, then

$$I = \int_o^\ell K^2(s,t)dt - \sum_{i=o}^{n} \phi_i^2(s)/\lambda_i^2 \geq 0\ . \tag{8.4.9}$$

This shows that the series

$$T(s) = \sum_{i=o}^{\infty} \frac{\phi_i^2(s)}{\lambda_i^2} \tag{8.4.10}$$

converges absolutely. Its integral gives

$$\int_o^\ell T(s)ds = \sum_{i=o}^{\infty} \frac{1}{\lambda_i^2} \leq \int_o^\ell \int_o^\ell K^2(s,t)dsdt\ . \tag{8.4.11}$$

This inequality implies that only a finite number of the λ_i can be equal and that $\lim_{i\to\infty} \lambda_i = \infty$. ■

This theorem also implies that an eigenvalue can have only a finite multiplicity.

It was noted in Theorem 8.4.3 that a degenerate symmetric kernel may be expanded as a bilinear series in its eigenfunctions. In general it is *not* true that a symmetric kernel can be expanded in the form

$$K(s,t) = \sum_{i=o}^{\infty} \frac{\phi_i(s)\phi_i(t)}{\lambda_i}, \tag{8.4.12}$$

but the following results *do* hold.

Theorem 8.4.5. If $h(t)$ is piecewise continuous on $[0,\ell]$ and

$$g(s) = \int_o^\ell K(s,t)h(t)dt\ , \tag{8.4.13}$$

then

$$g(s) = \sum_{i=o}^{\infty} g_i\phi_i(s)\ ,\quad g_i = (g,\phi_i) = (h,\phi_i)/\lambda_i\ . \tag{8.4.14}$$

The series converges uniformly and absolutely.

Theorem 8.4.6. The series (8.4.12) converges *in the mean* to $K(s,t)$, i.e.,

$$\lim_{n\to\infty}\int_o^\ell \left[K(s,t)-\sum_{i=o}^{n}\frac{\phi_i(s)\phi_i(t)}{\lambda_i}\right]^2 dt=0\ . \tag{8.4.15}$$

Theorem 8.4.7. Let $K^{(r)}(s,t)$ denote the *iterated kernel*, i.e., $K^{(1)}(s,t)=K(s,t)$ and

$$K^{(r+1)}(s,t)=\int_o^\ell K^{(r)}(s,u)K(u,t)du\ , \tag{8.4.16}$$

then

$$K^{(r)}(s,t)=\sum_{i=o}^{\infty}\frac{\phi_i(s)\phi_i(t)}{\lambda_i^r} \tag{8.4.17}$$

when $r>1$, all these expansions converge absolutely and uniformly both in s and in t, and uniformly in s and t together. (We note that the $T(s)$ of equation (8.4.10) is $K^{(2)}(s,s)$).

Introduce the bilinear form

$$J(\phi,\psi)=\int_o^\ell\int_o^\ell K(s,t)\phi(s)\psi(t)dsdt\ . \tag{8.4.18}$$

The symmetric kernel $K(s,t)$ is said to be *positive definite* if

$$J(\phi,\phi)>0 \quad \text{whenever } (\phi,\phi)>0\ .$$

Theorem 8.4.8. A symmetric kernel $K(s,t)$ is positive definite if and only if it has positive eigenvalues.

Proof:

If K is positive definite, then

$$(\phi_i,\phi_i)=\lambda_i J(\phi_i,\phi_i)\ , \tag{8.4.19}$$

so that $\lambda_i>0$. If $(\lambda_i)_1^\infty>0$ then Theorem 8.4.5 gives for any piecewise continuous $\phi(t)$

$$\int_o^\ell K(s,t)\phi(t)dt=\sum_{i=o}^{\infty}\frac{c_i\phi_i(s)}{\lambda_i}\ ,\quad c_i=(\phi,\phi_i) \tag{8.4.20}$$

and thus

$$J(\phi,\phi)=\sum_{i=o}^{\infty}\frac{c_i(\phi_i,\phi)}{\lambda_i}=\sum_{i=o}^{\infty}\frac{c_i^2}{\lambda_i}>0\ .\ \blacksquare \tag{8.4.21}$$

Theorem 8.4.9. If $K(s,t)$ is positive definite and continuous then $K(s,s)\geq 0$ for all s in $[0,\ell]$.

Proof:

Suppose that there is an s_o in $0 \le s_o \le \ell$ such that $K(s_o,s_o)<0$. Then there is a region $|s-s_o|<\varepsilon$, $|t-s_o|<\varepsilon$ (one-sided if $s_o=0$ or $s_o=\ell$) such that $K(s,t)<0$ everywhere in the region. But then by choosing $\phi(t)$ to be the piecewise continuous function such that $\phi(t)=1$ in the region and $\phi(t)=0$ elsewhere we will have

$$J(\phi,\phi) \equiv \int_o^\ell \int_o^\ell K(s,t)\phi(s)\phi(t)dsdt<0 \tag{8.4.22}$$

contrary to the positive definiteness of K. ∎

Theorem 8.4.10. If $K(s,t)$ is positive definite and continuous then the expansion (8.4.12) is valid and converges absolutely and uniformly

Proof:

Apply Theorem 8.4.9 to the positive definite kernel

$$H(s,t)=K(s,t)-\sum_{i=o}^{n}\frac{\phi_i(s)\phi_i(t)}{\lambda_i} . \tag{8.4.23}$$

We obtain

$$K(s,s)-\sum_{i=o}^{n}\frac{[\phi_i(s)]^2}{\lambda_i}\ge 0 , \tag{8.4.24}$$

so that the series

$$\sum_{i=o}^{\infty}\frac{[\phi_i(s)]^2}{\lambda_i} , \tag{8.4.25}$$

all of whose terms are positive (because $\lambda_i>0$) converges for all s. Because of the Schwarz inequality

$$\left(\sum_{i=m}^{n}\frac{\phi_i(s)\phi_i(t)}{\lambda_i}\right)^2 \le \left(\sum_{i=m}^{n}\frac{[\phi_i(s)]^2}{\lambda_i}\right)\left(\sum_{i=m}^{n}\frac{[\phi_i(t)]^2}{\lambda_i}\right) \tag{8.4.26}$$

the series (8.4.12) also converges absolutely. Because of this inequality the uniformity of the convergence of (8.4.12) follows from that of (8.4.25). The uniformity of the latter follows from the fact that $K(s,s)$ is continuous. ∎

Examples 8.4

1. Adapt the analysis of Theorem 8.4.9 to show that if $K(s,t)$ is positive definite, and $\mathbf{K}$ is the $n\times n$ matrix with elements $k_{ij}=K(s_i,s_j)$, and then, in the notation of equation (5.2.2),

$$K\begin{pmatrix} s_1,s_1,\ldots,s_n \\ s_1,s_2,\ldots,s_n \end{pmatrix} \geq 0$$

for all $(s_i)_1^n$ satisfying $0 \leq s_2 < s_2 < \ldots \ < s_n \leq \ell$.

8.5 Oscillatory properties of Sturm-Liouville kernels

It was shown in Section 8.3 that when the kernel $K(x,s)$ is symmetric, then the eigenvalues of the integral equation

$$y(x) = \lambda \int_o^\ell K(x,t)\rho(s)y(s)ds \tag{8.5.1}$$

are *real*, and that when K is positive definite, then the eigenvalues are *positive.* However, for a positive Sturm-Liouville system, it was proved that the eigenvalues, in addition to being real and positive, are *simple*, i.e., *distinct*. This implies that the Green's function kernel must have some additional properties which lead to this distinctness. We shall now discuss these properties.

We recall that a positive Sturm-Liouville system is governed by the equation

$$Ly \equiv -(py')' + qy = \lambda \rho y \ , \tag{8.5.2}$$

where p, p', q, ρ are continuous in $[0,\ell]$, $p>0$, $q \geq 0$, $\rho > 0$ and y satisfies the end conditions

$$y'(0) - hy(0) = 0 = y'(\ell) + Hy(\ell) \ , \tag{8.5.3}$$

where h, $H \geq 0$, $h+H>0$. We define the interval I as follows

$$\begin{aligned} I &= [0,\ell] \text{ if } h, \ H \text{ are finite} \\ &= (0,\ell] \text{ if } h=\infty, \ H \text{ is finite} \\ &= [0,\ell) \text{ if } h \text{ is finite, } H=\infty \\ &= (0,\ell) \text{ if } h=\infty=H \ . \end{aligned} \tag{8.5.4}$$

Thus when an end is fixed, e.g., $x=\ell$ when $H=\infty$, then that point is excluded from I; we may say that I is the set of *movable* points in $[0,\ell]$.

Definition 8.5.1 A function $f(x)$ is said to have *fixed sign* in an interval I if *either* $f(x) \geq 0$ for all x in I *or* $f(x) \leq 0$ for all x in I. It is said to have *strictly fixed sign* if *either* $f(x)>0$ for all x in I *or* $f(x)<0$ for all x in I.

Theorem 8.5.1. For a positive Sturm-Liouville system the function $\phi(x)$ of equation (8.3.7) has strictly fixed sign in $(0,\ell]$, and in $[0,\ell]$ if h is finite. Similarly $\psi(x)$ given in equation (8.3.8) has strictly fixed sign in $[0,\ell)$, and in $[0,\ell]$ if H is finite. The function $\phi(x)\psi(x)$ is strictly positive for x in I.

Proof:

Suppose $\phi(x_o)=0$ for some x_o in $(0,\ell]$, then as in Theorem 8.2.3,

$$0=\int_o^{x_o}\phi L\,\phi dx=[-p\phi'\phi]_o^{x_o}+\int_o^{x_o}p\phi'^2+q\phi^2dx\ . \tag{8.5.5}$$

If h is finite, then $\phi(0)\neq 0$; for if $\phi(0)=0$, then $\phi'(0)=0$ and $\phi(x)\equiv 0$. Therefore $[-p\phi'\phi]_o^{x_o}=p(0)\phi(0)\phi'(0)>0$. Thus the right hand side of (8.5.5) is positive, which provides a contradiction; hence $\phi(x)\neq 0$ for x in $[0,\ell]$. Since $\phi(x)$ is continuous, $\phi(x)$ has strictly fixed sign in $[0,\ell]$. If $h=\infty$ then $\phi(0)=0$, so that $\phi(x)$ has strictly fixed sign in $(0,\ell]$. We may argue similarly concerning $\psi(x)$, and conclude that $\phi(x)\psi(x)$ has strictly fixed sign in I. Theorem 8.2.3 shows that the eigenvalues of the system are positive, Theorem 8.4.8 shows that G is positive definite, and Theorem 8.4.9 that $G(x,x) = \phi(x)\psi(x)\geq 0$. Hence $\phi(x)\psi(x)>0$ for x in I. Without loss of generality we may assume that $\phi(x)>0$, $\psi(x)>0$ in the stated intervals. ■

Theorem 8.5.2. The Green's function of the rod satisfies

$$G(x,s)>0 \text{ for } x,s \text{ in } I$$

Proof:

This follows directly from equation (8.3.11) and Theorem 8.5.1. ■

Definition 8.5.2. A kernel $K(x,s)$ is said to be *oscillatory* if the following three conditions are satisfied:

1) $K(x,s)>0$ for x,s in I,

2)

$$K\begin{pmatrix}x_1,x_2,\ldots,x_n\\ s_1,s_2,\ldots,s_n\end{pmatrix}\geq 0 \quad\text{for}\quad 0<\begin{matrix}x_1<x_2<\ldots<x_n\\ s_1<s_2<\ldots<s_n\end{matrix}<\ell\ ,$$

3)

$$K\begin{pmatrix}x_1,x_2,\ldots,x_n\\ x_1,x_2,\ldots,x_n\end{pmatrix}>0 \quad\text{for}\quad 0<x_1<x_2<\ldots<x_n<\ell\ .$$

Here

$$K\begin{pmatrix}x_1,x_2,\ldots,x_n\\ s_1,s_2,\ldots,s_n\end{pmatrix}=\begin{vmatrix}K(x_1,s_1) & K(x_1,s_2) & \cdots & K(x_1,s_n)\\ K(x_2,s_1) & K(x_2,s_2) & \cdots & K(x_2,s_n)\\ \cdot & \cdot & \cdot & \cdot\\ K(x_n,s_1) & K(x_n,s_2) & \cdots & K(x_n,s_n)\end{vmatrix} \tag{8.5.6}$$

and note Ex. 8.5.1.

Theorem 8.5.3 The kernel $K(x,s)$ is an oscillatory kernel if and only if the matrix $\mathbf{K}^o \equiv (k_{ij}) = (K(s_i,s_j))$ is an oscillatory matrix for any choice of $(s_i)_1^n$ in I such that $0 \le s_1 < s_2 < ... < s_n \le \ell$, $(n=1,2,...)$, and there is at least one internal point, (a stipulation that is necessary only when $n=2$).

Proof:

If the kernel is oscillatory then condition (1) implies $k_{i,i+1} = K(s_i,s_{i+1}) > 0$, while condition (2) implies

$$K^o \begin{pmatrix} i_1,i_2,...,i_p \\ k_1,k_2,...,k_p \end{pmatrix} \ge 0\ , \quad 1 \le \begin{matrix} i_1<i_2<...<i_p \\ j_1<j_2<...<j_p \end{matrix} \le n$$

since

$$K^o \begin{pmatrix} i_1,i_2,...,i_p \\ k_1,k_2,...,k_p \end{pmatrix} = K \begin{pmatrix} x_1^o,x_2^o,...,x_p^o \\ s_1^o,s_2^o,...,s_p^o \end{pmatrix} \ge 0$$

where $x_j^o = s_{i_j}$, $s_j^o = s_{k_j}$. Condition (3) implies that the matrix $\mathbf{K}^o$ is non-singular. Hence $\mathbf{K}^o$ is an oscillatory matrix. All these implications apply in reverse also. ■

Theorem 8.5.1 shows that $G(x,s)$ has property 1. We shall now show that it has properties 2 and 3 also, and is therefore an oscillatory kernel. We introduce:

Definition 8.5.3. A matrix $\mathbf{G} = (g_{ij})$ is called a *Green's matrix* if

$$g_{ij} = \begin{cases} a_i b_j, & i \le j, \\ a_j b_i, & i \ge j, \end{cases} \tag{8.5.7}$$

where $(a_i)_1^n$ and $(b_i)_1^n$ are real constants.

Theorem 8.5.4 If $\mathbf{G}$ is a Green's matrix then

$$G \begin{pmatrix} i_1,i_2,...,i_p \\ j_1,j_2,...,j_p \end{pmatrix} = a_{k_1} \prod_{r=2}^{p} \begin{vmatrix} a_{k_r} & a_{\ell_{r-1}} \\ b_{k_r} & b_{\ell_{r-1}} \end{vmatrix} \cdot b_{\ell_p} \tag{8.5.8}$$

where $k_m = \min(i_m,j_m)$, $\ell_m = \max(i_m,j_m)$, provided $i_m,j_m < i_{m+1},j_{m+1}$ $(m=1,2,...,p-1)$. In all other cases the minor is zero.

Proof:

If $i_1 < i_2 \le j_1$ then the first two rows are

$$\begin{matrix} g_{i_1,j_1} & g_{i_1,j_2} & \cdots & g_{i_1,j_p} \\ g_{i_2,j_1} & g_{i_2,j_2} & \cdots & g_{i_2,j_p} \end{matrix}$$

but these are

$$a_{i_1}b_{j_1} \;\; a_{i_1}b_{j_2} \;\; \cdots \;\; a_{i_1}b_{j_p}$$

$$a_{i_2}b_{j_1} \;\; a_{i_2}b_{j_2} \;\; \cdots \;\; a_{i_2}b_{j_p}$$

and are thus proportional, so that the minor is zero. Similarly, if $j_1 < j_2 \le i_1$ the first two columns will be proportional and the minor zero. Thus we may assume that $\max(i_1, j_1) < \min(i_2, j_2)$. Suppose further, for definiteness that $i_2 \le j_2$ (otherwise the argument proceeds with the first two columns), then the first two rows are

$$a_{k_1}b_{\ell_1} \;\; a_{i_1}b_{j_2} \;\; \cdots \;\; a_{i_1}b_{j_p}$$

$$a_{j_1}b_{i_2} \;\; a_{i_2}b_{j_2} \;\; \cdots \;\; a_{i_2}b_{j_p} \, .$$

Multiplying row 2 by $-a_{i_1}/a_{i_2}$ and adding to the first, we find the first term

$$a_{k_1}b_{\ell_1} - a_{j_1}b_{i_2}a_{i_1}/a_{i_2} \equiv a_{k_1}b_{\ell_1} - a_{k_1}b_{k_2}a_{\ell_1}/a_{k_2}$$

$$= \frac{a_{k_1}}{a_{k_2}} \begin{vmatrix} a_{k_2} & a_{\ell_1} \\ b_{k_2} & b_{\ell_1} \end{vmatrix} ,$$

so that

$$G\begin{pmatrix} i_1, i_2, \ldots, i_p \\ j_1, j_2, \ldots, j_p \end{pmatrix} = a_{k_1} \begin{vmatrix} a_{k_2} & a_{\ell_1} \\ b_{k_2} & b_{\ell_1} \end{vmatrix} \cdot \frac{1}{a_{k_2}} G\begin{pmatrix} i_2, i_3, \ldots, i_p \\ j_2, j_3, \ldots, j_p \end{pmatrix} , \tag{8.5.9}$$

from which the theorem follows by induction. ■

Theorem 8.5.5. The Green's matrix **G** is completely non-negative if and only if all $(a_i)_1^n$, $(b_i)_1^n$ have the same strict sign and

$$\frac{a_1}{b_1} \le \frac{a_2}{b_2} \le \cdots \le \frac{a_n}{b_n} . \tag{8.5.10}$$

Moreover **G** will be oscillatory if and only if all $(a_i)_1^n$, $(b_i)_1^n$ have the same strict sign and

$$\frac{a_1}{b_1} < \frac{a_2}{b_2} < \cdots < \frac{a_n}{b_n} . \tag{8.5.11}$$

Proof:

There is no loss in generality in assuming that all a_i, b_i are positive. It was shown in the proof of Theorem 8.5.4 that a minor is zero unless

$$i_1, j_1 < i_2, j_2 < \cdots < i_p, j_p \, . \tag{8.5.12}$$

Each second order determinant in (8.5.8) is non-negative if and only if

$$\frac{a_{\ell_{r-1}}}{b_{\ell_{r-1}}} \le \frac{a_{k_r}}{b_{k_r}} \qquad r=1,2,\dots,p \; . \tag{8.5.13}$$

This is exactly the condition (8.5.10). **G** is completely non-negative and $g_{i,i+1}$ and $g_{i+,i}$ are positive so that the only condition which must be fulfilled for it to be oscillatory is that it must be non-singular. Thus each second order determinant must be positive, which is (8.5.11). ■

Corollary. Let $\phi(x)$, $\psi(x)$ be continuous in $[0,\ell]$ and

$$K(x,s)=\begin{cases}\phi(x)\psi(s) & 0\le x\le s\le \ell, \\ \phi(s)\psi(x) & 0\le s\le x\le \ell.\end{cases} \tag{8.5.14}$$

If $\phi(x)\psi(x)>0$ in $(0,\ell)$ and $\phi(x)/\psi(x)$ is an increasing function for x in $(0,\ell)$, then

$$K\begin{pmatrix} x_1,x, \dots ,x_n \\ s_1,s_2, \dots ,s_n\end{pmatrix} \ge 0 \quad \text{for} \quad 0< \begin{matrix} x_1<x_2< \cdots <x_n \\ s_1<s_2< \cdots <s_n\end{matrix} <\ell \; . \tag{8.5.15}$$

If $\phi(x)\psi(x)>0$ and $\phi(x)/\psi(x)$ is a strictly increasing function for x in I (see equation (8.5.4)), then

$$K\begin{pmatrix} x_1,x_2,\dots,x_n \\ s_1,s_2,\dots,s_n\end{pmatrix} >0 \tag{8.5.16}$$

if and only if $(x_i)_1^n$, $(s_i)_1^n$ are in I and

$$x_1,s_1<x_2,s_2< \cdots <x_n,s_n \; . \tag{8.5.17}$$

Theorem 8.5.6. The Green's function of the rod is oscillatory, and the minor

$$G\begin{pmatrix} x_1,x_2,\dots,x_n \\ s_1,s_2,\dots,s_n\end{pmatrix} >0 \tag{8.5.18}$$

if and only if $(x_i)_1^n$, $(s_i)_1^n$ are in I and

$$x_1,s_1<x_2,s_2< \cdots <x_n,s_n \tag{8.5.19}$$

Proof:

Because of (8.3.11) the Green's function satisfies the conditions of the Corollary to Theorem 8.5.5. The function $\phi(x)/\psi(x)$ is strictly increasing since (8.3.10) gives

$$p(x)\frac{d}{dx}\left(\frac{\phi(x)}{\psi(x)}\right)=\frac{1}{[\psi(x)]^2}>0 \quad ■ \tag{8.5.20}$$

In order to ascertain the physical meaning of the concept of an oscillatory kernel, consider a string or rod under the action of n concentrated forces $(F_i)_1^n$ (normal to the string, longitudinal for the rod (8.1.3), torsional for (8.1.5)), applied at points $(s_i)_1^n$ where $0 \le s_1 \le s_2 < \cdots < s_n \le \ell$. The displacement is

$$u(x)=\sum_{i=1}^{n} F_i G(x,s_i) . \qquad (8.5.21)$$

Thus condition (1) of the definition of an oscillatory kernel (see also Theorem 8.5.2) states that the displacement due to a single force occurs in the same 'direction' or sense as the force.

To see the meaning of condition (3) we note that the strain energy of the system under the action of the $(F_i)_1^n$ is

$$U=\frac{1}{2}\sum_{j=1}^{n} u(s_j)F_j=\frac{1}{2}\sum_{i=1}^{n}\ \sum_{i=1}^{n} G(s_i,s_j)F_iF_j , \qquad (8.5.22)$$

so that condition (3) (see also Theorem 8.5.6) states that U is positive definite provided that the $(F_i)_1^n$ are applied at movable points.

In Section 8.8, as a result of analysis developed in Section 8.7 we will show that condition (2) is closely related to the following result.

Theorem 8.5.7. Under the action of n forces $(F_i)_1^n$ the displacement $u(x)$ can change its sign no more than $n-1$ times.

Proof:

Suppose that forces $(F_i)_1^n$ are applied at points $(s_i)_1^n$, where $0 \le s_1 < s_2 < \ldots < s_n \le \ell$. If $s_1 > 0$, then

$$u(x)=\phi(x)\sum_{i=1}^{n} F_i\psi(s_i) , \quad 0 \le x \le s_1 , \qquad (8.5.23)$$

so that $u(x)$ is of one sign and can be zero only at 0, or is identically zero if

$$\sum_{i=1}^{n} F_i\psi(s_i)=0 .$$

In the interval $[s_j,s_{j+1}]$ $(j=1,2,\ldots,n-1)$

$$u(x)=\psi(x)\sum_{i=1}^{j} F_i\phi(s_i)+\phi(x)\sum_{i=j+1}^{n} F_i\psi(s_i) . \qquad (8.5.24)$$

Here $u(x)$ cannot be identically zero, and there can be at most one zero in $[s_j,s_{j+1}]$ because if there were two, namely ξ, η, such that $s_j \le \xi < \eta \le s_{j+1}$ then the determinant $\psi(\xi)\phi(\eta)-\psi(\eta)\phi(\xi)$ would be zero contradicting equation (8.5.20). Finally, if $s_n \le x \le \ell$ then

$$u(x)=\psi(x)\sum_{i=1}^{n}F_i\phi(s_i) , \tag{8.5.25}$$

so that again $u(x)$ has one sign and can be zero only at $x=\ell$ or is identically zero if

$$\sum_{i=1}^{n}F_i\phi(s_i)=0 .$$

We conclude that $u(x)$ has at most $n-1$ zeros in (s_1,s_n). If it is zero in either $[0,s_1]$ or $[s_n,\ell]$ then it is identically zero in that interval. ■ *Examples 8.5*

1. Continuity of the minor in (2) of Definition 8.5.2 shows that it will be non-negative for x_i, s_i satisfying the inequalities $0\le x_1<x_2<\cdots<x_n\le\ell$, $0\le s_1<s_2<\cdots<s_n\le\ell$. Use Theorem 5.5.4 to show that conditions (1)-(3) necessarily imply that (3) holds for all $x_1,x_2,...,x_n$ in I, not just those in $(0,\ell)$.
2. If concentrated loads $(F_i)_1^n$ are applied to the rod at points $(s_i)_1^n$, $s_1<s_2<\cdots<s_n$, in I, then the displacement at x is

   $$u(x)=\sum_{i=1}^{n}F_iG(x,s_i) .$$

   Show that if the forces are considered as being built up from zero to their final values, the strain energy stored in the rod is

   $$U=\frac{1}{2}\sum_{i=1}^{n}F_iu(s_i)=\frac{1}{2}\sum_{i=1}^{n}\sum_{j=1}^{n}G(s_i,s_j)F_iF_j .$$

   Hence deduce that the matrix $(g_{ij})\equiv(G(s_i,s_j))$ is positive definite provided that $h+H>0$.

8.6 Completeness

Let $\phi_o(s)$, $\phi_1(s)$,... be an orthonormal sequence of functions on $(0,\ell)$, i.e., for some $\rho(s)\neq 0$

$$(\rho\phi_i,\phi_j)\equiv\int_o^\ell\rho(s)\phi_i(s)\phi_j(s)ds=\delta_{ij} , \tag{8.6.1}$$

then the numbers

$$c_i=(\rho\phi,\phi_i) \tag{8.6.2}$$

are called the expansion coefficients of $\phi(s)$ w.r.t. $\{\phi_i(s)\}_o^\infty$. For any piecewise continuous function $\phi(s)$, the inequality

$$e_n\equiv\int_o^\ell\rho(s)\{\phi(s)-\sum_{i=o}^{n}c_i\phi_i(s)\}^2ds\ge 0 , \tag{8.6.3}$$

yields

$$e_n=(\rho\phi,\phi)-2\sum_{i=o}^{n}c_i(\rho\phi,\phi_i)+\sum_{i=o}^{n}c_i^2\equiv(\rho\phi,\phi)-\sum_{i=o}^{n}c_i^2\geq 0\ , \tag{8.6.4}$$

so that the series $\sum_{i=1}^{\infty}c_i^2$ converges,

$$\sum_{i=o}^{\infty}c_i^2\leq(\rho\phi,\phi)\ , \tag{8.6.5}$$

and $e_0,e_1,\ldots$ is a monotonically decreasing sequence. This leads to

Definition 8.6.1. The orthonormal sequence $\{\phi_i(s)\}_o^\infty$ is said to be *complete* on $(0,\ell)$ if, for any piecewise continuous function $\phi(s)$, the sequence $e_0,e_1,\ldots$ tends to zero. In this case therefore

$$\sum_{i=o}^{\infty}c_i^2=(\rho\phi,\phi)\ , \tag{8.6.6}$$

and the sequence

$$\{\sum_{i=o}^{n}c_i\phi_i(s)\}_o^\infty$$

is said to converge *in the mean* to $\phi(s)$.

Theorem 8.4.5 states that if $K(s,t)$ is a nondegenerate symmetric kernel, then any function $\phi(s)$ that may be expressed in the form

$$\phi(s)=\int_o^\ell K(s,t)\psi(t)dt\ , \tag{8.6.7}$$

where $\psi(t)$ is piecewise continuous, may be expanded in the form

$$\psi(s)=\sum_{i=o}^{\infty}c_i\phi_i(s)\ , \tag{8.6.8}$$

where the series converges uniformly and absolutely, and therefore also in the mean, to $\phi(s)$. Suppose that $K(s,t)$ is the Green's function of a Sturm-Liouville equation; Theorem 8.3.1 shows that, to be so expressed, $\phi(s)$ must satisfy the end conditions, have a continuous first derivative and a piecewise continuous second derivative. But since any piecewise continuous function $\phi(s)$ may be approximated in the mean to any desired accuracy by a sequence of such functions, it follows that the sequence of eigenfunctions $\{\phi_i(s)\}_o^\infty$ is complete. We now prove

Theorem 8.6.1. Let $K(s,t)$ be a continuous symmetric kernel. Then the maximum value of the integral form

$$J(\phi,\phi)\equiv\int_o^\ell\int_o^\ell K(s,t)\phi(s)\phi(t)dsdt$$

for all piecewise continuous $\phi(s)$ satisfying

$$(\phi,\phi)=1$$

is $1/\lambda_o$, where λ_o is the smallest positive eigenvalue; this maximum is attained when $\phi(s)=\phi_o(s)$.

Proof:

We shall prove this theorem for two classes of kernels only.

(i) $K(s,t)$ is degenerate. Theorem 8.4.3 yields

$$K(s,t)=\sum_{i=o}^{n}\frac{\phi_i(s)\phi_i(t)}{\lambda_i}\;, \tag{8.6.9}$$

so that

$$J(\phi,\phi)=\sum_{i=o}^{n}\frac{c_i^2}{\lambda_i}\;,\quad c_i=(\phi,\phi_i)\;. \tag{8.6.10}$$

Now let

$$\psi(s)=\sum_{i=o}^{n}c_i\phi_i(s)\;, \tag{8.6.11}$$

then

$$J(\psi,\psi)=J(\phi,\phi)\;, \tag{8.6.12}$$

while

$$(\phi-\psi,\phi-\psi)\geq 0\;, \tag{8.6.13}$$

yields

$$(\phi,\phi)-\sum_{i=o}^{n}c_i^2\equiv(\phi,\phi)-(\psi,\psi)\geq 0\;, \tag{8.6.14}$$

so that

$$\frac{J(\phi,\phi)}{(\phi,\phi)}\leq\frac{J(\psi,\psi)}{(\psi,\psi)}=\frac{\sum_{i=o}^{n}c_i^2/\lambda_i}{\sum_{i=o}^{n}c_i^2}\leq\frac{1}{\lambda_o}\;. \tag{8.6.15}$$

This shows that $1/\lambda_o$ is an upper bound for the ratio; it is achieved when $\phi(s)=\phi_o(s)$.

(ii) $K(s,t)$ has a complete sequence of eigenfunctions $\{\phi_i(s)\}_o^\infty$. Theorem 8.4.5 shows that

$$\int_o^\ell K(s,t)\phi(t)dt=\sum_{i=o}^{\infty} c_i\phi_i(s)/\lambda_i\ , \tag{8.6.16}$$

where $c_i=(\phi,\phi_i)$. Then

$$J(\phi,\phi)=\sum_{i=o}^{\infty} c_i^2/\lambda_i\ ,\quad (\phi,\phi)=\sum_{i=o}^{\infty} c_i^2\ , \tag{8.6.17}$$

and the result follows as before.

The analysis used here may clearly be used to give iterative and independent definitions of eigenvalues which are the analogues of those found for discrete systems in Section 2.9. See Courant and Hilbert (1953), p. 132.

8.7 Nodes and zeros

Let $\phi(x)$ be a function defined and continuous on $[a,b]$. For the purposes of our analysis we shall assume that $\phi(x)=0$ at a finite set of discrete points $(\xi_i)_1^n$ in $[a,b]$ and/or a finite set of continuous intervals $([\eta_i,\xi_i])_1^m$ within $[a,b]$. (We deliberately neglect the case where $\phi(x)=0$ at an infinite number of discrete points, etc.). Each point ξ_i is called a *zero*, each interval $[\eta_o,\zeta_i]$ a *zero interval*; either is called a *zero place* of $\phi(x)$.

The separate zeros of $\phi(x)$ in $[a,b]$ may be divided into *end zeros*, at a to t; *nodes*, where $\phi(x)$ crosses the axis, as in Figure 8.7.1(a), and null anti-nodes, where it touches the axis, as in Figure 8.7.1(b). In any two-sided vicinity of a node ξ there are points x_1,x_2 such that $x_1<\xi<x_2$ and

$$\phi(x_1)\phi(x_2)<0\ .$$

If ξ is a null anti-node of $\phi(x)$, then in any sufficiently small two-sided vicinity of $\xi,\phi(x)$ maintains its sign and is not identically zero.

The usual form of Rolle's theorem is

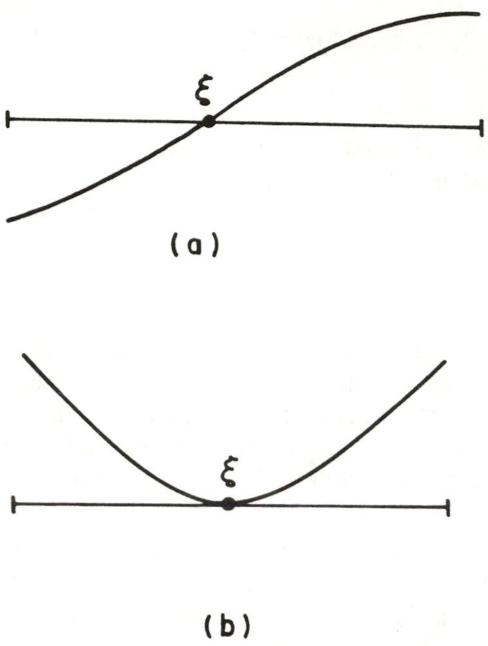

Figure 8.7.1 – *a)* $\phi(x)$ *has a note at* ξ *b)* $\phi(x)$ *has a null anti-node at* ξ

Theorem 8.7.1. (Rolle). Let $\phi(x)$ be continuous in $[a,b]$ and differentiable in (a,b). If $\phi(a)=0=\phi(b)$ then there is at least one point c in (a,b) at which $\phi'(c)=0$, i.e., $\phi'(x)$ has a zero place in (a,b). We need the following requirement.

Theorem 8.7.2. Let $\phi(x)$, $\phi'(x)$ be continuous in $[a,b]$, $\phi(x)$ be not identically zero and $\phi(a)=0=\phi(b)$, then $\phi'(x)$ has a *nodal* place in (a,b).

Theorem 8.7.3. Let $\phi'(x)$ be continuous in $[a,b]$ and have n nodes $(\xi_i)_1^n$ such that $a<\xi_1<\xi_2<...<\xi_n<b$, then the function $\phi(x)$ has at most one zero place in each of the intervals $[a,\xi_1]$, $[\xi_1,\xi_2]$,...,$[\xi_n,b]$, $n+1$ in all. If $\phi(a)\phi'(a)>0$ then $\phi(x)$ has no zero in $[a,\xi_1]$, while if $\phi(b)\phi'(b)<0$ it has no zero in $[\xi_n,b]$. The satisfaction of each of these inequalities thus reduces the number of zero places of $\phi(x)$ by 1.

Proof:

The first part follows directly from Theorem 8.7.2. Suppose $\phi(a)\phi'(a)>0$, then $\phi(a)\phi'(x)\geq 0$ for x in $[a,\xi_1)$ so that, by the mean value theorem, for every x in $[a,\xi_1]$ there is an x_o such that $a<x_o<x$ and

$$\phi(a)\phi(x)=\phi(a)[\phi(a)+(x-a)\phi'(x_o)]>0\,. \tag{8.7.1}$$

Similarly if $\phi(b)\phi'(b)<0$, then $\phi(b)\phi'(x)\leq 0$ for x in $(\xi_n,b]$, and for every x in $[\xi_n,b]$ there is an x_1 such that $x<x_1<b$ and

$$\phi(b)\phi(x)=\phi(b)[\phi(b)+(x-b)\phi'(x_1)]>0 \quad \blacksquare \tag{8.7.2}$$

Corollary. If instead of being continuous and having nodes at $(\xi_i)_1^n$, $\phi'(x)$ is continuous and of one sign in each of the intervals $[a,\xi_1),(\xi_1,\xi_2),\ldots,(\xi_n,b]$, so that it has jumps and may change sign only at $(\xi_i)_1^n$, then the results concerning $\phi(x)$ still hold.

We conclude this section by using the above analysis to provide an alternative proof of Theorem 8.5.7.

Theorem 8.7.4. Under the action of n forces $(F_i)_1^n$ acting at $(s_i)_1^n$, where $0\leq s_1<s_2<\ldots<s_n\leq\ell$, the displacement of a rod can reverse its sign at most $n-1$ times.

Proof:

First assume that h,H are finite and positive, $s_1>0$ and $s_n<\ell$. The displacement of the rod is

$$v(x)=\sum_{i=1}^{n}F_iG(x,s_i) \tag{8.7.3}$$

and because of (8.3.6) it satisfies

$$p(x)v'(x)=\begin{cases} c\ , & 0\leq x<s_1 \\ c_i\ , & s_i<x<s_{i+1}\quad (i=1,2,\ldots,n-1) \\ c_n\ , & s_n<x\leq\ell \end{cases} \tag{8.7.4}$$

where

$$c_i=c-\sum_{j=1}^{i}F_j\ ,\quad (i=1,2,\ldots,n)\,. \tag{8.7.5}$$

Thus $v'(x)$ has the property stated in the Corollary to Theorem 8.7.3, so that $v(x)$ has at most $n+1$ zeros, at most one in each of $[0,s_1]$, $[s_1,s_2],\ldots,[s_n,\ell]$. But

$$v(0)v'(0)=h[v(0)]^2\geq 0 \text{ and } v(\ell)v'(\ell)=-H[v(\ell)]^2\leq 0\,. \tag{8.7.6}$$

If $v(0)>0$ then Theorem 8.7.3 states that $v(x)$ has no zero in $[0,s_1]$; if $v(0)=0$ then $v'(0)=0$ and $v(x)$ is identically zero in $[0,s_1]$. In either case $v(x)$ does not change sign in $[0,s_1]$. Similarly $v(x)$ does not change sign in $[s_n,\ell]$. Therefore $v(x)$ has at most $n-1$ nodal places, one in each of $(s_1,s_2]$, $[s_2,s_3],\ldots,[s_{n-1},s_n)$.

Suppose $h=0$, then $H>0$ and equation (8.2.3) shows that $v'(x)\equiv 0$, so that $v(x)$ is constant, and therefore cannot change sign, in $[0,s_1]$. If $h=\infty$ then $v(0)=0$, and since $v'(x)$ has constant sign in $[0,s_1)$, $v(x)$ cannot change sign in $[0,s_1]$. If $s_1=0$ then $v(x)$ cannot change sign in $[s_1,s_2]$ so that it has at most $n-2$ changes of sign. Applying similar arguments to the cases $H=0$, ∞ and $s_n=\ell$ we see that the stated result holds in all cases ■

8.8 Oscillatory systems of functions

In this section we shall derive some basic results which are needed to establish further properties of the eigensolutions.

Let $\{\phi_{(x)}\}_o^n$ be a sequence of functions defined on an interval I, (closed, open or semi-closed) with ends $0,\ell$.

Theorem 8.8.1. The necessary and sufficient condition for the functions $\{\phi_i(x)\}_o^n$ to be linearly dependent is that

$$\Phi\begin{pmatrix} x_o,x_1,\ldots,x_n \\ 0,1,\ldots,n \end{pmatrix}\equiv\begin{vmatrix} \phi_o(x_o) & \phi_o(x_1) & \cdots & \phi_o(x_n) \\ \phi_1(x_o) & \phi_1(x_1) & \cdots & \phi_1(x_n) \\ \cdots & \cdots & \cdots & \cdots \\ \phi_n(x_o) & \phi_n(x_1) & \cdots & \phi_n(x_n) \end{vmatrix}=0$$

for any $(x_r)_o^n$ in the interval.

Proof:

The condition is necessary. For if the $\{\phi_i(x)\}_o^n$ are linearly dependent then there are constants $(c_i)_o^n$ not all zero such that

$$\sum_{i=o}^{n} c_i\phi_i(x)\equiv 0\,. \tag{8.8.1}$$

In this case there is a linear combination of rows of Φ which is zero so that $\Phi=0$.

We prove sufficiency by induction. If $n=0$, then $\Phi=0$ states that $\phi_o(x_o)=0$ so that $\phi_o(x)\equiv 0$.

Suppose therefore that

$$\Phi\begin{pmatrix} x_0,x_1,\ldots,x_n \\ 0,1,\ldots,n \end{pmatrix}=0\ , \tag{8.8.2}$$

for all $(x_r)_o^n$ in I. We need to prove that the $\{\phi_i(x)\}_o^n$ are linearly dependent. Assume that the $\{\phi_i(x)\}_o^{n-1}$ are linearly independent (for if they were dependent then so would the $\{\phi_i(x)\}_o^n$ be), then there are $\{x_r^o\}_o^{n-1}$ in I such that

$$\Phi\begin{pmatrix} x_0^o,x_1^o,\ldots,x_{n-1}^o \\ 0,1,\ldots,n-1 \end{pmatrix}\neq 0\ . \tag{8.8.3}$$

But then, for all x in I.

$$\Phi\begin{pmatrix} x_0^o,x_1^o,\ldots,x_{n-1}^o x \\ 0,1,\ldots,n-1,n \end{pmatrix}=0\ . \tag{8.8.4}$$

Expand this determinant along its last column: the result is (8.8.1) where c_n, being the determinant (8.8.3), is not zero ■

Definition 8.8.1. A sequence of continuous functions $\{\phi_i(x)\}_o^n$ is said to constitute a *Chebyshev* system on I if, for any set of real constants $(c_i)_o^n$ not all zero, the function

$$\phi(x)=\sum_{i=o}^{n} c_i\phi_i(x) \tag{8.8.5}$$

does not vanish more than n times on I.

Theorem 8.8.2. The necessary and sufficient conditions for $\{\phi_i(x)\}_o^n$ to form a Chebyshev system on I is that

$$\Phi\equiv\Phi\begin{pmatrix} x_o,x_1,\ldots x_n \\ 0,1,\ldots,n \end{pmatrix}$$

maintain strictly fixed sign for all $(x_r)_o^n$ in I such that $x_0<x_1<\cdots<x_n$.

Proof:

If $\Phi=0$ for certain $(x_r)_o^n$ such that $x_0<x_1<\cdots<x_n$, then, and only then, will the equations

$$\sum_{i=o}^{n} c_i\phi_i(x_r)=0\ ,\quad (r=0,1,\ldots,n) \tag{8.8.6}$$

have a non-zero solution $(c_i)_o^n$, i.e., the function $\phi(x)$ will have n zeros. Since the set of n-dimensional points $(x_0,x_1,\ldots,x_n)$ such that $x_0<x_1<\cdots<x_n$ and x_r in I is a connected set, and Φ is a continuous function, the fact that $\Phi\neq 0$ implies that Φ has strictly fixed sign. Note that without loss of generality we may take $\Phi>0$ ■

Gantmakher and Krein (1950) prove the following

Corollary: If the functions $\{\phi_i(x)\}_o^n$ form a Chebyshev system on I and the function

$$\phi(x)=\sum_{i=o}^{n} c_i\phi_i(x)\ , \quad \sum_{i=o}^{n} c_i^2>0\ , \tag{8.8.7}$$

has r zeros in I, including q null-antinodes, then

$$r+q\leq n\ ,$$

i.e., in the estimate of the number of zeros, each null anti-node may be counted twice.

Definition 8.8.2. A sequence of functions $\{\phi_i(x)\}_o^\infty$ will be called a Markov sequence in I, if for each $n=0,1,2,...$ the sequence $\{\phi_i(x)\}_o^n$ is a Chebyshev sequence.

Theorem 8.8.2 shows that the necessary and sufficient condition for $\{\phi_i(x)\}_o^\infty$ to be a Markov sequence is that the determinants

$$\Phi\begin{pmatrix} x_0,x_1,...,x_n \\ 0,1,...,n \end{pmatrix} \quad 0<x_0<x_1<\cdots<x_n<\ell\ ,$$

should have strictly the same sign ε_n. (Note that the sign need not be the same for all n).

We now prove a fundamental theorem.

Theorem 8.8.3. If the orthonormal sequence of functions $\{\phi_i(x)\}_o^\infty$ is a Markov sequence in $(0,\ell)$, i.e., for some $\rho(x)>0$

$$(\rho\phi_i,\phi_j)\equiv\int_o^\ell \rho(x)\phi_i(x)\phi_j(x)dx=\delta_{ij}\ , \tag{8.8.8}$$

then

(1) The function $\phi_o(x)$ has no zeros in $(0,\ell)$.

(2) The function $\phi_i(x)$ has i nodes and no other zeros in $(0,\ell)$.

(3) The function

$$\phi(x)=\sum_{i=k}^{m} c_i\phi_i(x)\ , \quad 0\leq k\leq m\ , \quad \sum_{i=k}^{m} c_i^2>0\ , \tag{8.8.9}$$

has not less than k nodes and not more then m zeros in $(0,\ell)$. (And in this case, according to the corollary to Theorem 8.8.2, each null anti-node counts as two zeros). In particular, if $\phi(x)$ has m different zeros in $(0,\ell)$ then all these zeros are nodes.

(4) The nodes of $\phi_i(x)$ and $\phi_{i+1}(x)$ interlace.

Proof:

Note that (1) and (2) are particular cases of (3), and that all that is left to be proved in (3) is that $\phi(x)$ has not less than k nodes.

Let $(\xi_i)_1^p$, where $\xi_1<\xi_2<\cdots<\xi_p$ be the nodes of $\phi(x)$. Define

$$\psi(x)=\Phi\begin{pmatrix}\xi_1, & \xi_2 & ,\ldots, & \xi_{p-1}, & x\\ 0, & 1 & ,\ldots, & p-1, & p\end{pmatrix}. \tag{8.8.10}$$

Definition 8.8.2 and Theorem 8.8.2 show that if $x\neq\xi_i$ for all $i=1,\ldots,p$, then $\psi(x)\neq 0$. Since the determinants are of strictly fixed sign, $\psi(x)$ changes sign at each ξ_i. Therefore $\psi(x)$ has just p zeros, the nodes $(\xi_i)_1^p$. These are the same nodes as those of $\phi(x)$, therefore

$$(\rho\phi,\psi)\neq 0\,. \tag{8.8.11}$$

But $\phi(x)$ is a combination of $(\phi_i(x))_k^m$ while $\psi(x)$ is a combination of $(\phi_i(x))_0^p$. The inequality (8.8.11) means that these two combinations must overlap, i.e.,

$$p\geq k\,. \tag{8.8.12}$$

In order to prove (4) we note that (3) shows that any combination

$$\phi(x)=c_i\phi_i(x)+c_{i+1}\psi_{i+l}(x)\,,\quad c_i^2+c_{i+1}^2>0\,, \tag{8.8.13}$$

has either i or $i+1$ zeros in $(0,\ell)$, and all these zeros are nodes. Suppose the zeros of $\phi_{i+1}(x)$ are $(\alpha_j)_1^{i+1}$, where $0=\alpha_0<\alpha_1<\cdots<\alpha_{i+1}<\alpha_{i+2}=\ell$. Consider the function

$$\psi(x)=\phi_i(x)/\phi_{i+1}(x)\,. \tag{8.8.14}$$

In each of the $i+2$ intervals (α_j,α_{j+1}), $j=0,1,\ldots,i+1$ the function $\psi(x)$ is continuous, since the denominator $\phi_{i+1}(x)$ has no zero there. We now show that $\psi(x)$ is *monotonic* in each of these intervals.

Suppose, if possible, that $\psi(x)$ were not monotonic in an interval (α_j,α_{j+1}). Then there would exist points x_1,x_2,x_3 such that $\alpha_j\leq x_1<x_2<x_3\leq\alpha_{j+1}$ and

$$(\psi(x_2)-\psi(x_1))(\psi(x_3)-\psi(x_2))<0\,. \tag{8.8.15}$$

Without loss of generality we may assume that $\psi(x_2)>\psi(x_1)$, then $\psi(x_2)>\psi(x_3)$. The function $\psi(x)$ reaches it maximum on the closed interval $[x_1,x_3]$ at a point x_0. This point will be an interior point since $\psi(x_1),\psi(x_3)$ are both less than $\psi(x_2)$. Therefore

$$\psi(x)-\psi(x_0)\leq 0 \text{ for all } x \text{ in } [x_1,x_3]\,, \tag{8.8.16}$$

and thus

$$\phi(x) \equiv \phi_{i+1}(x)[\psi(x) - \psi(x_0)] \equiv \phi_i(x) - \psi(x_0)\phi_{i+1}(x) \tag{8.8.17}$$

retains its sign in the neighbourhood of its zero x_0. This contradicts that statement that $\phi(x)$ can have nodes only. Hence $\psi(x)$ is monotonic in each interval (α_j, α_{j+1}), $j=0,1,...,i$.

We now show that in each of the i interior intervals (α_j, α_{j+1}) $j=1,2,...,i$, $\psi(x)$ takes all values from $-\infty$ to ∞, and so must have a node. This means that $\phi_i(x)$ will have just one node in each of these interior intervals.

We consider the behaviour of $\psi(x)$ near one of the nodes α_k $(k=1,2,...,i)$. Since $\psi(x)$ is monotonic in each of the $i+2$ intervals (α_j, α_{j+1}), the limits

$$\lim_{x \to \alpha_k -} \psi(x) = L_1 , \quad \lim_{x \to \alpha_k +} \psi(x) = L_2 \tag{8.8.18}$$

will exist; they may be finite or infinite. A limit can be finite only if $\phi_i(\alpha_k)=0$, and in this case, since α_k will be a node of $\phi_i(x)$, L_1 and L_2 will at least have the same sign; they need not be equal, and the other limit may be infinite.

Suppose, without loss of generality, that $\psi(x)$ is monotonic increasing in (α_{k-1}, α_k), and if possible, that L_1 is finite. In (α_k, α_{k+1}), $\psi(x)$ may be monotonic decreasing with (a) $L_2=\infty$, (b) $L_2=L_1$, (c) $L_2 \neq L_1$, as shown in Figure 8.8.1, or monotonic increasing with (d) $L_2=L_1$, (e) $L_2 \neq L_1$, as shown in Figure 8.8.2. In all but case (b) there is a line $y=h$, shown, such that $\psi(x)$ crosses this line as x passes through α_k. Thus $\psi(x)-h$ changes sign and

$$\phi(x) = \phi_{i+1}(x)(\psi(x)-h) \equiv \phi_i(x) - h\phi_{i+1}(x) \tag{8.8.19}$$

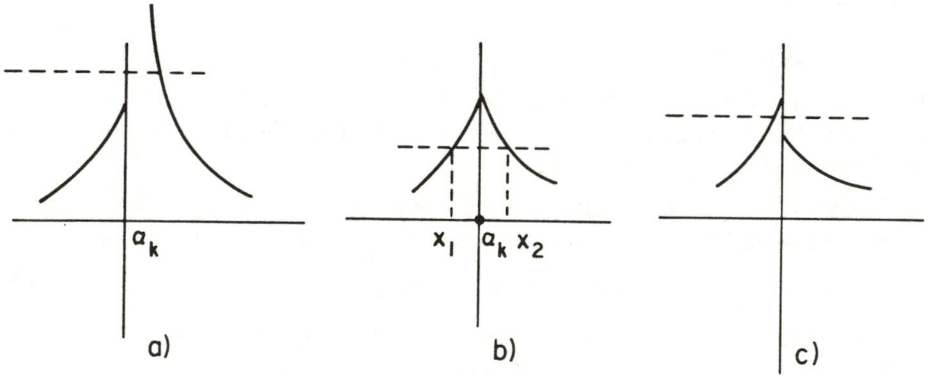

Figure 8.8.1 – Conceivable forms of $\psi(x)$

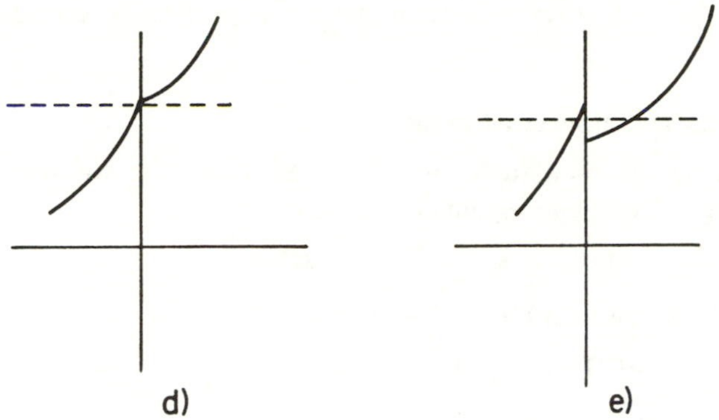

Figure 8.8.2 – *Conceivable forms of* $\psi(x)$

retains its sign as x passes through the zero α_k; this contradicts (3). In case (b), take a line $y=h$ below the common limit, such that the points x_1, x_2 satisfy $\alpha_{k-1}<x_1<\alpha_k<k_2<\alpha_{k+1}$. The function

$$\phi(x)=\phi_{i+1}(x)(\psi(x)-h)\equiv\phi_i(x)-h\phi_{i+1}(x) \qquad (8.8.20)$$

has nodes at x_1, x_2 because $\psi(x)-h$ changes sign while $\phi_{i+1}(x)$ retains its sign, and a node at α_k because $\phi_{i+1}(x)$ changes sign while $\phi(x)-h$ retains its sign. Thus $\phi(x)$ has two more nodes than $\phi_{i+1}(x)$, which again contradicts (3). We conclude that all the limits L_1, L_2 for $k=1,2,...,i$ are infinite. Thus $\psi(x)$ has the form shown in Figure 8.8.3 and has a node, where $\phi_i(x)=0$, at just one point in each of the interior intervals: $\psi(x)$ cannot have a zero in $(0,\alpha_1)$ or (α_{i+1},ℓ), since $\phi_i(x)$ has only i zeros ■

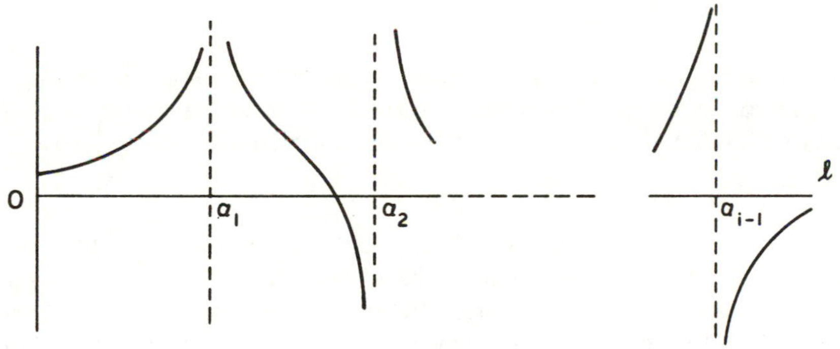

Figure 8.8.3 – *The form of* $\psi(x)$

In Section 8.9, we shall prove that the eigenfunctions of an oscillatory kernel form a Markov sequence; they will then have the properties stated in Theorem 8.8.3.

8.9 Perron's theorem and associated kernels

As with the analysis of oscillatory *matrices*, (Section 5.4) we first establish a theorem concerning a particular eigenvalue. The result is

Theorem 8.9.1. If the continuous kernel $K(x,s)$ satisfies

$$K(x,s)\geq 0\ ,\ \ K(x,x)>0\ ,\ \ 0<x,s<\ell\ ,$$

then the eigenvalues λ_1 of the integral equation

$$\phi(x)=\lambda\int_o^\ell K(x,s)\phi(s)ds \tag{8.9.1}$$

which has smallest absolute value, is positive and simple; the corresponding eigenfunction $\phi_0(x)$ has no zero in $(0,\ell)$.

Proof:

In Section 5.4 Perron's theorem was proved for an *arbitrary*, i.e., not necessarily symmetric, positive matrix. For *symmetric* positive matrices, for which the eigenvalues are known to be real, the argument given below may be adapted to yield a simple proof of the matrix theorem. On the other hand, the argument given in Section 5.4 may be adapted to prove the present theorem for arbitrary continuous kernels, but only by using mathematics that is beyond the scope of this book.

Theorem 8.7.1 shows that the maximum value of $J(\phi,\phi)$ on the 'sphere' $(\phi,\phi)=1$ is $1/\lambda_0$, where λ is the smallest positive eigenvalue. The maximum is achieved for $\phi(x)\ =\phi_0(x)$, where

$$\phi_o(x)=\lambda_o\int_o^\ell K(x,s)\phi_o(s)ds\ . \tag{8.9.2}$$

Now consider $\psi_o(x)=|\phi_o(x)|$. Clearly $(\psi_o,\psi_o)=1$ and $J(\psi_o,\psi_o)\geq J(\phi_o,\phi_o)$, but since $J(\phi_o,\phi_o)$ is the maximum value of $J(\phi,\phi)$, we must have $J(\psi_o,\psi_o)=J(\phi_o,\phi_o)$, which implies that $\psi_o(x)$ also is an eigenfunction corresponding to λ_o, i.e.,

$$|\phi_o(x)|=\lambda_o\int_o^\ell K(x,s)|\phi_o(s)|\,ds\ . \tag{8.9.3}$$

Suppose that $\phi_o(x)$ had an isolated zero, ξ, satisfying $0<\xi<\ell$. On the basis of $K(\xi,\xi)>0$ and the continuity of K we have $K(\xi,s)>0$, $|\phi_o(s)|>0$ in some interval $(\xi,\xi+\varepsilon)$. But this leads to a contradiction for, when $x=\xi$, the left hand side of equation (8.9.3) is zero, while the right is positive. A zero interval in $\phi_o(x)$ may be ruled out similarly. This shows that any eigenfunction corresponding to λ_o retains the same sign in $(0,\ell)$; there can therefore not be two orthogonal eigenfunctions corresponding to λ_o, so that i must be simple.

The maximal principal (Theorem 8.6.1) shows that λ_o is less than all the other positive eigenvalues of (8.9.1). We need therefore only prove that if λ is a negative eigenvalue of (8.9.1), then

$$\lambda_o < |\lambda| \ . \tag{8.9.4}$$

Let $\psi(x)$ be a normalized eigenfunction corresponding to λ, so that

$$\psi(x) = \lambda \int_o^\ell K(x,s)\psi(s)ds \ , \tag{8.9.5}$$

and therefore

$$|\psi(x)| \leq |\lambda| \int_o^\ell K(x,s)\,|\psi(s)|\,ds \ . \tag{8.9.6}$$

The function $\psi(x)$, being orthogonal to $\phi_0(x)$, cannot retain one sign in $(0,\ell)$ so that there must always be strict inequality in equation (8.9.6). Therefore

$$|\psi(x)| < |\lambda| \int_o^\ell K(x,s)\,|\psi(s)|\,ds \ , \tag{8.9.7}$$

and

$$(\psi,\psi) < |\lambda|\, J(|\psi|,|\psi|) \ , \tag{8.9.8}$$

i.e.,

$$|\lambda| > (\psi,\psi)/J(|\psi|,|\psi|) > \lambda_o \ . \ \blacksquare \tag{8.9.9}$$

We now define the associated kernels for the kernel $K(x,s)$.

Definition 8.9.1. The associated kernel $\underline{K}_p$ (X,S) is defined by the equation

$$\underline{K}_p(X,S) = K\begin{pmatrix} x_1,x_2,\ldots,x_p \\ s_1,s_2,\ldots,s_p \end{pmatrix} . \tag{8.9.10}$$

Each of the points $X=(x_1,x_2,\ldots,x_p)$, $S=(s_1,s_2,\ldots,s_p)$ runs through the p-dimensional simplex M^p defined by the inequalities

$$0 \leq x_1 \leq x_2 \leq \cdots \leq x_p \leq \ell \ . \tag{8.9.11}$$

If X is an internal point of M^p, then

$$0 < x_1 < x_2 < \cdots < x_p < \ell \ . \tag{8.9.12}$$

The place of the Cauchy-Binet theorem is taken by

Theorem 8.9.2. If three kernels $K(x,s)$, $L(x,s)$, $N(x,s)$ $(0 \leq x,s \leq \ell)$ are related by

$$N(x,s) = \int_o^\ell K(x,t)L(t,s)dt \ , \tag{8.9.13}$$

then

$$N\begin{pmatrix} x_1,x_2,\ldots,x_p \\ s_1,s_2,\ldots,s_p \end{pmatrix} = \int_o^{\ell}\int_o^{t_{p-1}} \cdots \int_o^{t_2} K\begin{pmatrix} x_1,x_2,\ldots,x_p \\ t_1,t_2,\ldots,t_p \end{pmatrix} L\begin{pmatrix} t_1,t_2,\ldots,t_p \\ s_1,s_2,\ldots,s_p \end{pmatrix} dt_1 dt_2 \ldots dt_p, \quad (8.9.14)$$

i.e.,

$$\underline{N}_p(X,S) = \int_{M^p} \underline{K}_p(X,T)\underline{L}_p(T,S)dT \ . \quad (8.9.15)$$

Proof:

The proof follows immediately from substituting

$$\psi_i(t) = K(x_i,t) \ , \ \chi_i(t) = L(t,s_i)$$

into the integral identity

$$\int_{M^p} \Delta\begin{pmatrix} \psi_1,\psi_2,\ldots,\psi_p \\ t_1,t_2,\ldots,t_p \end{pmatrix} \Delta\begin{pmatrix} \chi_1,\chi_2,\ldots,\chi_p \\ t_1,t_2,\ldots,t_p \end{pmatrix} dt_1 dt_2 \ldots dt_p =$$

$$\frac{1}{p!}\int_o^{\ell}\int_o^{\ell} \cdots \int_o^{\ell} \Delta\begin{pmatrix} \psi_1,\psi_2,\ldots,\psi_p \\ t_1,t_2,\ldots,t_p \end{pmatrix} \Delta\begin{pmatrix} \chi_1,\chi_2,\ldots,\chi_p \\ t_1,t_2,\ldots,t_p \end{pmatrix} dt_1 dt_2 \ldots dt_p \ . \ \blacksquare \quad (8.9.16)$$

Corollary. If $L=K$, then N is called the 2-iterated kernel of K, thus

$$K^{(2)}(x,s) = \int_o^{\ell} K(x,t)K(t,s)dt \ . \quad (8.9.17)$$

The kernel on the left of equation (8.9.15) is then the p-associated kernel of $K^{(2)}(x,s)$, namely $\{\underline{K}^{(2)}(X,T)\}_p$, while the integral on the right of (8.9.15) is the 2-iterate of $\underline{K}_p(X,T)$, i.e., $\{\underline{K}_p(X,T)\}^{(2)}$. Thus

$$\{\underline{K}^{(2)}(X,S)\}_p = \{\underline{K}_p(X,S)\}^{(2)} \ , \quad (8.9.18)$$

so that both may be written $\underline{K}_p^{(2)}(X,T)$. It is shown in Ex. 8.9.1 that

$$\{\underline{K}^{(q)}(X,S)\}_p = \{\underline{K}_p(X,S)\}^{(q)} = \underline{K}_p^{(q)}(X,S) \ . \quad (8.9.19)$$

Perron's Theorem 8.9.1 was stated for the integral equation (8.9.1), on the line $0 \le x \le \ell$, but it also holds on the simplex M^p for the equation

$$\Phi(X) = \Lambda\int_{M^p} \underline{K}_p(X,S)\Phi(S)dS \ . \quad (8.9.20)$$

The eigenvalues and eigenfunctions of this equation are related to those of equation (8.9.1) by

Theorem 8.9.3 If $(\lambda_i)_o^{\infty}$ and $\{\phi_i(x)\}_o^{\infty}$ are eigenvalues and corresponding eigenfunctions of equation (8.9.1), then the eigenvalues and eigenfunctions of equation (8.9.20) are

$$\Lambda=\lambda_{i_1}\lambda_{i_2}\cdots\lambda_{i_p}\ ,\quad \Phi=\Phi\begin{pmatrix}x_1,x_2,\ldots,x_p\\ i_1,i_2,\ldots,i_p\end{pmatrix}$$

where $0\leq i_1<i_2<\cdots<i_p$.

Proof:

Equation (8.9.1) shows that

$$\Phi\equiv\Phi\begin{pmatrix}x_1,x_2,\ldots,x_p\\ i_1,i_2,\ldots,i_p\end{pmatrix}=\lambda_{i_1}\lambda_{i_2}\cdots\lambda_{i_p}\int_{M^p}K\begin{pmatrix}x_1,x_2,\ldots,x_p\\ s_1,s_2,\ldots,s_p\end{pmatrix}\Delta\begin{pmatrix}\phi_{i_1},\phi_{i_2},\ldots,\phi_{i_p}\\ s_1,s_2,\ldots,s_p\end{pmatrix}ds_1ds_2\ldots ds_p\ ,$$

i.e.,

$$\Phi(X)=\lambda_{i_1}\lambda_{i_2}\cdots\lambda_{i_p}\int_{M^p}K(X,S)\Phi(S)dS\ .\ \blacksquare \tag{8.9.21}$$

We may now prove

Theorem 8.9.4 If the continuous kernel $K(x,s)$ satisfies

$$K\begin{pmatrix}x_1,x_2,\ldots,x_p\\ s_1,s_2,\ldots,s_p\end{pmatrix}\geq 0\ ,\quad K\begin{pmatrix}x_1,x_2,\ldots,x_p\\ x_1,x_2,\ldots,x_p\end{pmatrix}>0\ .$$

for

$$0<\begin{matrix}x_1<x_2<\ldots<x_p\\ s_1<s_2<\ldots<s_p\end{matrix}<\ell\ ,$$

then all the eigenvalues of the equation (8.9.1) are positive and simple, i.e., $0<\lambda_o<\lambda_1<\cdots$, and the corresponding eigenfunctions form a Markov sequence within $(0,\ell)$.

Proof:

Order the eigenvaues of equation (8.9.1) so that $|\lambda_o|\leq|\lambda_1|\leq\cdots$. Then the eigenvalue of equation (8.9.20) having smallest absolute values is $\lambda_0\lambda_1\cdots\lambda_{p-1}$. Thus Perron's theorem applied to equation (8.9.20) states that (a) $\lambda_0\lambda_1\cdots\lambda_{p-1}>0$, (b) $\lambda_0\lambda_1\cdots\lambda_{p-1}<|\lambda_0\lambda_1\cdots\lambda_{p-2}\lambda_p|$ for all $p=1,2,\ldots$. Thus $0<\lambda_0<\lambda_1<\cdots$. Theorem 8.9.3 shows that the eigenfunction

$$\Phi\begin{pmatrix}x_1,x_2,\ldots,x_p\\ 0,1,\ldots,p-1\end{pmatrix}$$

corresponding the the lowest eigenvalue $\lambda_0\lambda_1\cdots\lambda_{p-1}$ has no zeros in $(0,\ell)$. According to Theorem 8.8.2, this is the necessary and sufficient condition for $\{\phi_i(x)\}_o^\infty$ to form a Markov sequence in $(0,\ell)$. ■

At this point we may summarize some of the important results that have been found. Theorem 8.5.6 shows that the Green's function for the rod or string under the end conditions (8.1.2) is oscillatory. Theorem 8.9.4 shows that the eigenvalues are simple and that the eigenfunctions form a Markov sequence, with the properties stated in Theorem 8.8.3. The typical form of the mode shapes, for a cantilever rod with end conditions

$$u(0)=0=u'(\ell)\ , \qquad (8.9.22)$$

is shown in Figure 8.9.1.

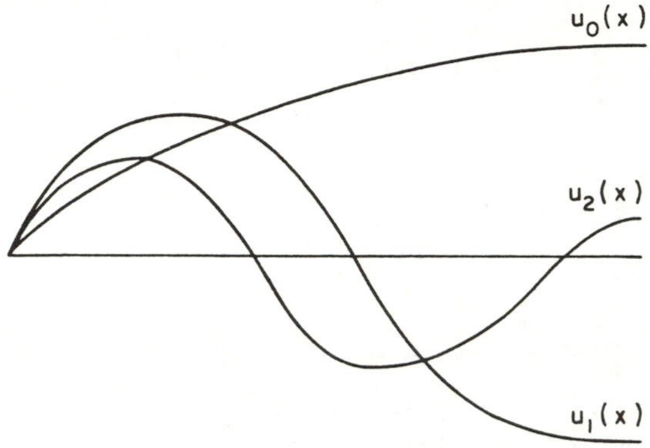

Figure 8.9.1. – Modes of a cantilever rod

It was shown in Ex. 8.3.4 that for the end conditions (8.9.22) the slope $v(x)=u'(x)$ satisfies the integral equation

$$v(x)=\lambda\int_o^\ell L(x,s)B(s)v(s)ds\ , \qquad (8.9.23)$$

where

$$L(x,s)=\int_{x^+}^\ell A(t)dt\ .$$

This integral equation may be written in the symmetric form

$$\phi(x)=\lambda\int_o^\ell K(x,s)\phi(s)ds\ , \qquad (8.9.24)$$

with

$$\phi(x)=[B(x)]^{1/2}v(x)\ ,\ \ K(x,s)=[B(x)B(s)]^{1/2}L(x,s)\ , \qquad (8.9.25)$$

and $K(x,s)$ is an oscillatory kernel. Thus the eigenfunctions $\phi_i(x)=[B(x)]^{1/2}u'_n(x)$ form a Markov sequence, and will have the form shown in Figure 8.9.2.

8.10 The interlacing of eigenvalues

Consider equation (8.1.1), namely

$$u''(x)+\lambda\rho(x)u(x)=0\ , \qquad (8.10.1)$$

subject to the end conditions

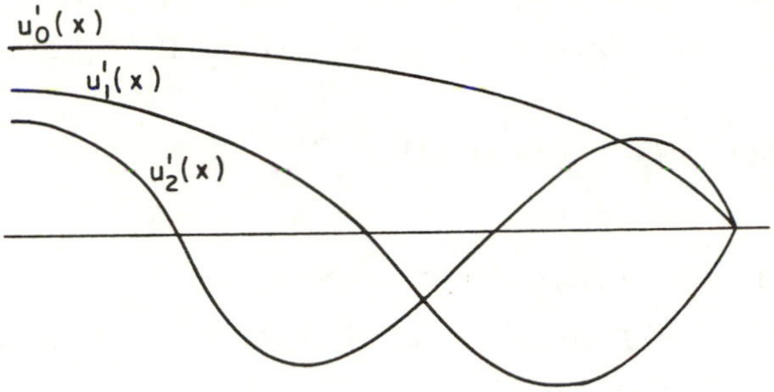

Figure 8.9.2 – The derivatives of cantilever rod modes

$$u'(0)-hu(0)=0=u'(\ell)+Hu(\ell)\ . \tag{8.10.2}$$

Using the analysis of Section 8.6, we may phrase the eigenvalue problem (8.10.1), (8.10.2) as a variational problem for the functional $J(u,u)$ subject to the condition $(u,u)=1$; the kernel $K(s,t)$ will be the Green's function $G(s,t)$. In this section, following Courant and Hilbert (1953, p. 398) we shall give an alternative variational formulation. This will enable us to find how the eigenvalues of (8.10.1), (8.10.2) vary with h,H.

The eigenvalue problem (8.10.1), (8.10.2) is equivalent to the problem of finding the stationary values of

$$J(u)\equiv\int_o^\ell [u'(x)]^2dx+hu^2(0)+Hu^2(\ell)\ , \tag{8.10.3}$$

subject to

$$(\rho u,u)\equiv\int_o^\ell \rho u^2(x)dx=1\ . \tag{8.10.4}$$

The following argument may be made rigorous.

We introduce a Lagrange parameter λ and consider

$$G(u)\equiv J(u)-\lambda(\rho u,u)\ . \tag{8.10.5}$$

Then

$$\lim_{\varepsilon\to o}\frac{G(u+\varepsilon\eta)-G(u)}{2\varepsilon}=\int_o^\ell u'\eta'dx+hu(0)\eta(0)+ + Hu(\ell)\eta(\ell)-\lambda\int_o^\ell \rho u\eta dx\ . \tag{8.10.6}$$

Integrate the first term by parts, rearrange the terms, and equate the whole to zero; then

$$-\int_o^\ell (u''+\lambda\rho u)\eta dx-[u'(0)-hu(0)]\eta(0)+[u'(\ell)+Hu(\ell)]\eta(\ell)=0\ . \tag{8.10.7}$$

This will be zero for all variations $\eta(x)$ only if $u(x)$ satisfies (8.10.1), (8.10.2).

Now, in order to show the dependence of $J(u)$ and the eigenvalue λ_n on h,H we write

$$J(u)\equiv J(u,h,H)\ ,\ \ \lambda_n\equiv\lambda_n(h,H)\ . \tag{8.10.8}$$

The lowest eigenvalue, $\lambda_o(h,H)$ is

$$\lambda_o(h,H)=\min J(u,h,H) \text{ for } (\rho u,u)=1\ , \tag{8.10.9}$$

so that the explicit form of J, given in (8.10.3), shows that $\lambda_o(h,H)$ is an increasing function of h and of H. In particular,

$$\lambda_0(0,0)<\lambda_0(h,0)<\lambda_0(h,H)<\lambda_0(h,\infty)<\lambda_0(\infty,\infty)\ . \tag{8.10.10}$$

The inequalities are strict. (Ex. 8.10.1).

In order to compare the $\lambda_n(h,H)$ for different pairs of values of h and H, we must use the independent definition of eigenvalues (Courant and Hilbert, 1953, p. 405). Thus

$$\lambda_n=\max_{v_0,v_1,\ldots,v_{n-1}}\ \min_{\substack{(u,v_m)=0\\ m=0,1,\ldots,n-1}} J(u,h,H) \ \text{ for } (\rho u,u)=1 \tag{8.10.11}$$

and since $J(u,h,H)$ increases with h,H, we may conclude that λ_n is an increasing function of h,H. Thus if $h,H>0$ then

$$\lambda_n(0,0)<\lambda_n(h,0)<\lambda_n(h,H)<\lambda_n(h,\infty)<\lambda_n(\infty,\infty)\ . \tag{8.10.12}$$

On the other hand, $\lambda_n(h,H)$ is the maximum value that the minimum of the functional $J(u,h,0)$ can take when subjected to $n+1$ constraints, one of which is $u'(\ell)+Hu(\ell)=0$. This maximum must be less than the maximum of the minimum subjected to *any* $n+1$ constraints, i.e.,

$$\lambda_n(h,H)<\lambda_{n+1}(h,0)\ . \tag{8.10.13}$$

From the inequalities (8.10.11), (8.10.12) we conclude that, if $H<H'$ then

$$\lambda_n(h,H)<\lambda_n(h,H')<\lambda_{n+1}(h,H)\ , \tag{8.10.14}$$

i.e., the spectra $\{\lambda_n(h,H)\}_o^\infty$ and $\{\lambda_n(h,H')\}_o^\infty$ interlace. Clearly, if $h<h'$, then

$$\lambda_n(h,H)<\lambda_n(h',H)<\lambda_{n+1}(h,H)\ , \tag{8.10.15}$$

so that $\{\lambda_n(h,H)\}_o^\infty$, $\{\lambda_n(h',H)\}_o^\infty$ interlace.

There is an alternative way of analysing the interlacing. Suppose $\{\lambda_n,u_n(x)\}_o^\infty$ are the eigenvalues and eigenfunctions of equations (8.10.1), (8.10.2), normalized so that

$$(\rho u_n, u_m) = \delta_{mn} . \tag{8.10.16}$$

Then,

$$\int_o^\infty u_n'(x)u_m'(x)dx = [u_n(x)u_m'(x)]_o^\ell + \lambda_m \delta_{mn}$$
$$= -hu_n(0)u_m(0) - Hu_n(\ell)u_m(\ell) + \lambda_m \delta_{mn} . \tag{8.10.17}$$

Now consider the variational problem for the equation (8.10.1) under the end conditions

$$u'(0) - h^*u(0) = 0 = u'(\ell) + Hu(\ell) . \tag{8.10.18}$$

This is the problem of finding the stationary values of

$$J^*(u) = \int_o^\ell [u'(x)]^2 dx + h^*u^2(0) + Hu^2(\ell) , \tag{8.10.19}$$

subject to (8.10.4). Since the $\{u_n(x)\}_o^\infty$ form a complete set on $(0,\ell)$ we may write

$$u(x) = \sum_{m=o}^\infty c_m u_m(x) , \tag{8.10.20}$$

and use equation (8.10.17) to find

$$J^*(u) = \sum_{m=o}^\infty c_m^2 \lambda_m + (h^* - h)u^2(0) , \tag{8.10.21}$$

so that

$$\lambda_m c_m + (h^* - h)u_m(0)u(0) - \lambda c_m = 0 , \quad m = 0,1,... , \tag{8.10.22}$$

i.e.,

$$c_m = (h^* - h)\frac{u_m(0)u(0)}{\lambda - \lambda_m} , \tag{8.10.23}$$

so that the condition $u(0) = \sum_{m=o}^\infty c_m u_m(0)$ gives

$$1 = (h^* - h) \sum_{m=o}^\infty \frac{[u_m(0)]^2}{\lambda - \lambda_m} . \tag{8.10.24}$$

This is the analogue of equation (4.3.39), and it immediately yields the inequalities (8.10.15).

We note that if the eigenvalues $(\mu_n)_o^\infty$ of equation (8.10.1) for the end conditions (8.10.18) are known, then $\{[u_m(0)]^2\}_o^\infty$ may be determined apart from a factor. For then

$$1-(h^*-h)\sum_{m=o}^{\infty}\frac{[u_m(0)]^2}{\lambda-\lambda_m}=C\frac{\prod_{m=o}^{\infty}(1-\frac{\lambda}{\mu_m})}{\prod_{m=o}^{\infty}(1-\frac{\lambda}{\lambda_m})}\ , \tag{8.10.25}$$

and since λ_m, $\mu_m=0(m^2)$, both products converge. Thus

$$(h^*-h)[u_m(0)]^2=C\lambda_n\prod_{m=o}^{\infty}(1-\frac{\lambda_n}{\mu_m})/\prod_{m=o}^{\infty}{}'(1-\frac{\lambda_n}{\lambda_m})\ , \tag{8.10.26}$$

where

$$C\prod_{m=o}^{\infty}(\frac{\lambda_m}{\mu_m})=1\ .$$

An alternative derivation of equation (8.10.25) may be found in Levitan (1964a). (Note that his function $\phi(x)$ is normalized so that $\phi(0)=1$, rather than by taking $\int_o^{\pi}\phi^2(x)dx=1$. It is therefore necessary to compare his $\phi^2(\pi,\lambda_n)/\int_o^{\pi}\phi^2(\pi,\lambda_n)$, with our equation (8.10.25).

8.11 Asymptotic behaviour of eigenvalues and eigenfunctions

For the solution of inverse problems in Chapter 9, we shall need to know the behaviour of the eigenvalues λ_n and eigenfunctions $\phi_n(x)$ for large n. To examine this behaviour it is convenient to consider the Sturm-Liouville equation in the normal form on the interval $(0,\pi)$, thus

$$u''(x)+(\lambda-q(x))u(x)=0\ , \tag{8.11.1}$$

$$u'(0)-hu(0)=0=u'(\pi)+Hu(\pi)\ . \tag{8.11.2}$$

Write $\lambda=\omega^2$, then equation (8.11.1) may be written

$$u''(x)+\omega^2u(x)=q(x)u(x)\ . \tag{8.11.3}$$

Treating the right hand side as a forcing function, we may write the solution in the form

$$u(x)=A\cos\omega x+B\sin\omega x+\omega^{-1}\int_o^x\sin\omega(x-t)q(t)u(t)dt\ . \tag{8.11.4}$$

If h is finite, we may chooose $u(x)$ so that

$$u(0)=1\ ,\ u'(0)=h\ , \tag{8.11.5}$$

then

$$u(x)=\cos\omega x+h\omega^{-1}\sin\omega x+\omega^{-1}\int_o^x\sin\omega(x-t)q(t)u(t)dt\ , \tag{8.11.6}$$

so that

$$u(x)=\cos\omega x\{1-\omega^{-1}q_1(x)\}+\omega^{-1}\sin\omega x\{h+q_2(x)\}\ , \tag{8.11.7}$$

$$u'(x)=\cos\omega x\{h+q_2(x)\}-\omega\sin\omega x\{1-\omega^{-1}q_1(x)\}\ , \tag{8.11.8}$$

where

$$q_1(x)=\int_o^x \sin\omega t\ q(t)u(t)dt\ ,\ \ q_2(x)=\int_o^x \cos\omega t\ q(t)u(t)dt\ . \tag{8.11.9}$$

Thus, when H is finite, $u(x)$ will satisfy the second end condition (8.11.2) if

$$\tan\omega\pi=\frac{\dfrac{h+q_2(\pi)}{\omega}+\dfrac{H}{\omega}-\dfrac{Hq_1(\pi)}{\omega^2}}{1-\dfrac{(h+q_2(\pi))H}{\omega^2}-\dfrac{q_1(\pi)}{\omega}}\ . \tag{8.11.10}$$

The case $q_1(\pi)=0=q_2(\pi)$ was encountered in Ex. 8.2.2. We may establish the result (iv) in a more general case.

Equation (8.11.6) shows that

$$u(x)=\cos\omega x+\omega^{-1}\alpha(\omega,x)\ , \tag{8.11.11}$$

where $\alpha(\omega,x)$ is bounded. Therefore

$$q_1(x)=\frac{1}{2}\int_o^x \sin 2\omega t\ q(t)dt+\omega^{-1}\int_o^x \sin\omega t\ \alpha(\omega,t)q(t)dt\ , \tag{8.11.12}$$

$$q_2=\frac{1}{2}\int_o^x q(t)dt+\frac{1}{2}\int_o^x \cos 2\omega t\ q(t)dt + \omega^{-1}\int_o^x \cos\omega t\ \alpha(\omega,t)q(t)dt\ . \tag{8.11.13}$$

If $q(x)$ is continuous, then $\alpha(\omega,t)$ will be continuous and all except the first integral in $q_2(t)$ will be $0(\omega^{-1})$. Thus

$$q_1(x)=0(\omega^{-1})\ ,\ \ q_2(x)=\frac{1}{2}\int_o^x q(t)dt+0(\omega^{-1})\ , \tag{8.11.14}$$

and on substituting these into equation (8.11.7) we find

$$u(x)=\cos\omega x\{1+0(\omega^{-2})\}+\sin\omega x\{\omega^{-1}Q(x)+0(\omega^{-2})\}\ , \tag{8.11.15}$$

where

$$Q(x)=h+\frac{1}{2}\int_o^x q(t)dt\ . \tag{8.11.16}$$

The eigenvalue equation (8.11.10) becomes

$$\tan\omega\pi=\frac{c\pi\omega^{-1}+0(\omega^{-2})}{1+0(\omega^{-2})}\ , \tag{8.11.17}$$

where

$$c=\frac{h+H}{\pi}+\frac{1}{2}m\ ,\quad m=\frac{1}{\pi}\int_o^\pi q(t)dt\ . \tag{8.11.18}$$

Equation (8.11.10) will have just one root in each interval $(n,n+1)$, so that putting

$$\omega\pi=n\pi+\theta\ , \tag{8.11.19}$$

we find

$$\tan\omega\pi=\tan\theta=\frac{\frac{c\pi}{n}+0(n^{-2})}{1+0(n^{-2})}\ , \tag{8.11.20}$$

so that

$$\theta=\frac{c\pi}{n}+0(n^{-2})\ , \tag{8.11.21}$$

and

$$\omega_n=n+\frac{c}{n}+0(n^{-2})\ . \tag{8.11.22}$$

This gives the asymptotic form of the eigenvalues. For the eigenfunctions, we first approximate $\cos\omega_n x$ and $\sin\omega_n x$; we have

$$\cos\omega_n x=\cos nx\{1+0(n^{-2})\}-sinnx\{cn^{-1}x+0(n^{-2})\}\ , \tag{8.11.23}$$

$$\sin\omega_n=\sin nx\{1+0(n^{-2})\}+\cos nx\{cn^{-1}x+0(n^{-2})\}\ , \tag{8.11.24}$$

so that, on substituting into (8.11.15) we find

$$u_n(x)=\cos nx\{1+0(n^{-2})\}+\sin nx\{n^{-1}[Q(x)-cx]+0(n^{-2})\}\ , \tag{8.11.25}$$

and

$$\int_o^\pi u_n^2(x)dx=\frac{\pi}{2}+0(n^{-2})\ . \tag{8.11.26}$$

For the next stage of the asymptotic development we note that if $q(t)$ has a continuous derivative on $[0,\pi]$, then

$$\int_o^\pi \cos 2\omega t\ q(t)dt=\frac{1}{2}\omega^{-1}\sin(2\omega\pi)q(\pi)+0(\omega^{-2})\ , \tag{8.11.27}$$

and the eigenvalue equation (8.11.20) becomes

$$\tan\omega\pi=\frac{\frac{c\pi}{\omega}+\frac{1}{2\omega^2}\sin(2\omega\pi)q(\pi)+0(\omega^{-3})}{1+0(\omega^{-2})}\ . \tag{8.11.28}$$

Again using equations (8.11.19), (8.11.21), we find that

$$\frac{1}{2\omega^2}\sin(2\omega\pi)=\frac{1}{2\omega^2}\sin(2\theta)=0(n^{-2}) , \tag{8.11.29}$$

so that

$$\tan\theta=\frac{\frac{c\pi}{n}+0(n^{-3})}{1+0(n^{-2})} , \tag{8.11.30}$$

and

$$\omega_n=n+\frac{c}{n}+0(n^{-3}) . \tag{8.11.31}$$

The most extensive study of asymptotic estimates of the Sturm-Liouville spectrum has been carried out by Hochstadt (1961). He supposes that the mean value of $q(t)$, given in equation (8.11.18), is zero. Equation (8.11.1) may be reduced to this form by writing it as

$$u''(x)+(\lambda^*-q^*(x))u(x)=0 , \tag{8.11.32}$$

where

$$\lambda^*=\lambda-m , \quad q^*(x)=q(x)-m . \tag{8.11.33}$$

When h,H, are finite and $q(x)$ is continuously differentiable, he shows that

$$(\omega_n^2-m)^{1/2}=n+\frac{h+H}{\pi n}+\frac{d}{n^3}+0(n^{-4}) , \tag{8.11.34}$$

where $d=d_2-d_1$ and

$$d_1=(\frac{h+H}{\pi})^2+\frac{1}{3}(\frac{h^3+H^3}{\pi}) , \tag{8.11.35}$$

$$d_2=\frac{1}{8\pi}\{\int_o^\pi[q^*(t)]^2dt+q'(0)-q'(\pi)-4hq^*(0)-4Hq^*(\pi)\} . \tag{8.11.36}$$

Equation (8.11.34) is consistent with (8.11.31). Note that when $q(x)\equiv 0$, then equation (8.11.34) shows that

$$\omega_n=n+\frac{h+H}{n}-d_1n^{-3}+0(n^{-4}) . \tag{8.11.37}$$

In special case $h=0=H$, $q(x)\equiv 0$, there is a zero eigenvalue. This case, considered in Ex. 8.3.2, is excluded in (8.11.37). Other special cases are the subject of Ex. 8.11.1 and Ex. 8.11.2. See also Fix (1967).

Examples 8.11

1. Show that when $h=\infty$ then

$$(\omega_n^2-m)^{1/2}=n+\frac{1}{2}+\frac{H}{\pi(n+1/2)}+\frac{d}{n^3}+0(n^{-2})$$

where $d=d_2-d_1$ and

$$d_1=\left(\frac{H}{\pi}\right)^2+\frac{1}{3}\frac{H^3}{\pi}, \quad d_2=\frac{1}{8\pi}\{\int_o^\pi[q^*(t)]^2dt-q'(0)-q'(\pi)-4Hq^*(\pi)\}.$$

2. Show that when $h=\infty=H$, then

$$(\omega_n^2-m)^{1/2}=n+d_2n^{-3}+0(n^{-4})$$

where

$$d_2=\frac{1}{8\pi}\{\int_o^\pi[q^*(t)]^2dt+q'(\pi)-q'(0)\}.$$

8.12 Impulse responses

Consider a rod of density ρ, Young's modulus E, cross-section $A(x)$ and length ℓ, free at $x=0$ and fixed at $x=\ell$. Suppose that at time $t'=0$ the rod is at rest, and is then set in motion by a force $G(t')$ applied at the end $x=0$. The governing equations are

$$\rho A(x)\frac{\partial^2 u}{\partial t'^2}=\frac{\partial}{\partial x}(EA(x)\frac{\partial u}{\partial x}), \tag{8.12.1}$$

$$EA\frac{\partial u}{\partial x}\,|_{x=o}=g(t'), \tag{8.12.2}$$

$$u(\ell,t')=0, \quad t'\geq 0, \tag{8.12.3}$$

$$u(x,0)=0=\frac{\partial u}{\partial t}(x,0), \quad 0\leq x\leq \ell. \tag{8.12.4}$$

Instead of real time t' we use the scaled time $t=ct'$, $c=\sqrt{E/\rho}$, and put $g(t)=g(t')/E$. We may replace the end force $g(t)$ by a distributed loading

$$g(x,t)=\lim_{\varepsilon\to o}\begin{cases} g(t)/\varepsilon, & 0<x<\varepsilon \\ 0, & \varepsilon\leq x\leq \ell \end{cases}=g(t)\delta(x) \tag{8.12.5}$$

so that equation (8.12.1) becomes

$$A(x)\frac{\partial^2 u}{\partial t^2}=\frac{\partial}{\partial x}(A(x)\frac{\partial u}{\partial x})+g(t)\delta(x). \tag{8.12.6}$$

Take the Laplace transform of this equation and put

$$U(x,s)=\int_o^\infty e^{-st}u(x,t)dt, \quad G(s)=\int_o^\infty e^{-st}g(t)dt, \tag{8.12.7}$$

to obtain

$$s^2 A(x)U(x,s) = \frac{\partial}{\partial x}\left(A(x)\frac{\partial U}{\partial x}\right) + G(s)\delta(x) . \tag{8.12.8}$$

The solution of this equation which satisfies the end condition

$$U(\ell,s) = 0 \tag{8.12.9}$$

may be written

$$U(x,s) = K(x,s)G(s) , \tag{8.12.10}$$

so that, by the convolution theorem

$$u(x,t) = \int_o^t k(x,t-\tau)g(\tau)d\tau . \tag{8.12.11}$$

Here $k(x,t)$ is the inverse Laplace transform of $K(x,s)$, i.e.,

$$k(x,t) = \frac{1}{2\pi i}\int_\Gamma K(x,s)e^{st}ds , \tag{8.12.12}$$

where Γ is a line $(\gamma - i\infty,\ \gamma + i\infty)$ lying to the right of the singularities of $K(x,s)$. The function $k(x,t)$ is called the (displacement) impulse response function. Clearly, when $g(\tau)$ is a unit impulse, i.e., $g(\tau) = \delta(\tau)$, then equation (8.12.11) shows that $u(x,t) = k(x,t)$.

If $\{\omega_n^2, u_n(x)\}_o^\infty$ are the (scaled) eigenvalues and normalized eigenfunctions of the free-fixed rod, i.e.,

$$A(x)\omega_n^2 u_n(x) = -\frac{d}{dx}\left(A(x)\frac{du_n}{dx}\right) \tag{8.12.13}$$

$$u'_n(0) = 0 = u_n(\ell) , \tag{8.12.14}$$

then we may expand $U(x,s)$ in the form

$$U(x,s) = \sum_{n=o}^{\infty} \alpha_n(s)u_n(x) , \tag{8.12.15}$$

so that equation (8.12.8) becomes

$$\sum_{n=o}^{\infty} (s^2 + \omega_n^2)A(x)\alpha_n(s)u_n(x) = G(s)\delta(x) . \tag{8.12.16}$$

Multiplying through by $u_m(x)$, integrating from 0 to ℓ, using orthogonality and the result

$$\int_o^\ell u_n(x)\delta(x)dx = u_n(o) , \tag{8.12.17}$$

we obtain

$$(s^2 + \omega_n^2)\alpha_n(s) = u_n(o)G(s) , \tag{8.12.18}$$

and

$$K(x,s)=\sum_{n=o}^{\infty}\frac{u_n(o)u_n(x)}{s^2+\omega_n^2}\ , \tag{8.12.19}$$

for which the inverse is

$$k(x,t)=\begin{cases}\displaystyle\sum_{n=o}^{\infty}\frac{u_n(o)u_n(x)}{\omega_n}\ \sin(\omega_n t) & ,\ t>0\\ 0 & ,\ t\le 0\end{cases} \tag{8.12.20}$$

For a *uniform* rod

$$u_n(x)=\left(\frac{2}{\ell}\right)^{1/2}\cos[\frac{(2n+1)\pi x}{2\ell}]\ ,\ \omega_n=\frac{(2n+1)\pi}{2\ell}\ . \tag{8.12.21}$$

so that

$$k(x,t)\ =\ \frac{4}{\pi}\sum_{n=o}^{\infty}\frac{\cos[\frac{(2n+1)\pi x}{2\ell}]\sin[\frac{(2n+1)\pi t}{2\ell}]}{(2n+1)}\ ,$$

i.e.,

$$k(x,t)\ =\ \frac{1}{2}\{S[\frac{\pi(x+t)}{2\ell}]-S[\frac{\pi(x-t)}{2\ell}]\}\ , \tag{8.12.22}$$

where

$$S(x)=\frac{4}{\pi}\sum_{n=o}^{\infty}\frac{\sin(2n+1)x}{2n+1}\ . \tag{8.12.23}$$

Now $S(x)$ is discontinuous at 0, $\pm\pi$, $\pm 2\pi$,... and

$$S(x)=sign(x)\ \ ,\ \ -\pi<x<\pi\ , \tag{8.12.24}$$

as shown in Figure 8.12.1 (Gradshteyn and Ryzhik (1965), p. 38).

From equation (8.12.22) we may deduce the behaviour of the rod subjected to an impulse at $t=0$. Thus if $x>t$, then $x+t<2x<2\ell$, so that

$$S[\frac{\pi(x+t)}{2\ell}]=1\ \ ,\ \ S[\frac{\pi(x-t)}{2\ell}]=1\ , \tag{8.12.25}$$

and $k(x,t)=0$. This may be interpreted as showing that the effect of the impulse moves along the rod with scaled speed 1, i.e., real speed c, and the rod is at rest for $x>t$. Analysis of the partial differential equation (8.12.1) shows that this result is true even when $A(x)$ is not uniform (Courant and Hilbert, Vol. II, 1962). For the uniform rod, behind the initial disturbance, i.e., for $x<t$, $x+t<2\ell$, we have $k(x,t)=1/2$. When the disturbance reaches the end $x=\ell$ and starts to return we have

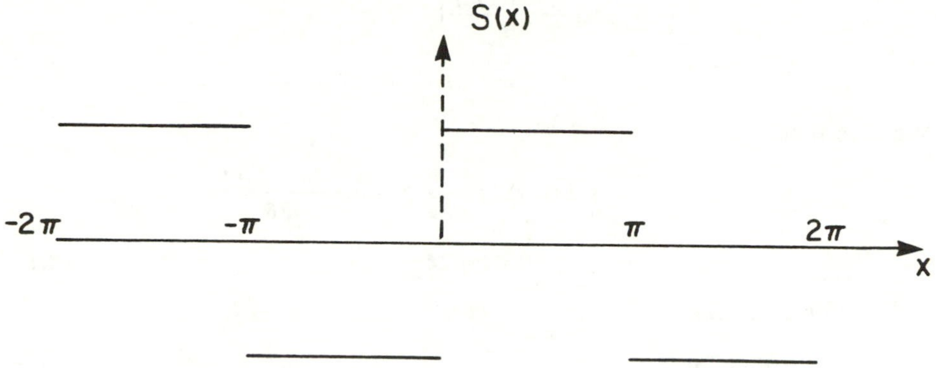

Figure 8.12.1 – The function $S(x)$ has jumps at $\pm n\pi$

$$S[\frac{\pi(x+t)}{2\ell}]=-1 \quad S[\frac{\pi(x-t)}{2\ell}]=-1 \tag{8.12.26}$$

so the step 1/2 which had stretched from $x=o$ to $x=\ell$ is annihilated, starting from $x=\ell$. So the process continues indefinitely.

Sometimes it is convenient to use velocity and (scaled) stress as variables, i.e.,

$$v(x,t)=\frac{\partial u}{\partial t}\ , \ p(x,t)=A(x)\frac{\partial u}{\partial x}\ , \tag{8.12.27}$$

then equation (8.12.1) may be written

$$A(x)\frac{\partial v}{\partial t}=\frac{\partial p}{\partial x}\ , \ A(x)\frac{\partial v}{\partial x}=\frac{\partial p}{\partial t}\ , \tag{8.12.28}$$

and the velocity $v(x,t)$ is given by

$$v(x,t)=\int_o^t \hat{h}(x,t-\tau)g(\tau)d\tau\ , \tag{8.12.29}$$

where

$$\hat{h}(x,t-\tau)=\frac{\partial k}{\partial t}(x,t) \tag{8.12.30}$$

must be interpreted as a generalized function.

Equation (8.12.20) shows that $\hat{h}(x,t)=0$ when $t<o$, and

$$\hat{h}(x,t)=\sum_{n=o}^{\infty} u_n(o)u_n(x)\cos(\omega_n t)\ , \quad t\geq 0\ , \tag{8.12.31}$$

and thus

$$\hat{h}(0,t)=\sum_{n=0}^{\infty}[u_n(0)]^2\cos(\omega_n t)\ , \quad t\geq 0\ . \tag{8.12.32}$$

For the uniform rod, therefore

$$\hat{h}(0,t)=\begin{cases}\frac{2}{\ell}\sum_{n=o}^{\infty}\cos[\frac{(2n+1)\pi t}{2\ell}] & , \ t\geq 0\\ 0 & , \ t<0\ ,\end{cases} \tag{8.12.33}$$

We note that

$$\int_{-\infty}^{t}\hat{h}(0,\tau)d\tau = \int_{o}^{t}\hat{h}(0,\tau)d\tau=\frac{4}{\pi}\sum_{n=o}^{\infty}\sin[\frac{(2n+1)\pi t}{2\ell}]$$

$$= 1 \quad , \quad 0<t<2\ell \tag{8.12.34}$$

so that, for $0<t<2\ell$,

$$\hat{h}(0,t)=\delta(t)\ . \tag{8.12.35}$$

For larger values of t, $\hat{h}(0,t)$ may be evaluated by using its periodicity. For a non-uniform rod it can be shown that (Courant and Hilbert (1962), Vol II, p. 512ff)

$$\hat{h}(0,t)=\delta(t)+h(t)\ , \tag{8.12.36}$$

where $h(t)$ is continuously differentiable (Ex. 8.12.2).

Examples 8.12

1. Show that $S(x+\pi)=-S(x)$, $S(x+2\pi)=S(x)$, and hence show that

$$S(x)=(-)^{|n|} \quad , \quad n\pi<x<(n+1)\pi\ ,$$

2. Show that if the rod is such that its eigenvalues ω_m and eigenfunctions $u_m(x)$ satisfy

$$\omega_m=\frac{(2m+1)\pi}{2\ell} \quad , \quad [u_m(0)]^2=\frac{2}{\ell} \quad , \quad m=n+1,n+2,\ldots\ ,$$

then its impulse response function may be written in the form (8.12.36), where

$$h(t)=\sum_{m=o}^{n}\{[u_m(o)]^2\cos(\omega_m t)-\frac{2}{\ell}\cos[\frac{(2m+1)\pi t}{2\ell}]\}\ .$$

CHAPTER 9

INVERSION OF CONTINUOUS SECOND-ORDER SYSTEMS

If there were no obscurity, man would not be sensible of his corruption; if there were no light, man would not hope for a remedy.

Pascal's Pensées

9.1 Introduction

In Chapter 8 we analysed the Sturm-Liouville equation

$$y''+[\lambda-q(x)]y=0\ , \tag{9.1.1}$$

or the equivalent equations (8.1.1), (8.1.3), subject to the end conditions

$$y'(0)-hy(0)=0=y'(\ell)+Hy(\ell)\ . \tag{9.1.2}$$

In this chapter we will consider whether and how the function $q(x)$, and possibly the constants h, H, may be recovered from information about the solutions of the equation.

In Section 9.2, we shall give a brief historical overview. The researcher coming to the literature will find that it is extensive; it relates to a variety of problems which, though superficially similar, are in fact very different; it employs a variety of mathematical concepts, many of which are deep.

In this section we shall therefore point out a fundamental distinction which, when understood, will allow the reader to select the research material that he needs: the distinction is that between a *continuous* and a *discrete* spectrum. In this book, we shall be concerned exclusively with problems having a discrete spectrum, but it is worthwhile to point out the two concepts. For a discussion of these matters in the context of the vibration nodes of a sphere (the Earth) see Sabatier (1978).

Consider the simple equation

$$y''+\lambda y=0 \tag{9.1.3}$$

on the infinite half-line $(0,\infty)$, and with the end condition

$$y'(0)=0\ . \tag{9.1.4}$$

If $\lambda<0$, we may put $\lambda=-\xi^2$; the solution is then

$$y(x)=A\sinh\xi x+B\cosh\xi x \tag{9.1.5}$$

and $y'(0)=0$ implies $A=0$. Thus

$$y(x)=B\cosh\xi x \tag{9.1.6}$$

and there is *no* solution other than $y\equiv 0$, i.e., $B=0$, which is bounded on $(0,\infty)$. If $\lambda\geq 0$ we may put $\lambda=\xi^2$ and find

$$y(x)=A\sin\xi x+B\cos\xi x\ , \tag{9.1.7}$$

so that the solution satisfying $y'(0)=0$ is

$$y(x)=B\cos\xi x\ . \tag{9.1.8}$$

For all values of ξ, i.e., all non-negative λ, this solution is bounded on $(0,\infty)$. We say that, on $(0,\infty)$, equations (9.1.3), (9.1.4) have the continuous spectrum $\lambda\geq 0$.

Now consider the same equation (9.1.3) on the finite interval $(0,\ell)$ and with the end conditions

$$y'(0)=0=y(\ell)\ . \tag{9.1.9}$$

If $\lambda<0$ then $y(x)$ given by equation (9.1.6) will satisfy the second end condition if

$$B\cosh\xi\ell=0\ , \tag{9.1.10}$$

i.e., $B=0$. On the other hand, if $\lambda\geq 0$, then $y(x)$ given by equation (9.1.8) will satisfy $y(\ell)=0$ if

$$B\cos\xi\ell=0\ , \tag{9.1.11}$$

that is, if $\xi=\xi_n$, where

$$\xi_n\ell=(n+1/2)\pi\ ,\quad n=0,1,2,...\ . \tag{9.1.12}$$

We say that equations (9.1.3), (9.1.9) have the discrete spectrum $\{\lambda_n\}_o^\infty$, where

$$\lambda_n=\{(n+1/2)\pi/\ell\}^2\ ,\quad n=0,1,2,...\ . \tag{9.1.13}$$

The distinction between a continuous and a discrete spectrum is intimately connected to the distinction between a Fourier *integral* and a Fourier *series*. Provided $f(x)$ satisfies certain conditions, e.g., it is piecewise continuous and $\int_o^\infty [f(x)]^2$ exists, or more generally $f(x)$ is in $L^2\,(0,\infty)$, the integral

$$F_N(\xi)=\int_o^N f(x)\cos(x\xi)dx\ , \qquad (9.1.14)$$

converges in the mean to a function $F(\xi)$ in $L^2\ (0,\infty)$ and

$$f_N(x)=\frac{2}{\pi}\int_o^N F(\xi)\cos(x\xi)d\xi \qquad (9.1.15)$$

converges in the mean to $f(x)$, i.e.,

$$\lim_{N\to\infty}\int_o^\infty |f_N(x)-f(x)|^2dx=0\ . \qquad (9.1.16)$$

$F(\xi)$ is called the Fourier (cosine) integral of $f(x)$, and Parseval's equality holds, i.e.,

$$\int_o^\infty [f(x)]^2dx=\frac{2}{\pi}\int_o^\infty [F(\xi)]^2d\xi\ . \qquad (9.1.17)$$

On the other hand, if $f(x)$ is defined and, say, piecewise continuous in $(0,\ell)$, we may form the Fourier coefficients

$$c_n=\int_o^\ell f(x)\cos(\xi_n x)dx \qquad (9.1.18)$$

where ξ_n is given by equation (9.1.12). The sum

$$f_N(x)=\frac{2}{\ell}\sum_{n=o}^{N} c_n\cos(\xi_n x), \qquad (9.1.19)$$

converges in the mean to $f(x)$, i.e.,

$$\lim_{N\to\infty}\int_o^\ell |f_N(x)-f(x)|^2dx=0\ , \qquad (9.1.20)$$

and there is the Parseval identity

$$\int_o^\ell [f(x)]^2dx=\frac{2}{\ell}\sum_{n=o}^{\infty} c_n^2\ . \qquad (9.1.21)$$

Equations (9.1.14), (9.1.18) involve the eigenfunctions of the equation (9.1.3); in (9.1.14) the parameter ξ, and therefore $\lambda=\xi^2$, runs through the infinite interval $(0,\infty)$, while in equation (9.1.18) it takes discrete values. One of the major problems connected with Sturm-Liouville equations is the formation and justification of such eigenfunction expansions for the general equation (9.1.1), see Titchmarsh (1962).

We have already encountered a generalisation of (9.1.21). For

$$\phi_n(x)=\left(\frac{2}{\ell}\right)^{1/2}\cos(\xi_n x)\ , \quad n=0,1,2,... \qquad (9.1.22)$$

are the normalized eigenfunctions of equations (9.1.3), (9.1.9). It was shown in Section 8.6 that the eigenfunctions of (9.1.1), (9.1.2) are *complete* in the sense of Definition 8.6.1; equation (9.1.21) is a particular case of equation (8.6.6).

The generalization of equations (9.1.14), (9.1.17) involves the idea of a *spectral function* $\sigma(\lambda)$, with the following properties. It is a monotonic function, bounded on each finite interval, such that for any square integrable function $f(x)$ we have

$$\int_o^\infty [f(x)]^2 dx = \int_{-\infty}^\infty E^2(\lambda) d\sigma(\lambda) . \tag{9.1.23}$$

Here $E(\lambda)$ is the Fourier transform of the function $f(x)$, i.e.,

$$E(\lambda) = \int_o^\infty f(x)\phi(x,\lambda) dx \tag{9.1.24}$$

where $\phi(x,\lambda)$ is the solution of equation (9.1.1) on $(0,\infty)$ with the end conditions

$$y(0)=1 \quad , \quad y'(0)=h . \tag{9.1.25}$$

The integral in equation (9.1.23) is taken in the sense of Stieltjes. By comparing equations (9.1.17), (9.1.23) we see that for the equation (9.1.3) the spectral function is

$$\sigma(\lambda) = \begin{cases} \frac{2}{\pi}\sqrt{\lambda} \ , & \lambda \geq 0 \\ 0 \ , & \lambda < 0 \end{cases} . \tag{9.1.26}$$

An infinite series such as that in equation (9.1.21) may be written as a Stieltjes integral. We may write

$$c(\lambda) = \int_o^\ell f(x)\cos(\sqrt{\lambda}x) dx , \tag{9.1.27}$$

and define $\sigma(\lambda)$ to be the step function with jumps $2/\ell$ at $\lambda_0, \lambda_1, \ldots$, i.e.,

$$\sigma(\lambda) = \frac{2}{\ell}\sum_{n=1}^\infty H(\lambda - \lambda_n) , \tag{9.1.28}$$

where $H(x)$ is the Heaviside function defined by

$$H(x) = \begin{cases} 1 \ , & x \geq 0 \\ 0 \ , & x < 0 \end{cases} . \tag{9.1.29}$$

Then

$$\int_o^\infty c^2(\lambda) d\sigma(\lambda) = \frac{2}{\ell}\sum_{n=o}^\infty c_n^2 . \tag{9.1.30}$$

For a discrete spectrum there are two ways of scaling the eigenfunctions. They may be normalized, like (9.1.22), so that $(\phi_n,\phi_n)=1$, or taken so that $\phi_n(0)=1$, like $\cos(\xi_n x)$. If they are not scaled at all, the spectral function $\sigma(\lambda)$ will have jumps

$$\sigma_n = [\phi_n(0)]^2/(\phi_n,\phi_n) \tag{9.1.31}$$

at λ_n, and

$$\sigma(\lambda)=\sum_{n=o}^{\infty}\sigma_n H(\lambda-\lambda_n)\ . \tag{9.1.32}$$

9.2 A historical overview

It was shown in Section 8.1 that the Sturm-Liouville problem can appear in three different forms: the one preferred by mathematicians seems to be (8.1.14), namely

$$y''(x)+\{\lambda-q(x)\}y(x)=0\ . \tag{9.2.1}$$

In vibration problems, equation (8.1.1), namely

$$u''(x)+\lambda\rho(x)u(x)=0 \tag{9.2.2}$$

is encountered in the lateral vibration of a taut non-uniform string, while

$$(A(x)v'(x))'+\lambda A(x)v(x)=0 \tag{9.2.3}$$

occurs in the longitudinal or torsional vibration of a rod of cross-section $A(x)$.

As with all inverse problems (see Parker (1977)), the introduction of Newton (1983), or Sabatier (1978, 1985)) there are three aspects to the inverse problem:

i) *existence*; i.e., mathematically, is there a function $q(x)$ (or $\rho(x)$, $A(x)$), or physically, is there a vibrating system, with the required properties?

ii) *uniqueness*; i.e., is there only one system with these properties?

iii) *reconstruction*; how can we construct a (or the) system from the given data?

These questions, which are closely related, have been gradually elucidated over the past forty years.

Ambarzumian (1929) considered the question of uniqueness in a special case. He considered equation (9.2.1) with the end conditions

$$y'(0)=0=y'(\pi)\ , \tag{9.2.4}$$

and the equation

$$y''(x)+\lambda y(x)=0 \tag{9.2.5}$$

with the same end conditions. He showed that if the two systems have the same spectrum, $((\lambda_n)_o^\infty = (n^2)_o^\infty)$, then $q(x)$ is identically zero. Notice that he considered symmetrical end conditions, so that only one spectrum is needed. His proof has a defect in that it relies on a perturbation method which requires $q(x)$ to be small.

The fundamental paper on the inverse problem for the equation (9.2.1) is Borg (1946). He showed that if $q(x)$ is symmetric, i.e.,

$$q(x)=q(\pi-x)\ , \tag{9.2.6}$$

then the spectrum of equation (9.2.1) corresponding to the end conditions (9.2.4) or to the end conditions

$$y(0)=0=y(\pi) \tag{9.2.7}$$

determines $q(x)$ uniquely. This validates Ambarzumian's earlier result. (See also Hochstadt and Kim (1970)).

Borg also considered equation (9.2.1) for two sets of end conditions, of the form

$$\cos\alpha\, y(0)+\sin\alpha\, y'(0)=0=\cos\beta\, y(\pi)+\sin\beta\, y'(\pi) \tag{9.2.8}$$

$$\cos\alpha\, y(0)+\sin\alpha\, y'(0)=0=\cos\gamma\, y(\pi)+\sin\gamma\, y'(\pi)\,. \tag{9.2.9}$$

If $\sin\alpha=0=\sin\gamma$, so that (9.2.9) is equivalent to (9.2.7), and $\sin\beta\neq0$, then two interlacing spectra (see Section 8.10) determine a unique nonsymmetric function $q(x)$. If $\sin\alpha\sin\beta\neq0$, then $q(x)$ is uniquely determined by two spectra which are short of the first eigenvalue λ_o of the first spectrum, corresponding to (9.2.8).

Borg's results were extended and simplified by Levinson (1949). He proved that if the spectra of (9.2.1) for each of the end conditions (9.2.8), (9.2.9) are given, and if $\sin(\gamma-\beta)\neq0$, that is if (9.2.8), (9.2.9) are not identical, then $q(x)$ is uniquely determined. This result was extended by Hochstadt (1973, 1975a) who considered the extent to which $q(x)$ was determined when some eigenvalues λ_n, μ_n were unknown; see also Hald (1978a) and further references given there.

For the symmetrical case Levinson showed if it is known that (9.2.6) holds almost everywhere in $(0,\pi)$, and if $\alpha+\beta=\pi$, i.e., $h=H$, then $q(x)$ is uniquely determined by the spectrum for the end conditions (9.2.8). This result includes Borg's results for (9.2.4), (9.2.7) as special cases.

Marchenko (1952, 1953) made these results a little sharper. He showed that if

$$\int_o^\pi |q(x)|\,dx<\infty \tag{9.2.10}$$

and $sin(\alpha-\beta)\neq0$, then the spectra of (9.2.1) corresponding to (9.2.8) and (9.2.9) determine $q(x)$ and $\tan\alpha$, $\tan\beta$, $\tan\gamma$, uniquely. A full account of the uniqueness theorem may be found in Levitan (1964a). Further results may be found in Hochstadt (1973, 1975b, 1976, 1977) Hochstadt and Lieberman (1978), Hald (1984), Sabatier (1979a, 1979b), McLaughlin (1986).

The results listed above are all concerned with *uniqueness*. Basically, they all state that it is not possible to find more than one function $q(x)$ corresponding to two spectra. However, it was shown in Chapter 8 that the eigenvalues of (9.2.1), (9.2.8) have a number of specific properties, e.g., they interlace according to the rule (8.10.14) and they have the asymptotic form (8.11.31). The question is therefore, what are sufficient conditions for two sets of numbers $(\lambda_n)_o^\infty$, $(\mu_n)_o^\infty$ to be the spectra of an equation of the form (8.2.1) with a function $q(x)$ which is k-times continuously differentiable?

To answer this question we recall some results from Chapter 8. Suppose $(\lambda_n)_o^\infty$, $(\mu_n)_o^\infty$ are to be the spectra of equation (9.2.1) corresponding to the end conditions

$$y'(0)-hy(0)=0=y'(\pi)+Hy(\pi)\ , \tag{9.2.11}$$

$$y'(0)-hy(0)=0=y'(\pi)+H^1y(\pi)\ . \tag{9.2.12}$$

If $H<H^1$ then equation (8.10.13) show that the two spectra must interlace according to the rule

$$\lambda_o<\mu_o<\lambda_1<\mu_1<\cdots\ . \tag{9.2.13}$$

This holds even when $h=\infty=H^1$. If h, H, H^1 are finite, and $q'(x)$ is continuous on $[0,\ell]$, then equation (8.11.31) shows that, for large n,

$$\sqrt{\lambda_n}=n+\frac{c_0}{n}+\frac{c_1}{n^3}+\ \cdots \tag{9.2.14}$$

$$\sqrt{\mu_n}=n+\frac{c_o'}{n}+\frac{c_1'}{n^3}+\ \cdots\ , \tag{9.2.15}$$

and equation (8.11.18) shows that

$$c_0'-c_0=\frac{1}{\pi}(H^1-H)\ . \tag{9.2.16}$$

Levitan (1964b) proved the following theorem.

Let $(\lambda_n)_o^\infty$, $(\mu_n)_o^\infty$ be sets of numbers satisfying (9.2.13) - (9.2.15), where $c_0 \neq c_0'$. Then there exists an equation of the form (9.2.1) with a continuous real valued function $q(x)$ and real numbers h, H, H^1 such that the sequence $(\lambda_n)_o^\infty$ is the spectrum of the problem (9.2.1), (9.2.11), the sequence $(\mu_n)_o^\infty$ is the spectrum of the problem (9.2.1), (9.2.12), and equation (9.2.16) holds. If in the asymptotic expansions (9.2.14), (9.2.15) there are k exact terms (not counting the first), then the function $q(x)$ is continuously differentiable $(k-2)$-times. In particular, an infinite classical asymptotic expansion for the numbers $\sqrt{\lambda_n}$ and $\sqrt{\mu_n}$ exists if and only if the function $q(x)$ is infinitely differentiable.

Levitan's theorem is a refinement of results contained in the final section of Gel'fand and Levitan (1951). In that paper the data for the inverse problem was taken to be the spectral function described in Section 9.1; for problems with a discrete spectrum the data is therefore $(\lambda_n)_o^\infty$, $(\sigma_n)_o^\infty$. They showed that provided λ_n, σ_n had the correct asymptotic form, $\sqrt{\lambda_n}$ given by equation (9.2.14), and σ_n by

$$\sigma_n=\frac{2}{\pi}+\frac{e_0}{n^2}+0\left(\frac{1}{n^4}\right)\ , \tag{9.2.17}$$

then there exists a function $q(x)$ for which $(\lambda_n,\sigma_n)_o^\infty$ are the spectra.

For the question of reconstruction, the basic paper is Gel'fand and Levitan (1951). They show that the function $q(x)$ and the constants h, H, are uniquely determined by the spectral function $\rho(\lambda)$. For problems with a discrete spectrum this function is given by (9.1.32). They show that $q(x)$,h and H are uniquely determined from the spectrum $(\lambda_n)_o^\infty$, the constants $(\sigma_n)_o^\infty$ and the asymptotic expansions for λ_n and σ_n for large n. Gel'fand and Levitan develop a procedure for reconstructing $q(x)$ which was introduced by Marchenko (1950); a modified form of this procedure will be described in Section 9.3. Levitan (1964b) established necessary and sufficient conditions for two sequences $(\lambda_n)_o^\infty$, $(\mu_n)_o^\infty$ to be the spectra of equations (9.1.1), (9.1.2) corresponding to two constants H and H^1, where $q(x)$ is k times differentiable. In so doing he derives the spectral constants σ_n and their asymptotic expansion.

It was shown in Section 8.10 that the spectral constants may be determined, apart from a common factor, by two spectra corresponding to two different end conditions, at one end or the other. We shall return to this matter in Section 9.6, when considering the details of the reconstruction of a system from two spectra.

Three papers by Krein (1951a,b, 1952) answered questions i)-iii) for a nonhomogeneous taut string (equation (8.1.1)) by using his theory of the extension of positive definite functions. His results were generally stated without proof, and his methods have been used only by a few later authors; see Gopinath and Sondhi (1971), and Landau (1983).

Gopinath and Sondhi (1970), in considering the determination of the shape of the human vocal tract from acoustical measurements, encountered Webster's horn equation (8.12.1), and devised two methods for its inversion. The first is in the spirit of Gel'fand and Levitan, and can be replaced by the analysis of Section 9.3. The second is set in the time domain and relies on the impulse response described in Section 8.12. This formulation was improved and extended by Sondhi and Gopinath (1970) and is described in Section 9.7. A recent review of the vocal tract inverse problem was given by Sondhi (1984). The interconnections between all the various procedures for the inversion of second-order problems have been analysed by Burridge (1980); he pays particular attention to the case in which the cross-sectional area function $A(x)$ is discontinuous. See also Hald (1984).

9.3 The reconstruction procedure

The basic idea for the reconstruction procedure was introduced by Marchenko (1950), and developed by Gel'fand and Levitan (1951) and others. Suppose that we wish to construct a function $q(x)$ such that the equation

$$y''(x)+(\lambda-q(x))y(x)=0 , \tag{9.3.1}$$

under end conditions of the form

$$y'(0)-hy(0)=0=y'(\pi)+Hy(\pi) , \tag{9.3.2}$$

has certain eigenproperties. First we note that for given values of λ, ξ there is a unique function $y(x)\equiv y(x;\lambda,\xi)$ which is the solution of equation (9.3.1) satisfying

$$y(0)=\xi \ , \ y'(0)=h\xi , \tag{9.3.3}$$

i.e., the left hand end condition. We now introduce an *auxiliary* equation of the same form as (9.3.1), say

$$z''(x)+(\lambda-q_o(x))z(x)=0 \tag{9.3.4}$$

where $q_o(x)$ is a known function; and a new pair of end-conditions

$$z'(0)-h_o z(0)=0=z'(\pi)+H_o z(\pi)\ . \tag{9.3.5}$$

Again, for given λ, ξ, there is a unique solution $z(x)\equiv z(x;\lambda,\xi)$ of equation (9.3.4) saitsfying

$$z(0)=\xi\ \ ,\ \ z'(0)=h_o\xi\ , \tag{9.3.6}$$

i.e., the left hand end-condition. Equation (9.3.4) has a solution satisfying both end-condtions (9.3.5) only for certain eigenvalues $(\lambda_i^o)_o^\infty$; if the corresponding eigenfunctions are $\{z_i^o(x)\}_o^\infty$, and we shall suppose them to be normalized so that

$$\int_o^\pi z_i^o(x)z_j^o(x)dx=\delta_{ij}\ , \tag{9.3.7}$$

then clearly

$$z(x;\lambda_i^o,z_i^o(0))=z_i^o(x)\ . \tag{9.3.8}$$

We now introduce the fundamental step. We do not attempt to construct $q(x)$ directly, instead we construct the *solutions* of equation (9.3.1). Specifically we suppose that, *for all values of* λ, the solutions $y(x)\equiv y(x;\lambda,\xi)$ of equation (9.3.1) are related to the solutions $z(x)\equiv z(x;\lambda,\xi)$ of equation (9.3.4) by an equation of the form

$$y(x)=z(x)+\int_o^x K(x,s)z(s)ds\ . \tag{9.3.9}$$

The kernel $K(x,s)$ is determined by the condition that the normalized *eigenfunctions* $y_i(x)\equiv y(x;\lambda_i,y_i(0))$ of equation (9.3.1) subject to the end-conditions (9.3.2), with h, H at first undetermined, should form a complete orthogonal set.

In the form that has been stated above, the procedure is due in part to McLaughlin (1976); see also McLaughlin (1978, 1984a). She introduced it in connection with fourth-order inverse problems, but realised that the (integral) equation satisfied by $K(x,s)$ does not depend on the equation, second or fourth-order or whatever, satisfied by the function $z(x)$.

Since the reconstruction procedure is long, we shall state the major results in this section, and leave their derivation to the following two sections.

The first result is the (integral) equation satisfied by the kernel $K(x,s)$. We suppose that, from some value of $n+1$ on the given eigenvalues are the same as those for the auxiliary problem, i.e.,

$$\lambda_i=\lambda_i^o\ \ ,\ \ i=n+1,n+2,... \tag{9.3.10}$$

This assumption has the effect of ensuring that the eigenvalues have the appropriate asymptotic form, see equation (8.11.34). Secondly, we assume that we know the end values $y(0,\lambda_i)\equiv y_i(0)$ for the (as yet unknown) eigenfunctions scaled so that

$$\int_o^\pi y_j(x)y_i(x)dx=\delta_{ij} \tag{9.3.11}$$

and again we assume that these end values are, from $n+1$ on, the same as those for the auxiliary problem, i.e.,

$$y_i(0)=z_i^o(0) \quad i=n+1,n+2,... \tag{9.3.12}$$

The effect of equations (9.3.10), (9.3.12) is that one constructs a system that is *similar* to the auxiliary system, in the sense that it has similar high frequency behaviour. Note that the system that is constructed, i.e., $q(x)$ and the constants h, H, will change with changing $q_o(x)$, h_o, H_o. This is in accordance with the uniqueness theorem which states that $q(x)$ is determined by the eigenvalues $(\lambda_i)_o^\infty$, (i.e., $(\lambda_i)_o^n$ and $(\lambda_i^o)_{n+1}^\infty$), and by the end values of the normalized eigenfunctions, i.e., $(y_i(0))_o^\infty$, (i.e., $(y_i(0))_o^n$ and $(z_i^o(0))_{n+1}^\infty$.

The first theorem is:

Theorem 9.3.1. Let $z_i(x)=z(x;\lambda_i,y_i(0))$, $z_i^o(x)$ and $y_i(x)$ be defined as above for $i=0,1,2,...$, and $0\leq x\leq\pi$. Suppose in addition that $K(s,t)$ is continuous in $0\leq t\leq s\leq\pi$. Then $\{y_i(s)\}_o^\infty$ is a complete orthogonal set (c.o.s.) on $0\leq s\leq\pi$, with eigenvalues λ_i and end values $y_i(0)$ satisfying equations (9.3.10), (9.3.12) if and only if $K(s,t)$ satisfies the integral equation

$$F(x,s)+\int_o^x K(x,t)F(t,s)dt+K(x,s)=0\ , \tag{9.3.13}$$

where

$$F(s,t)=\sum_{i=o}^n\{z_i(s)z_i(t)-z_i^o(s)z_i^o(t)\}\ . \tag{9.3.14}$$

Note that although equations (9.3.10), (9.3.12) imply

$$z_i(x)=z_i^o(x) \quad i=n+1,n+2,...\ , \tag{9.3.15}$$

the eigenfunctions for the auxilairy system $(z_i^o(x))$ are not the same as the eigenfunctions of the constructed system, even for $i\geq n+1$. For all values of i the eigenfunctions of the latter are given by

$$y_i(x)=z_i(x)+\int_o^x K(x,s)z_i(s)ds\ . \tag{9.3.16}$$

Equation (9.3.13) is a degenerate integral equation (equation (8.4.4.)) and its solution is given by:

Theorem 9.3.2. Assume that $K(s,t)$ is continuous in t, $0\leq t\leq s$, for each s, $0\leq s\leq\pi$. Then the solution of equation (9.3.13) has the form

$$K(x,s)=\sum_{i=o}^n\{F_i(x)z_i(s)-G_i(x)z_i^o(s)\}\ , \tag{9.3.17}$$

where $F_i(x)$, $G_i(x)$, $i=0,1,...,$, are solutions of the $2n+2$ linear equations

$$z_i(x)+F_i(x)+\sum_{j=o}^{n}\{b_{ij}F_j(x)-c_{ij}G_j(x)\}=0 \quad i=0,1,\ldots,n\ , \tag{9.3.18}$$

$$z_i^o(x)+G_i(x)+\sum_{j=o}^{n}\{c_{ji}F_j(x)-d_{ij}G_j(x)\}=0 \quad i=0,1,\ldots,n\ , \tag{9.3.19}$$

where

$$b_{ij}(x)=\int_o^x z_i(t)z_j(t)dt\ ,\quad c_{ij}(x)=\int_o^x z_i(t)z_j^o(t)dt\ , \tag{9.3.20}$$

$$d_{ij}(x)=\int_o^x z_i^o(t)z_j^o(t)dt\ . \tag{9.3.21}$$

Note that $b_{ij}=b_{ji}$, $d_{ij}=d_{ji}$, but $c_{ij}\neq c_{ji}$.

With this kernel $K(x,s)$, the function $y(x)$ satisfies the differential equation (9.3.1) with

$$q(x)=q_o(x)+\frac{2dK(x,x)}{dx} \tag{9.3.22}$$

and the boundary condition

$$y'(0)-hy(0)=0\ , \tag{9.3.23}$$

where

$$h=h_o+K(0,0)\ . \tag{9.3.24}$$

When $\lambda\neq\lambda_i$ for $i=0,1,\ldots$ the function $y(x,\lambda)$ does not satisfy any particular boundary condition at $x=\pi$, but $y(x,\lambda_i)\equiv y_i(x)$ does satisfy the condition

$$y_i'(\pi)+Hy_i(\pi)=0\ , \tag{9.3.25}$$

where

$$H=H_o-K(\pi,\pi)\ . \tag{9.3.26}$$

In brief therefore the reconstruction procedure consists in the following steps:

i) Choose $q_o(x)$, h_o, H_o. We may choose $q_o(x)\equiv 0$.

ii) Find the $(\lambda_i^o,z_i^o(x))_o^n$, and, with the given $(\lambda_i,y_i(0))_o^n$, find $(z_i(x))_o^n$ and hence find $F(x,s)$.

iii) Solve equations (9.3.18)-(9.3.19) and hence find $K(x,s)$.

iv) The function $q(x)$ is given by equation (9.3.22).

v) The end conditions are given by equations (9.3.24)-(9.3.26).

Examples 9.3

1. Use equations (9.3.9), (9.3.17)-(9.3.19) to show that

$$F_i(x)=-y_i(x)\ ,\quad G_i(x)=-y_i^o(x)\equiv -y(x;\lambda_i^o,z_i^o(0))$$

for $i=0,1,...,n$.

2. Show that, at least to $0(n^{-2})$, the asymptotic development of the eigenvalues of equation (9.3.1), with $q(x)$ given by equation (9.3.22) and the end-conditions given by equations (9.3.24), (9.3.26), is the same as that for equations (9.3.4), (9.3.5). Use equation (8.11.34) to give, for large n,

$$\lambda_n = n^2 + [n + 2\frac{(h_o+H_o)}{\pi}] + \frac{(h_o+H_o)^2}{\pi^2 n^2} + \frac{2d_o}{n^2} + \cdots$$

In fact, a detailed analysis shows that the two asymptotic developments are identical, as would be expected from equation (9.3.10).

9.4 The Gel'fand-Levitan integral equation

This section is devoted to the derivation of Theorems 9.3.1 and 9.3.2. First we present some integral identities, the proofs of which depend on formal properties of multiple integrals. We shall assume that all the functions introduced below are such that the various integrals exist. The domains of the integrals have been taken to be $(0,\pi)$; this choice is arbitrary – they could as easily have been taken to be $(0,1)$.

Lemma 9.4.1.

$$\int_o^\pi f(u)du \int_o^\pi g(v)dv = \int_o^\pi \int_o^u (f(u)g(v)+f(v)g(u))dvdu \ . \tag{9.4.1}$$

Proof:

Divide the square $0 \le u,\ v \le \pi$ into two triangles. The left hand side is

$$\int_o^\pi \int_o^\pi f(u)g(v)dudv = \int_o^\pi \int_o^u f(u)g(v)dvdu + \int_o^\pi \int_o^v f(u)g(v)dudv \ .$$

Now interchange the dummy variables u, v in the second integral. ■

Lemma 9.4.2.

$$H(u) = \int_u^\pi h(t)K(t,u)dt \ , \quad G(v) = \int_v^\pi g(t)K(t,v)dt \ , \tag{9.4.2}$$

then

$$\int_o^\pi H(u)g(u)du + \int_o^\pi G(v)h(v)dv = \int_o^\pi \int_o^u \{g(u)h(v)+g(v)h(u)\}K(u,v)dvdu \ . \tag{9.4.3}$$

Proof:

Consider just the first integral. It is

$$\int_o^\pi H(v)g(v)dv = \int_o^\pi \int_v^\pi h(u)K(u,v)dug(v)dv = \int_o^\pi \int_o^u h(u)g(v)K(u,v)dvdu \ . \ ■$$

Lemma 9.4.3. With the notation of Lemma 9.4.2,

$$\int_o^\pi G(u)H(u)du=\int_o^\pi\int_o^u(g(u)h(v)+g(v)h(u))\int_o^v K(u,s)K(v,s)dsdvdu\ . \qquad (9.4.4)$$

The proof follows as in Lemma 9.4.2.

Lemma 9.4.4. With the notation of Lemma 9.4.2,

$$\int_o^\pi G(u)w(u)du\int_o^\pi h(v)w(v)dv+\int_o^\pi H(v)w(v)dv\int_o^\pi g(u)w(u)du$$
$$=\int_o^\pi\int_o^u(g(u)h(v)+g(v)h(u))\int_o^u K(u,t)w(t)w(v)dtdvdu$$
$$+\int_o^\pi\int_o^v(g(u)h(v)+g(v)h(u))\int_o^u K(u,t)w(t)w(v)dtdudv\ . \qquad (9.4.5)$$

Lemma 9.4.5.

$$\int_o^\pi G(u)w(u)du\int_o^\pi H(v)w(v)dv=\int_o^\pi\int_o^v(g(u)h(v)+g(v)h(u))\int_o^u K(u,s)w(s)ds$$
$$\int_o^v K(v,t)w(t)dtdudv\ . \qquad (9.4.6)$$

The necessary and sufficient condition for $\{y_i(x)\}_o^\infty$ to form a complete orthogonal set on $(0,\pi)$, normalized so that

$$\int_o^\pi y_i(x)y_j(x)dx=\delta_{ij}\ , \qquad (9.4.7)$$

is that

$$S\equiv\sum_{i=o}^\infty\int_o^\pi g(u)y_i(u)du\int_o^\pi h(v)y_i(v)dv=\int_o^\pi g(u)h(u)du\ , \qquad (9.4.8)$$

for arbitrary $g(u)$, $h(u)$ in $L^2(0,\pi)$. Equation (9.4.8) is called Parseval's equality.

We now suppose that $\{z_i^o(x)\}_o^\infty\equiv\{z(x;\lambda_i^o,z_i^o(0)\}_o^\infty$ form a c.o.s. on $(0,\pi)$, while $\{y_i\}_o^\infty\equiv\{y(x;\lambda_i,y_i(0)\}_o^\infty$ given by equation (9.3.16) form a second c.o.s. Thus, in addition to (9.4.8), we have

$$S^o\equiv\sum_{i=o}^\infty\int_o^\pi g(u)z_i^o(u)du\int_o^\pi h(v)z_i^o(v)dv=\int_o^\pi g(u)h(u)du\ . \qquad (9.4.9)$$

Expand the integrals on the left hand side of equation (9.4.8) by using equation (9.3.16). The result is

$$S=S_1+S_2+S_3\ , \qquad (9.4.10)$$

where

$$S_1=\sum_{i=o}^\infty\int_o^\pi g(u)z_i(u)du\int_o^\pi h(v)z_i(v)dv\ . \qquad (9.4.11)$$

and $z_i(u)\equiv z(u;\lambda_i,y_i(0))$. Lemma 9.4.1 shows that

$$S_1=\sum_{i=o}^{\infty}\int_o^\pi\int_o^u (g(u)h(v)+g(v)h(u))z_i(u)z_i(v)dvdu\ , \tag{9.4.12}$$

so that

$$S_1-S^o=\int_o^\pi\int_o^u (g(u)h(v)+g(v)h(u))F(u,v)dvdu\ , \tag{9.4.13}$$

where

$$F(u,v)=\sum_{i=o}^{\infty}[z_i(u)z_i(v)-z_i^o(u)z_i^o(v)]\ . \tag{9.4.14}$$

The second term in the expansion of (9.4.8) is

$$\begin{aligned} S_2 = & \sum_{i=o}^{\infty}\{\int_o^\pi h(u)\int_o^u K(u,t)z_i(t)dtdu\int_o^\pi g(v)z_i(v)dv \\ & + \int_o^\pi g(v)\int_o^v K(v,t)z_i(t)dtdv\int_o^\pi h(u)z_i(u)du\}\ . \end{aligned} \tag{9.4.15}$$

In the notation of Lemma 9.4.2

$$\begin{aligned} \int_o^\pi h(u)\int_o^u K(u,t)z_i(t)dtdu &= \int_o^\pi z_i(u)\int_u^\pi K(t,u)h(t)dtdu \\ &= \int_o^\pi H(u)z_i(u)du\ , \end{aligned} \tag{9.4.16}$$

so that

$$\begin{aligned} S_2 = & \sum_{i=o}^{\infty}\{\int_o^\pi H(u)z_i(u)du\int_o^\pi g(v)z_i(v)dv \\ & + \int_o^\pi G(u)z_i(u)du\int_o^\pi h(v)z_i(v)dv\}\ . \end{aligned} \tag{9.4.17}$$

Now Parseval's equality applied first to $H(u)$, $g(v)$ and then to $G(u)$, $h(v)$ shows that S_2^o, obtained by replacing $z_i(u)$ by $z_i^o(u)$, is

$$S_2^o=\int_o^\pi H(u)g(u)du+\int_o^\pi G(u)h(u)du\ , \tag{9.4.18}$$

and by Lemma 9.4.2

$$S_2^o=\int_o^\pi\int_o^u (g(u)h(v)+g(v)h(u))K(u,v)dvdu\ . \tag{9.4.19}$$

Now use Lemma 9.4.4 on $S_2-S_2^o$ and write

$$S_2\equiv S_2-S_2^o+S_2^o\ , \tag{9.4.20}$$

to obtain

$$\begin{aligned} S_2 = & \int_o^\pi\int_o^u \{g(u)h(v)+g(v)h(u)\}\int_o^u K(u,t)F(t,v)dtdvdu \\ & + \int_o^\pi\int_o^v \{g(u)h(v)+g(v)h(u)\}\int_o^u K(u,t)F(t,v)dtdudv \end{aligned}$$

$$+\int_o^\pi\int_o^u \{g(u)h(v)+g(v)h(u)\}K(u,v)dvdu\ . \tag{9.4.21}$$

The last term in the expansion of (9.4.8) is

$$S_3=\sum_{i=o}^{\infty}\int_o^\pi h(u)\int_o^u K(u,t)z_i(t)dtdu\int_o^\pi g(v)\int_o^v K(v,s)z_i(s)dsdv$$

$$=\sum_{i=o}^{\infty}\int_o^\pi G(u)z_i(u)du\int_o^\pi H(v)z_i(v)dv\ . \tag{9.4.22}$$

Thus, by Lemma 9.4.5

$$S_3-S_3^o=\int_o^\pi\int_o^v \{g(u)h(v)+g(v)h(u)\}\int_o^u K(u,t)ds\int_o^v K(v,s)F(s,t)dsdudv\ . \tag{9.4.23}$$

Again Parseval's equality applied to $G(u)$, $H(u)$ shows that

$$S_3^o=\int_o^\pi G(u)H(u)du=\int_o^\pi\int_o^v \{g(u)h(v)$$

$$+\ g(v)h(u)\}\int_o^u K(u,t)K(v,t)dtdudv\ . \tag{9.4.24}$$

Collecting terms from equations we find that

$$S-S^o=S_1-S^o+S_2-S_2^o+S_2^o+S_3-S_3^o+S_3^o=0\ , \tag{9.4.25}$$

yields the equation

$$\int_o^\pi\int_o^u \{g(u)h(v)+g(v)h(u)\}\{F(u,v)$$

$$+\int_o^u K(u,t)F(t,v)dt+K(u,v)\}dvdu$$

$$+\int_o^\pi\int_o^v \{g(u)h(v)+g(v)h(u)\}\int_o^u K(u,t)[F(v,t)$$

$$+\int_o^v K(v,s)F(s,t)ds+K(v,t)]dtdudv=0\ . \tag{9.4.26}$$

Write

$$J(u,v)=F(u,v)+\int_o^u K(u,t)F(t,v)dt+K(u,v)\ , \tag{9.4.27}$$

then equation (9.5.26) may be written

$$\int_o^\pi\int_o^u \{J(u,v)+\int_o^v J(u,t)K(v,t)dt\}\{g(u)h(v)+g(v)h(u)\}dvdu=0\ . \tag{9.4.28}$$

Since $g(h)$, $h(u)$ are arbitrary functions in $L^2(0,\pi)$, we have

$$J(u,v)+\int_o^v J(u,t)K(v,t)dt=0\ ,\quad 0\le u,v\le\pi\ . \tag{9.4.29}$$

For fixed u, $0\le u\le\pi$, this is a homogeneous Volterra integral equation with kernal $K(v,t)$. The only solution is the zero solution (Yoshida (1960)). Hence, since u was arbitrary, $J(u,v)=0$ for $0\le v\le u\le\pi$, i.e., $K(u,v)$ must satisfy

$$F(u,v)+\int_o^u K(u,t)F(t,v)dt+K(u,v)=0\,. \tag{9.4.30}$$

The function $F(u,v)$ is defined in equation (9.4.14); it is an infinite sum. However, if for $i=n+1,n+2,...$ we have $\lambda_i=\lambda_i^o$ and $y_i(0)=z_i^o(0)$, then $z_i(u)\equiv z_i^o(u)$ and the sum becomes the *finite* sum (9.3.14), and we obtain Theorem 9.3.1.

We shall now prove that, when $F(u,v)$ is given by equation (9.3.14), equation (9.4.30) has at most one solution. Then we shall construct a solution, which must therefore be the solution.

Theorem 9.4.1. Let $F(u,v)$ be defined as in equation (9.3.14), and suppose that $K(u,v)$ is continuous in v, $0\le v\le u\le\pi$, for each fixed u, $0\le u\le\pi$. Then there exists at most one solution of the integral equation (9.4.30).

Proof:

For fixed u, equation (9.4.30) is a nonhomogeneous Fredholm equation for $K(u,v)$ with kernel $F(t,v)$. In order to show that (9.4.30) has at most one solution, one need only show that for each u the homogeneous equation

$$\int_o^u f(t)F(t,v)dt+f(v)=0\,,\quad 0\le v\le u\,, \tag{9.4.31}$$

has only the zero solution (i.e., $f(x)\equiv 0$). Therefore, fix u, assume that $f(t)$ is continuous in $0\le t\le u$ and satisfies (9.4.31). Multiply by $f(v)$ and integrate from 0 to u to obtain

$$\int_o^u [f(v)]^2dv+\int_o^u\int_o^u f(t)F(t,v)f(v)dtdv=0\,. \tag{9.4.32}$$

Since $\{z_i^o\}_o^\infty$ is a c.o.s., Parseval's equality (9.4.8) gives

$$\int_o^u [f(v)]^2dv=\sum_{i=o}^{\infty}[\int_o^u f(t)z_i^o(t)dt]^2\,. \tag{9.4.33}$$

On the other hand, equation (9.3.14) yields

$$\int_o^u\int_o^u f(t)F(t,v)f(v)dtdv=$$
$$=\sum_{i=o}^{n}\{[\int_o^u f(t)z_i(t)dt]^2-[\int_o^u f(t)z_i^o(t)dt]^2\}\,, \tag{9.4.34}$$

so that, recalling equation (9.3.15), we deduce that

$$\sum_{i=o}^{\infty}[\int_o^u f(t)z_i(t)dt]^2=0\,, \tag{9.4.35}$$

i.e.,

$$\int_o^u f(t)z_i(t)dt=0\,,\quad i=0,1,2,...\,. \tag{9.4.36}$$

We shall now show that $\{z_i(t)\}_o^\infty$ is a complete set in $0\le t\le\pi$; it will then be a complete set on $0 \leq t \leq u$. To do this we shall show that if $g(t)$ is in $L^2(0,\pi)$ then

$$\int_o^\pi g(t)z_i(t)dt=0\ , \quad i=0,1,2,... \quad \text{implies}$$

$$\int_o^\pi g(t)z_i^o(t)dt=0\ , \quad i=0,1,2,... . \tag{9.4.37}$$

Since $\{z_i^o(t)\}_o^\infty$ is a c.o.s. on $(0,\pi)$, this means that $g(t)=0$ almost everywhere. If $i \geq n+1$, then $z_i(t)=z_i^o(t)$, so that equation (9.4.37) holds immediately. Suppose therefore that $0 \leq i \leq n$, and that $\int_o^\pi g(t)z_j(t)dt=0$ for $j=0,1,2,...$. Then by Parseval's equality

$$0 = \int_o^\pi g(t)z_j(t)dt = \sum_{j=o}^{\infty} \int_o^\pi g(t)z_j^o(t)dt \int_o^\pi z_i(t)z_j^o(t)dt$$

$$= \sum_{j=o}^{n} \int_o^\pi g(t)z_j^o(t)dt \int_o^\pi z_i(t)z_j^o(t)dt\ . \tag{9.4.38}$$

If for some i, $\lambda_i=\lambda_k^o$ where $0 \leq k \leq n$, we have $z_i(t) \equiv z_k^o(t)$, then equation (9.4.38) is identically satisfied. Therefore, without loss of generality, we may assume that $\lambda_i \neq \lambda_j^o$ for all $i,j=0,1,...,n$. Equation (9.4.38) then yields n homogeneous equation for the determination of $\int_o^\pi g(t)z_j^o(t)dt$. We show that the determinant of coefficients is non-zero.

The equations

$$z_i''(x)+(\lambda_i-q_o(x))z_i(x)=0=z_j^{o''}(x)+(\lambda_j^o-q_o(x))z_j^o(x) \tag{9.4.39}$$

yield

$$(\lambda_i-\lambda_j^o)\int_o^\pi z_i(x)z_j^o(x)dx+[z_j^o(x)z_i'(x)-z_i(x)z_j^{o'}(x)]_o^\pi=0\ . \tag{9.4.40}$$

the bracketed quantity is

$$z_j^o(\pi)\{z_i'(\pi)+H_o z_i(\pi)\}-z_j^o(0)\{z_i'(0)-h_o z_i(0)\}$$

since $z_j^o(x)$ satisfies both of equations (9.3.5). But $z_i(x)$ satisfies the first of equations (9.4.5) so that

$$c_{ij} \equiv \int_o^\pi z_i(x)z_j^o(x)dx=-z_j^o(\pi)\frac{\{z_i'(\pi)+H_o z_i(\pi)\}}{(\lambda_i-\lambda_j^o)}\ , \tag{9.4.41}$$

and both $z_j^o(\pi)$ and $z_i'(\pi)+H_o z_i(\pi)$ will be non-zero. Equation (9.4.41) shows that the coefficient c_{ij} has the form

$$c_{ij}=b_i c_j/(\lambda_i-\lambda_j^o)\ , \tag{9.4.42}$$

so that

$$|\mathbf{C}|=det(c_{ij})=\det[(\lambda_i-\lambda_j^o)^{-1}]\prod_{i=o}^{n} b_i c_i\ . \tag{9.4.43}$$

This may easily be shown to be non-zero provided $\lambda_i \neq \lambda_j^o$ for all i,j, as we have assumed (Ex. 9.4.1). Thus $\{z_i(t)\}_o^\infty$ do form a complete set on $(0,\pi)$, and the only solution of equation (9.4.31) is $f(v)\equiv 0$. ■

The proof of Theorem 9.3.2 consists in assuming that $K(u,v)$ has the form (9.3.17) and substituting into the integral equation (9.3.13). We now show that the matrix of coefficients for equations (9.3.18), (9.3.19) is non-singular. Suppose that the matrix were singular for some $u=u_o$, then the homogeneous equations obtained by replacing the first terms, $z_i(u)$, $z_i^o(u)$, in equations (9.3.18), (9.3.19), by zeros, would have a non-zero solution. Multiply the first of these equations by $F_i(u)$, the second by $G_i(u)$ and add all $2n+2$ equations together; the result is

$$\sum_{i=o}^{n}\{\sum_{j=o}^{n}(\delta_{ij}+b_{ij})F_i(u)F_j(u)+\sum_{j=o}^{n}(\delta_{ij}-d_{ij})G_i(u)G_j(u)\}=0\,. \tag{9.4.44}$$

The matrix with coefficients $\delta_{ij}+b_{ij}$ is positive definite for all u. The second matrix has coefficients

$$\delta_{ij}-d_{ij}=\int_o^{\pi}z_i^o(t)z_j^o(t)dt-\int_o^{u}z_i^o(t)z_j^o(t)dt=\int_u^{\pi}z_i^o(t)z_j^o(t)dt\,. \tag{9.4.45}$$

This matrix is therefore positive definite for all $0\le u<\pi$, and identically zero if $u=\pi$, Ex. 9.4.2. Thus the only way for the homogeneous equations to be satisfied with non-trivial $F_i(u)$, $G_i(u)$ is for $u_o=\pi$, $F_i(u_o)=0$, $i=0,1,2,...,n$, and

$$\sum_{j=o}^{n}c_{ij}(\pi)G_j(\pi)=0\,. \tag{9.4.46}$$

But the matrix of coefficients in this equation is the one encountered in equation (9.4.42); it is non-singular. We conclude that equations (9.3.13), (9.3.19) have a unique solution for each u, $0\le u\le\pi$.

Examples 9.4

1. Show that if $\lambda_i\neq\lambda_j^o$ for all $i,j=0,1,...,n$, then the matrix (a_{ij}) defined by

   $$a_{ij}=(\lambda_i-\lambda_j^o)^{-1}\quad i,j=0,1,...,n$$

   is non-singular.
2. Show that the matrix $(\delta_{ij}-d_{ij})$ is positive definite. Hint: construct a positive quadratic form from the matrix.

9.5 Reconstruction of the differential equation

In Section 9.3 we showed that if $K(x,s)$ satisfied the Gel'fand-Levitan integral equation (9.3.13), then the set of functions $\{y_i(x)\}_o^{\infty}$ form a c.o.s. In this section we shall derive the differential equation for which these functions are the eigenfunctions. We first find the end-conditions and differential equations satisfied by the kernel $K(x,s)$. We note that if $q_o(x)\equiv 0$, or if $q_o(x)$ has derivatives of all orders, then the form (9.3.17) ensures that $K(x,s)$ is infinitely differentiable in both variables x,s such that $0\le s\le x\le\pi$.

We note that $z(x;\lambda,\xi)$ satisfies

$$z''(x)+(\lambda-q_o(x))z(x)=0\ , \tag{9.5.1}$$

and the end conditions

$$z'(0)-h_o z(0)=0\ , \tag{9.5.2}$$

while the eigenfunction $z(x;\lambda_i,x_i(0))\equiv z_i^o(x)$ satisfies

$$z'(\pi)+H_o z(\pi)=0 \tag{9.5.3}$$

in addition to equation (9.5.2). Clearly $F(u,v)$ satisfies

$$F_u(0,v)-h_o F(0,v)=0=F_v(u,0)-h_o F(u,0)\ , \tag{9.5.4}$$

where

$$F_u(0,v)\equiv\frac{\partial F}{\partial u}(u,v)\,|_{u=o}\ \text{etc.}, \tag{9.5.5}$$

and

$$F_{uu}(u,v)-q_o(u)F(u,v)=F_{vv}(u,v)-q_o(v)F(u,v)\ . \tag{9.5.6}$$

Equation (9.3.13) gives

$$K(0,v)+F(0,v)=0\ , \tag{9.5.7}$$

$$K_v(u,v)+\int_o^u K(u,t)F_v(t,v)dt+F_v(u,v)=0 \tag{9.5.8}$$

so that

$$K_v(u,0)-h_oK(u,0)=0\ . \tag{9.5.9}$$

On the other hand

$$K_u(u,v)+K(u,u)F(u,v)+\int_o^u K_u(u,t)F(t,v)dt+F_u(u,v)=0\ , \tag{9.5.10}$$

so that

$$K_u(0,v)+K(0,0)F(0,v)+F_u(0,v)=0\ , \tag{9.5.11}$$

and therefore, on using (9.5.7), (9.5.11), we obtain

$$K_u(0,v)-(h_o+K(0,0))K(0,v)=0\ . \tag{9.5.12}$$

In order to obtain the partial differential equation satisfied by $K(u,v)$, we differentiate (9.5.8) w.r.t. v and substitute for $F_{vv}(u,v)$ from equation (9.5.6). We find

$$\begin{aligned}K_{vv}(u,v)\ +\ &\int_o^u K(u,t)\{F_{tt}(t,v)+[q_o(v)-q_o(t)]F(t,v)\}dt+F_{uu}(u,v)\\ &+\ [q_o(v)-q_o(u)]F(u,v)=0\ .\end{aligned} \tag{9.5.13}$$

But

$$\int_o^u K(u,t)F_{tt}(t,v)dt = [K(u,t)F_t(t,v)-K_t(u,t)F(t,v)]_o^u + \int_o^u K_{tt}(u,t)F(t,v)dt \; . \tag{9.5.14}$$

Equations (9.5.4), (9.5.9) show that the integrated term is zero at the lower limit: thus

$$K_{vv}(u,v) + \int_o^u \{K_{tt}(u,t)+[q_o(v)-q_o(t)]K(u,t)\}F(t,v)dt + K(u,u)F_u(u,v)-K_t(u,t)|_{t=u}F(u,v)+F_{uu}(u,v) + [q_o(v)-q_o(u)]F(u,v)=0 \; . \tag{9.5.15}$$

Now differentiate equation (9.5.10) w.r.t. u; we find

$$K_{uu}(u,v) + \frac{dK(u,u)}{du}F(u,v)+K(u,u)F_u(u,v)+K_u(u,t)|_{t=u}F(u,v) + \int_o^u K_{uu}(u,t)F(t,v)dt+F_{uu}(u,v)=0 \; . \tag{9.5.16}$$

Put

$$K_{uu}(u,v)-K_{vv}(u,v)=U(u,v) \tag{9.5.17}$$

and subtract equation (9.5.15) from equation (9.5.16) to obtain

$$U(u,v)+\int_o^u U(u,t)F(t,v)dt+\frac{2dK(u,u)}{du}F(u,v)+W(u,v)=0 \; , \tag{9.5.18}$$

where

$$W(u,v)=[q_o(u)-q_o(v)]F(u,v)+\int_o^u [q_o(t)-q_o(v)]K(u,t)F(t,v)dt \; , \tag{9.5.19}$$

and we have used the fact that

$$K_u(u,t)|_{t=u}+K_t(u,t)|_{t=u}=\frac{dK(u,u)}{du} \; . \tag{9.5.20}$$

Now use equation (9.3.13) to rewrite $W(u,v)$ in the form

$$W(u,v)=[q_o(v)-q_o(u)]K(u,v)+\int_o^u [q_o(t)-q_o(u)]K(u,t)F(t,v)dt \; . \tag{9.5.21}$$

Now, putting

$$V(u,v)=U(u,v)+[q_o(v)-q_o(u)]K(u,v) \; , \tag{9.5.22}$$

we may write equation (9.5.18) in the form

$$V(u,v)+\int_o^u V(u,t)F(t,v)dt+\frac{2dK(u,u)}{du}F(u,v)=0 \; . \tag{9.5.23}$$

But, since the solution of equation (9.3.13) is unique we have

$$V(u,v)=\frac{2dK(u,u)}{du}K(u,v)\ , \tag{9.5.24}$$

i.e.,

$$K_{uu}(u,v)-K_{vv}(u,v)+[q_o(v)-q_o(u)]K(u,v)=\frac{2dK(u,u)}{du}K(u,v)\ . \tag{9.5.25}$$

This is the partial differential equation satisfied by $K(u,v)$.

Now we find the boundary conditions and differential equations satisfied by $y(x)$. Equation (9.3.9) gives

$$y'(x)=z'(x)+K(x,x)z(x)+\int_o^x K_x(x,s)z(s)ds\ . \tag{9.5.26}$$

Thus,

$$y'(0)=z'(0)+K(0,0)z(0)=[K(0,0)+h_o]y(0)\ , \tag{9.5.27}$$

and

$$\begin{aligned} y''(x) &= z''(x)+K(x,x)z'(x)+\frac{dK(x,x)}{dx}z(x) \\ &\quad + K_x(x,s)\,|_{s=x}z(x)+\int_o^x K_{xx}(x,s)z(s)ds\ . \end{aligned} \tag{9.5.28}$$

Now substitute for $K_{xx}(x,s)$ from equation (9.5.25) and integrate by parts twice to obtain

$$\begin{aligned} y''(x)&=z''(x)+K(x,x)z'(x)+\frac{dK(x,x)}{dx}z(x) \\ &+K_x(x,s)\,|_{s=x}z(x)-\int_o^x [q_o(s)-q_o(x)]K(x,s)z(s)ds \\ &+\frac{2dK(x,x)}{dx}\int_o^x K(x,s)z(s)ds+[K_s(x,s)z(s)-K(x,s)z'(s)]_o^x \\ &+\int_o^x K(x,s)z''(s)ds\ . \end{aligned} \tag{9.5.29}$$

Again the integrated term is zero at the lower limit and

$$\begin{aligned} y''(x)&=z''(x)+\frac{2dK(x,x)}{dx}\{z(x)+\int_o^x K(x,s)z(s)ds\} \\ &+\int_o^x K(x,s)\{z''(s)-[q_o(s)-q_o(x)]z(s)\}ds\ . \end{aligned} \tag{9.5.30}$$

Now use the fact that $z''(x)=(q_o(x)-\lambda)z(x)$ to obtain

$$y''(x)=\{q_o(x)+\frac{2dK(x,x)}{dx}-\lambda\}\{z(x)+\int_o^x K(x,s)z(s)ds\}\ , \tag{9.5.31}$$

i.e., $y(x)$ satisfies equation (9.3.1), where

$$q(x)=q_o(x)+\frac{2dK(x,x)}{dx}. \tag{9.5.32}$$

Finally, we find the end condition satisfied by $y_i(x)$ at $x=\pi$. Since $d_{ij}(\pi)=\delta_{ij}$, equation (9.3.19) for $u=\pi$ gives

$$z_i^o(\pi)+\sum_{j=o}^{n} c_{ji}(\pi)F_j(\pi)=0\ , \quad i=0,1,2,\ldots . \tag{9.5.33}$$

On differentiating equation (9.3.19) w.r.t. x we find

$$z_i^{o'}(u) + G_i'(u)+\sum_{j=o}^{n}\{c_{ji}F_j'(u)-d_{ij}'G_j'(u)\}$$
$$+\sum_{j=o}^{n}\{c_{ji}'F_j(u)-d_{ij}'G_j(u)\}=0\ . \tag{9.5.34}$$

But

$$\sum_{j=o}^{n}\{c_{ji}'F_j(u)-d_{ij}'G_j(u)\}=\sum_{j=o}^{n}z_i^o(u)\{z_j(u)F_j(u)-z_j^o(u)G_j(u)\}$$
$$=z_i^o(u)K(u,u)\ , \tag{9.5.35}$$

so that equation (9.5.34) with $u=\pi$ gives

$$z_i^{o'}(\pi)+z_i^o(\pi)K(\pi,\pi)+\sum_{j=o}^{n}c_{ji}(\pi)F_j'(\pi)=0\ , \quad i=0,1,2,\ldots . \tag{9.5.36}$$

Now combine equations (9.5.33), (9.5.36) and use the fact that $z_i^{o'}(\pi)+H_oz_i^o(\pi)=0$; the result is

$$\sum_{j=o}^{n}c_{ji}(\pi)\{F_j'(\pi)+(H_o-K(\pi,\pi))F_j(\pi)\}=0\ , \quad i=0,1,2,\ldots . \tag{9.5.37}$$

But the matrix of coefficients is non-singular so that the bracketed terms are all zero. Using the fact (Ex. 9.3.1) that $F_j(x)=-y_j(x)$, $j=0,1,2,\ldots,n$, we deduce that

$$y_j(\pi)+(H_o-K(\pi,\pi))y_j(\pi)=0\ , \quad j=0,1,\ldots,n\ . \tag{9.5.38}$$

Examples 9.5

1. Show that $y_j(x)$ satisfies the same boundary condition (9.5.38) even when $j=n+1,n+2,\ldots$.

9.6 The inverse problem for the vibrating rod

In Section 9.3 it was shown how the differential equation (9.3.1) and the end constants h, H can be reconstructed from the set of eigenvalues $(\lambda_i)_o^\infty$ and end values $\{y_i(0)\}_o^\infty$ assuming that, from some $n+1$ on, these quantities are the same as those for the known auxiliary problem – see equations (9.3.10), (9.3.12). (The analysis of Gel'fand and Levitan (1951) or McLaughlin (1978) may be adapted to lift this restriction, i.e., to construct $q(x)$ from two given sequences $(\lambda_i)_o^\infty$, $\{y_i(0)\}_o^\infty$, having the proper asymptotic form. The restricted problem is quite adequate for applications). The basic step in the reconstruction was the determination of the kernel $K(x,s)$ linking two c.o.s. For a rod in longitudinal (or torsional) vibration, the differential equation is equation (9.2.3), i.e.,

$$(A(x)u'(x))'+\lambda A(x)u(x)=0 \tag{9.6.1}$$

and the eigenfunctions are orthogonal w.r.t. the weighting function $A(x)\equiv a^2(x)$, i.e.,

$$\int_o^\ell a^2(x)u_i(x)u_j(x)dx=\delta_{ij} , \tag{9.6.2}$$

so that the functions

$$y_i(x)=a(x)u_i(x) \tag{9.6.3}$$

form a c.o.s. For the auxiliary problem we take

$$(A_o(x)v'(x))'+\lambda A_o(x)v(x)=0 , \tag{9.6.4}$$

where $A_o(x)=a_o^2(x)$. The function

$$z(x)=a_o(x)v(x) \tag{9.6.5}$$

then satisfies equation (9.3.4) with

$$q_o(x)=a_o''(x)/a_o(x) , \tag{9.6.6}$$

while $y(x)$ satisfies equation (9.3.1) with

$$q(x)=a''(x)/a(x) . \tag{9.6.7}$$

We now use the analysis of Section 9.3 to link $y(x)$ with $z(x)$, i.e., to link $u(x)$ to $v(x)$, via

$$a(x)u(x)=a_o(x)v(x)+\int_o^x K(x,s)a_o(s)v(s)ds , \tag{9.6.8}$$

where $K(x,s)$ satisfies

$$K(x,s)+\int_o^x K(x,t)F(t,s)dt+F(x,s)=0 \quad 0\le s\le x , \tag{9.6.9}$$

and

$$F(s,t)=a_o(s)a_o(t)\sum_{i=o}^n \{v_i(s)v_i(t)-v_i^o(s)v_i^o(t)\} . \tag{9.6.10}$$

The solution of equation (9.6.9) has the form

$$K(x,s)=a_o(x)a_o(s)\sum_{i=o}^n \{F_i(x)v_i(s)-G_i(x)v_i^o(s)\} \tag{9.6.11}$$

where $F_i(x)$, $G_i(x)$ satisfy

$$v_i(x)+F_i(x)+\sum_{j=o}^{n}\{b_{ij}F_j(x)-c_{ij}G_j(x)\}=0\ , \tag{9.6.12}$$

$$v_i^o(x)+G_i(x)+\sum_{j=o}^{n}\{c_{ji}F_j(x)-d_{ij}G_j(x)\}=0\ , \tag{9.6.13}$$

and

$$b_{ij}(x)=\int_o^x a_o^2(t)v_i(t)v_j(t)dt\ \ ,\ \ c_{ij}(x)=\int_o^x a_o^2(t)v_i(t)v_j^o(t)dt\ , \tag{9.6.14}$$

$$d_{ij}(x)=\int_o^x a_o^2(t)v_i^o(t)v_j^o(t)dt\ . \tag{9.6.15}$$

We note that $d_{ij}(\ell)=\delta_{ij}$.

Suppose that $z(x)$ satisfies the end condition

$$z'(0)-h_oz(0)=0\ , \tag{9.6.16}$$

then $v(x)$ satisfies

$$a_o(0)v'(0)+\{a_o'(0)-h_oa_o(0)\}v(0)=0\ . \tag{9.6.17}$$

Without loss of generality we may choose $a_o(x)$ so that

$$a_o'(0)-h_oa_o(0)=0\ ,\quad a_o(0)=1\ . \tag{9.6.18}$$

Then the auxiliary rod has a free end at $x=0$, i.e.,

$$v'(0)=0\ . \tag{9.6.19}$$

This means that the (scaled) solution of (9.6.4) for $\lambda=0$ which satisfies the end condition (9.6.19) is simply

$$v(x)=1\ . \tag{9.6.20}$$

The end condition satisfied by $y(x)$ is given by equation (9.3.23); it is

$$y'(0)-\{h_o+K(0,0)\}y(0)=0\ , \tag{9.6.21}$$

i.e.,

$$a(0)u'(0)+\{a'(0)-[h_o+K(0,0)]a(0)\}u(0)=0\ . \tag{9.6.22}$$

For given data $(\lambda_i,y_i(0))_o^\infty$, i.e., for given $K(x,s)$, there is a family of rods, each member of which has its own end condition at $x=0$. The simplest choice is to take $a(x)$ satisfying the end conditions

$$a'(0)=[h_o+K(0,0)]a(0)\ ,\ \ a(0)=1\ , \tag{9.6.23}$$

so that the reconstructed rod also has a free end at $x=0$, i.e.,

$$u'(0)=0\ . \tag{9.6.24}$$

In this case the (scaled) solution of (9.6.1) corresponding to $\lambda=o$ is

$$u(x)=1 \, , \tag{9.6.25}$$

and equation (9.6.8) for $\lambda=0$ yields the relation between $u(x)$ and $v(x)$ as

$$a(x)=a_o(x)+\int_o^x K(x,s)a_o(s)ds \, . \tag{9.6.26}$$

To verify that this gives the required cross-section $A(x)=a^2(x)$ we immediately see that $a(x)$ satisfies the end conditions (9.6.23). The differential equation satisfied by $a(x)$ is found to be

$$a''(x)=\{q_o(x)+\frac{2dK(x,x)}{dx}\}a(x)=q(x)a(x) \, . \tag{9.6.27}$$

Thus equation (9.3.1), i.e.,

$$y''(x)+[\lambda-q(x)]y(x)=0 \tag{9.6.28}$$

may be transformed, as in (8.1.13) into

$$(a^2(x)u'(x))'+\lambda a^2(x)u(x)=0 \, . \tag{9.6.29}$$

The end conditions satisfied by $v_i(x)$ and $u_i(x)$ at $x=\ell$ are

$$a_o(\ell)v_i^{o'}(\ell)+\{a_o'(\ell)+H_o a_o(\ell)\}v_i^o(\ell)=0 \, , \quad i=0,1,2,... \, , \tag{9.6.30}$$

$$a(\ell)u_i'(\ell)+\{a'(\ell)+[H_o-K(\ell,\ell)]a(\ell)\}u_i(\ell)=0 \, , \quad i=0,1,2,... \, . \tag{9.6.31}$$

If $H_o=\infty$ then both auxiliary and reconstructed rod are fixed at $x=\ell$, i.e.,

$$v_i^o(\ell)=0=u_i(\ell) \, , \quad i=0,1,2,... \, . \tag{9.6.32}$$

Before using the procedure outlined above to reconstruct a rod from two spectra, we must show how two spectra may be used to determine the end values of the normalized eigenvectors. Suppose that it is required to construct a rod such that its first $n+1$ free-fixed eigenvalues are $(\lambda_i)_o^n$ while the first $n+1$ fixed-fixed eigenvalues are $(\mu_i)_o^n$. If the (unknown) end values of the normalized eigenfunctions $u_i(x)$ are $u_i(0)$, then the analysis of Section 8.10 (which can be rephrased for a rod, rather than a string) shows that $(\mu_i)_o^n$ are the first $n+1$ zeros of

$$\sum_{i=o}^{\infty}\frac{[u_i(0)]^2}{\lambda_i-\lambda}=0 \, . \tag{9.6.33}$$

Now assume that

$$(\lambda_i,u_i(0))_{n+1}^{\infty}=(\lambda_i^o,u_i^o(0))_{n+1}^{\infty} \, , \tag{9.6.34}$$

where ^{o} refers now to the free-fixed auxiliary uniform bar. Then equation (9.6.33) can be rewritten

$$\sum_{i=o}^{n}\frac{[u_i(0)]^2}{\lambda_i-\lambda}+\sum_{i=n+1}^{\infty}\frac{[u_i^o(0)]^2}{\lambda_i^o-\lambda}=0 \, . \tag{9.6.35}$$

But, for the uniform bar, the $u_i^o(x)$ are the eigenfunctions of

$$u_i^{o''}(x)+\lambda_i^o u_i^o(x)=0 \tag{9.6.36}$$

$$u_i^{o'}(0)=0=u_i^o(\ell)\ , \tag{9.6.37}$$

so that $\lambda_i^o=(i+1/2)^2\pi^2/\ell^2$, $[u_i^o(0)]^2=2/\ell$, and

$$\sum_{i=n+1}^{\infty}\frac{[u_i^o(\ell)]^2}{\lambda_i^o-\lambda}=\sum_{i=o}^{\infty}\frac{[u_i^o(\ell)]^2}{\lambda_i^o-\lambda}-\sum_{i=o}^{n}\frac{[u_i^o(\ell)]^2}{\lambda_i^o-\lambda}\ . \tag{9.6.38}$$

Put $\theta^2=\ell^2\lambda$ and use entry 1.4211 of Gradshteyn and Ryzhik (1965), to write the infinite sum as

$$S=2\ell\sum_{i=o}^{\infty}\frac{1}{(i+1/2)^2\pi^2-\theta^2}=\ell\frac{\tan\theta}{\theta}\ . \tag{9.6.39}$$

The eigenvalue equation is now

$$\sum_{i=o}^{n}\frac{[u_i(0)]^2}{\lambda_i-\lambda}=\sum_{i=o}^{n}\frac{[u_i^o(0)]^2}{\lambda_i-\lambda}-\ell\frac{\tan\theta}{\theta}=f(\lambda) \tag{9.6.40}$$

from which $[u_i(0)]^2$ can be found by solving the set of equations

$$\sum_{i=o}^{n}\frac{[u_i(0)]^2}{\lambda_i-\mu_j}=f(\mu_j)\qquad j=0,1,...,n\ . \tag{9.6.41}$$

It may easily be verified that the matrix of coefficients will be non-singular provided that the $(\lambda_i)_o^n$ and $(\mu_i)_o^n$ are distinct, and that the $[u_i(0)]^2$ so found will be positive provided that

$$0<\lambda_o<\mu_o<\lambda_1<\mu_1<\cdots<\lambda_n<\mu_n<\lambda_{n+1}^o<\cdots\ . \tag{9.6.42}$$

Note that the inequality $\mu_n<\lambda_{n+1}^o$ provides a constraint on the length ℓ of the rod which is to be constructed.

This procedure may be used to construct a symmetrical fixed-fixed rod with given eigenvalues $(\lambda_i)_o^{2n-1}$. The eigenvalues $(\lambda_{2j})_o^{n-1}$ will be those for half the rod with the mid-point *free*, while $(\lambda_{2j-1})_1^n$ will be those for the mid-point fixed. The problem thus reduces to the earlier one.

The procedure may also be used to construct a free-fixed rod with a given response function at the free end. If a force F is applied at the free end then, in the usual notation

$$\frac{u(0)}{F}=\sum_{i=o}^{\infty}\frac{[u_i(0)]^2}{\lambda_i-\lambda}\ , \tag{9.6.43}$$

from which an initial set $(\lambda_i,u_i(0))_o^n$ may be identified; the remaining $(u_i(0),\lambda_i)_{n+1}^{\infty}$ may be taken from a known auxiliary problem.

It is important to note that the $a(x)$ generated by the construction procedure will always be positive. We show this by supposing that $a(x_o)=0$ for some x_o in $[0,\ell]$ and arriving at a contradiction. By analogy with Ex. 9.3.1 we have

$$a(x)u(x,\lambda_i)=-F_i(x)\ ,\quad i=0,1,...n \tag{9.6.44}$$

$$a(x)u(x,\lambda_i^o)=-G_i(x)\ ,\quad i=0,1,...n\ . \tag{9.6.45}$$

Thus $a(x_o)=0$ implies $F_i(x_o)=0=G_i(x_o)$, $i=0,1,...n$, and therefore, from equations (9.6.12), (9.6.13),

$$v_i(x_o)=0=v_i^o(x_o)\ \ ,\ i=0,1,2,...n\ . \tag{9.6.46}$$

But when $i=o$, $v_i^o(x)$ has *no* zero, except possibly at $x=\ell$ when $H=\infty$. In this case $v_i(\ell)\neq 0$, since otherwise λ_i would also be an eigenvalue of the auxiliary problem. This contradiction implies $a(x)\neq 0$ for x in $[0,\ell]$; since $a(0)=1$ and $a(x)$ is continuous, we must have $a(x)>0$.

Finally, we note that the inverse problems for the vibrating string may be solved by using the relationship between the string and rod contained in equations (8.1.7)-(8.1.8).

There have been a number of attempts to reduce the inverse problem for a continuous second-order system to one for a discrete system. Anderson (1970) replaced the differential equation by a difference equation, but did not have the results on the inversion of matrix systems at his disposal. Hald (1972) made a detailed study of the problem, later (Hald (1977)) paying particular attention to the Sturm-Liouville problem with symmetric potential. Hald (1978b) discussed the discrete system obtained by applying the Rayleigh-Ritz procedure to the continuous problem, and considered the limiting case in which the number of terms in the Fourier series expansion of $q(x)$ tends to infinity. Barcilon (1983) attempted to derive the (continuous) density of a string from the known solution to the inverse problem for the discrete system, but his procedure does not lend itself to computation.

In conclusion it may be stated that the procedure described in the main body of this section provides an efficient method for reconstructing the function $a(x)$. With this method available it is not necessary to reduce the problem to discrete form using difference equations or the Rayleigh-Ritz method.

Examples 9.6

1. Show that

$$a''(x)=a_o''(x)+\frac{2dK(x,x)}{dx}\{a_o(x)+\int_o^x K(x,s)a_o(s)ds\}$$
$$+\int_o^x K(x,s)a_o''(s)ds+\int_o^x [q_o(x)-q_o(s)]K(x,s)a_o(s)ds$$

   and hence deduce equation (9.6.27).

2. Use equations (8.1.7)-(8.1.8) to modify the reconstruct process to find the density of a taut string from eigenvalue data.

9.7 Reconstruction from the impulse response

In this section we shall describe analysis, derived by Sondhi and Gopinath (1970), by which $A(x)$ can be reconstructed from the impulse response $\hat{h}(0,t)$ of Section 8.12. Suppose a unit impulse if applied to the free end, $x=0$, of the rod at time $t=0$. It is intuitively clear that the response $\hat{h}(0,T)$ at the end of the rod at time T is independent of the shape of the rod for $x>L$, where $L=T/2$. This is because any effect on $\hat{h}(0,T)$ due to the shape for $x>T/L$ would not be felt until *after* time T, the time taken for a disturbance moving with (scaled) speed 1 to reach $x=L$ and return. Sondhi and Gopinath demonstrate the converse, namely that knowledge of $\hat{h}(t)$ for $t\leq T$ is sufficient (and necessary) for the determination of $A(x)$ for $0\leq x\leq L$.

The solution is based upon the following observation. Suppose the rod is at rest at time $t=t_o$, i.e. $v(x,t_o)=0=p(x,t_o)$ for $0\leq x\leq L$, and a force is applied at the end $x=0$. At time $t=t_o+a$ the rod will still be at rest for $x\geq a$, because the scaled wave speed is 1. Integrating the first of equations (8.12.28) we obtain

$$A(x)[v(x,t)]_{t_o}^{t_o+a} = A(x)v(x,t_o+a)=\int_{t_o}^{t_o+a}\frac{\partial p}{\partial x}dt \tag{9.7.1}$$

and on a second integration w.r.t. x we find

$$\int_o^a A(x)v(x,t_o+a)dx=\int_{t_o}^{t_o+a} p(0,t)dt\,. \tag{9.7.2}$$

If now for every a we could find a force $p(0,t)$ such that $v(x,t_o+a)=1$ for $0\leq x\leq a$ then, for that case, equation (9.7.2) would give

$$V(a)\equiv\int_o^a A(x)dx=\int_{t_o}^{t_o+a} p(0,t)dt\,. \tag{9.7.3}$$

Thus the volume $V(a)$, and hence $A(a)$ would be determined as a function of a. We now show that such a force does exist and can be determined from a knowledge of $\hat{h}(t)$.

If $v(x,t)$, $p(x,t)$ satisfy equations (8.12.28), then so do $v(x,-t)$, $-p(x,-t)$ and, by superposition

$$V(x,t)=v(x,t)+v(x,-t)\,, \tag{9.7.4}$$

$$P(x,t)=p(x,t)-p(x,-t)\,. \tag{9.7.5}$$

Trivially $V(x,t)=2$, $P(x,t)=0$ is such a solution. A well known theorem in the theory of partial differential equations (Cauchy's problem) (Garabedian (1964)) states that this is the unique solution in the triangular region R of Figure 9.7.1 which satisfies the conditions $V(0,t)=2$, $P(0,t)=0$ for $-a\leq t\leq a$. Thus if $p(0,t)$ is such that $P(0,t)=0$ and $V(0,t)=2$ for $-a\leq t\leq a$, then everywhere in the triangle R, $P(x,t)=0$, $V(x,t)=2$. In particular $v(x,0)=1$, $0\leq x\leq a$; this gives the $v(x,t)$ required in equation (9.7.2) if t_o is taken to be $-a$. To find the required pressure $p(0,t)$ we note that since the rod is at rest at $t=-a$, equation (8.12.29) gives

$$v(0,t)=\int_{-a}^{t}\hat{h}(0,t-\tau)p(0,\tau)d\tau \tag{9.7.6}$$

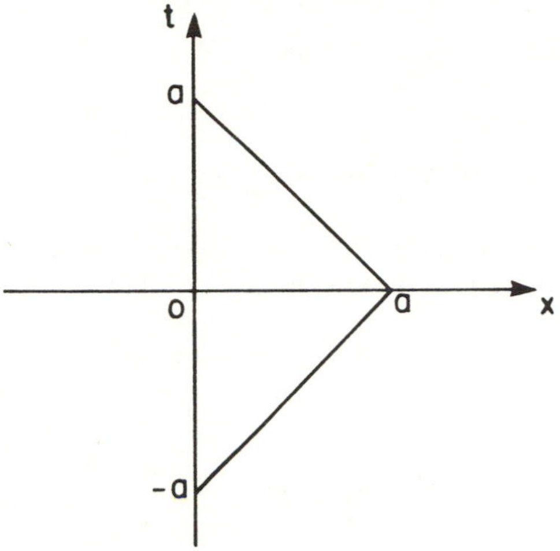

Figure 9.7.1. – The region R in the x,t plane

so that if $v(0,t)+v(0,-t)=2$ then

$$\int_{-a}^{t}\hat{h}(0,t-\tau)p(0,\tau)d\tau+\int_{-a}^{-t}\hat{h}(0,-t-\tau)p(0,\tau)d\tau=2\ . \tag{9.7.7}$$

The solution of this equation depends on a; we therefore write

$$p(0,\tau)=f(a,\tau)\ . \tag{9.7.8}$$

Now using the fact that $p(0,\tau)$ is even in τ, and that equation (8.12.36) yields

$$\hat{h}(t)=\delta(t)+h(t)\ , \tag{9.7.9}$$

we find

$$f(a,t)+\frac{1}{2}\int_{-a}^{a}h(|t-\tau|)f(a,\tau)d\tau=1\ . \tag{9.7.10}$$

Once $f(a,t)$ is known, equation (9.7.3) gives

$$\int_{o}^{a}A(x)dx=\int_{o}^{a}f(a,t)dt\ . \tag{9.7.11}$$

Equation (9.7.10) may be written in operator form

$$(I+H_a)f(a,t)=1\ . \tag{9.7.12}$$

Sondhi and Gopinath show that if $\hat{h}(t)$ is the impulse response of an actual rod, then the operator $I+H_a$ will be positive definite, so that equation (9.7.12) will have a unique solution. They show moreover that the corresponding $A(a)$ will be *positive*, provided that it is continuous. In addition, they show that if $I+H_a$ is positive definite, then there is a rod (i.e., an $A(x)$) which has this impulse response.

We now apply the procedure to the problem of Section 9.3, i.e., the reconstruction of a rod which has, from some index on, the same eigenvalues and end values of normalized eigenfunctions, as the uniform rod, i.e.,

$$\omega_i = \omega_i^o \ , \ \ u_i(0) = u_i^o(0) \ , \ \ i = n+1, n+2, \ldots \tag{9.7.13}$$

where

$$\omega_i^o = \frac{(2i+1)\pi}{2\ell} \ , \ \ [u_i^o(0)]^2 = \frac{2}{\ell} \ . \tag{9.7.14}$$

In this case $h(t)$ is given by Ex. 8.12.2 and the kernel $h\,|(t-\tau)|$ is degenerate. Since $F(a,t)$ is even in t, we may write equation (9.7.10) as

$$f(a,t) + \frac{1}{2}\int_o^a \{h(|t-\tau|) + h(t+\tau)\} f(a,\tau) d\tau = 1 \ , \ \ 0 \le t \le a \ . \tag{9.7.15}$$

Now the kernel is

$$\begin{aligned} H(t,\tau) &= \frac{1}{2}\{h(|t-\tau|) + h(t+\tau)\} \\ &= \sum_{i=o}^{n} \{[u_i(0)]^2 \cos\omega_i t \ \cos\omega_i \tau - [u_i^o(0)]^2 \cos\omega_i^o t \ \cos\omega_i^o \tau\} \ . \end{aligned} \tag{9.7.16}$$

Since the kernel is degenerate, the solution may be found by a straightforward matrix inversion. Thus equation (9.7.15) gives

$$f(a,t) = 1 + \sum_{i=o}^{n} a_i(a)[u_i(0)]^2 \cos\omega_i t - b_i(a)[u_i^o(0)]^2 \cos\omega_i^o t \ , \tag{9.7.17}$$

and on substituting this into equation (9.7.15) we find

$$a_i + \int_o^a \cos\omega_i \tau \ d\tau + \sum_{j=o}^{n} (b_{ij} a_j - c_{ij} b_j) = 0 \tag{9.7.18}$$

$$b_i + \int_o^a \cos\omega_i^o \tau \ d\tau + \sum_{j=o}^{n} (c_{ji} a_j - d_{ij} b_j) = 0 \tag{9.7.19}$$

where

$$b_{ij} = u_j^2(o)\int_o^a \cos\omega_i \tau \ \cos\omega_j \tau \ d\tau \ , \ \ c_{ij} = u_j^{o2}(0)\int_o^a \cos\omega_i \tau \ \cos\omega_j^o \tau \ d\tau \ , \tag{9.7.20}$$

$$d_{ij} = u_j^{o2}(0)\int_o^a \cos\omega_i^o \tau \ \cos\omega_j^o \tau \ d\tau \ . \tag{9.7.21}$$

Once $f(a,t)$ has been found, $A(x)$ may be computed from equation (9.7.11). This completes the inversion.

CHAPTER 10

THE EULER-BERNOULLI BEAM

The last process of reason is to recognise that there is an infinity of things which are beyond it. It is but feeble if it does not see so far as to know this.

Pascal's Pensées

10.1 Introduction

The free undamped infinitesimal transverse vibrations, of frequency ω^*, of a thin straight beam of length ℓ shown in Figure 10.1.1 are governed by the Euler-Bernoulli equation

$$\frac{d^2}{dx^2}\left(EI(x)\frac{d^2u(x)}{dx^2}\right)=A(x)\rho\omega^{*2}u(x)\ ,\quad 0\leq x\leq\ell\ . \tag{10.1.1}$$

Here E is Young's modulus, ρ is the density, both assumed constant; $A(x)$ is the cross-sectional area at section x, $I(x)$ is the second moment of this area about the axis through the centroid at right angles to the plane of vibration (the neutral axis). We put

$$x=\ell s\ ,\quad u(s)=u(x)\ ,\quad r(s)=\frac{I(x)}{I(x_o)}\ ,\quad a(s)=\frac{A(x)}{A(x_o)}\ ,$$

$$\lambda=\frac{A(x_o)\rho\ell^4\omega^{*2}}{EI(x_o)}\ , \tag{10.1.2}$$

where x_o is a chosen point in $[0,\ell]$. Equation (10.1.1) then becomes

$$(r(s)u''(s))''=\lambda a(s)u(s)\ , \quad 0\leq s\leq 1\ , \tag{10.1.3}$$

where $'\equiv d/ds$. (We shall now use x for the dimensionless variable). Both $r(x)$ and $a(x)$ are positive functions, i.e.,

$$r(x)>0\ , \ a(x)>0\ , \ 0\leq x\leq 1\ . \tag{10.1.4}$$

We shall assume throughout that $a(x)$ and $r(x)$ are twice continuously differentiable.

For a beam the most common end-conditions are

$$\text{free: } u''=0=u''\ , \tag{10.1.5}$$

$$\text{pinned: } u=0=u''\ , \tag{10.1.6}$$

$$\text{sliding: } u'=0=(r(x)u'')'\ , \tag{10.1.7}$$

$$\text{clamped: } u=0=u'\ . \tag{10.1.8}$$

There are some combinations of these end conditions which allow movement of the beam as a rigid body. These, together with the possible movements (so called rigid-body modes) are shown in Figure 10.1.2. These modes are eigenmodes of equation (10.1.3) for $\omega=0$ and the relevant end-conditions. Notice that the free-free beam has two. Although mode (b) shown in Figure 10.1.2 will not necessarily be orthogonal, with weight function $a(x)$, to (a), a combination of (a) and (b) can always be found which will be orthogonal to (a). (See Example 10.1.1).

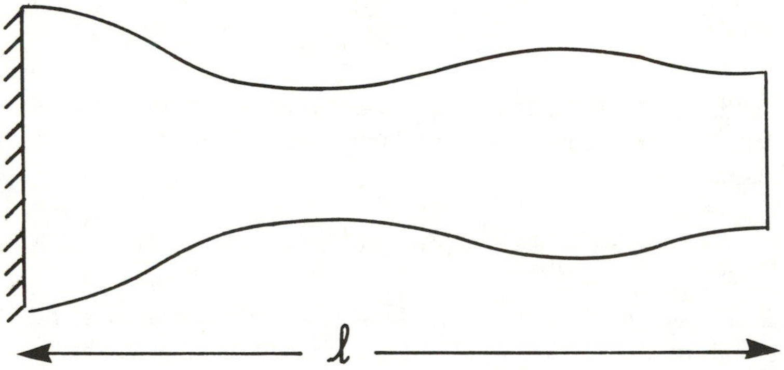

Figure 10.1.1 – A Non-Uniform Beam

The ends may be restrained by translational and rotational springs. In this case the end-conditions are

$$(r(x)u''(x))'_{x=o}+h_1u(0)=0=(r(x)u''(x))'_{x=1}-h_2u(1)\ , \tag{10.1.9}$$

$$r(0)u''(0)-k_1u'(0)=0=r(1)u''(1)+k_2u'(1)\ . \tag{10.1.10}$$

Here h_1, h_2 are the translational and k_1, k_2 the rotational stiffnesses. The

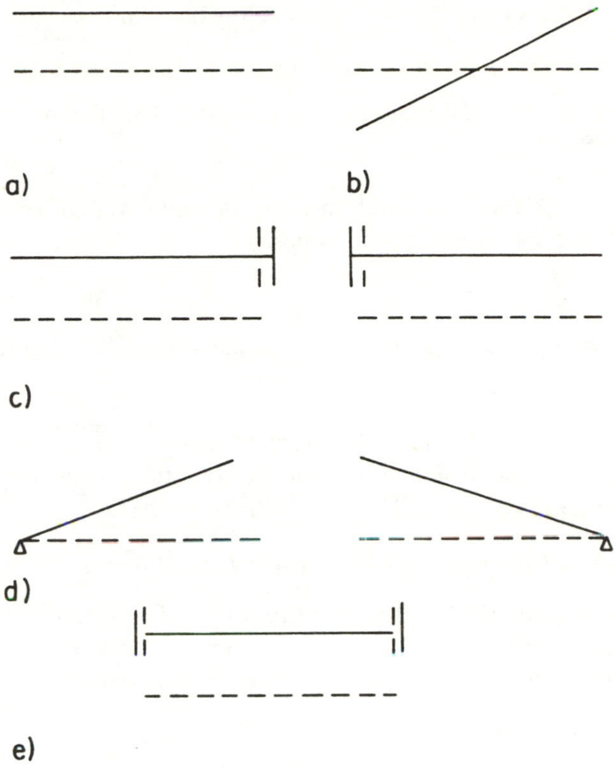

Figure 10.1.2 – Rigid-Body Modes of a Beam. a) Uniform translation and b) rotation of a free-free beam, c) Uniform translation of a free-sliding or sliding-free beam, d) Uniform translation of a sliding-sliding beam, e) Rotation of a pinned-free or free-pinned beam.

conditions (10.1.5)-(10.1.8) correspond respectively to $h=0=k$; $h=\infty, k=0$; $h=0$, $k=\infty; h=\infty=k$. We shall say that the system governed by equations (10.1.3), (10.1.9), (10.1.10) is *positive* if

$$h_1+h_2>0\ , \quad k_1+k_2>0\ . \tag{10.1.11}$$

Since h_1, h_2, k_1, $k_2 \geq 0$ this means that one of h_1, h_2 and one of k_1, k_2 must be non-zero; this rules out all of the rigid-body modes shown in Figure 10.1.2.

Theorem 10.1.1. The Euler-Bernoulli operator

$$B \equiv (r(x)u''(x))''$$

is self-adjoint, i.e., $(B\,u,v)=(u,B\,v)$, under the end-condition (10.1.9), (10.1.10).

Proof:

$$(Bu,v)-(u,Bv)=\int_o^1 \{(ru'')''v-(rv'')''u\}dx$$
$$=[(ru'')'v-ru''v'-(rv'')'u+rv''u']_o^1 . \qquad (10.1.12)$$

Under any of the end-conditions (10.1.9)-(10.1.10), the bracketed term is zero at each end. ■

Theorem 10.1.2. Eigenvalues of an Euler-Bernoulli system are non-negative, and are positive if and only if the system is positive.

Proof:

Suppose $u(x)$ is an eigenfunction of equation (10.1.3) corresponding to λ, then

$$Bu=\lambda au . \qquad (10.1.13)$$

Thus $(Bu,\bar{u})=\lambda(au,\bar{u})$. But $(Bu,\bar{u})= (B\bar{u},u) = \overline{(Bu,\bar{u})} = \bar{\lambda}\overline{(au,\bar{u})} = \bar{\lambda}(au,\bar{u})$ since B is self adjoint and $a(x)$ is a real function. Thus $\lambda=\bar{\lambda}$ and λ is real. Now $(Bu,u) = [(ru'')'u-ru''u']_o^1+ \int_o^1 r(u'')^2dx$ so that

$$\lambda(au,u)=h_1u^2(0)+h_2u^2(1)+k_1[u'(0)]^2+k_2[u'(1)]^2+\int_o^1 r(u'')^2dx . \qquad (10.1.14)$$

If $u(x)$ is not identically zero, then $(au,u)>0$. There can be a zero eigenvalue only if the integral on the right is zero, i.e., $u''(x)\equiv0$, or $u(x)=cx+d$, *and* each of the other terms in the right of equation (10.1.14) is separately zero, i.e.,

$$h_1d=0=h_2(c+d)=k_1c=k_2c . \qquad (10.1.15)$$

Suppose h_1, h_2, k_1, k_2 are *finite*. Equation (10.1.15) implies that either $k_1=0=k_2$, in which case the system is not positive, or $c=0$. If $c=0$, then either $h_1=0=h_2$, in which case the system is not positive, or $d=0$. Thus, if the system is positive, then $\lambda>0$. Conversely, if the system is not positive, then one or both of c, d may be taken to be non-zero and there is a rigid-body mode. The cases when an h or a k can be infinite are covered in Figure 10.1.2.

Before introducing the Green's function in general, we consider a special case, the cantilever beam, i.e., a beam clamped at $x=0$, free at $x=1$. (In the reconstruction of the beam, in Section 10.6, we shall find it convenient to reverse the end-conditions; here however the clamped-free configuration is the more convenient). If a unit concentrated load (made dimensionless as in (10.1.2)) is applied to the beam at $x=s$ $(0<s\le1)$ then the deflection $u(x)$ and its first two derivatives will be continuous in [0,1], while its third derivative will have a jump discontinuity at $x=s$. Equilibrium demands

$$[(r(x)u''(x))']_{x=s-}^{x=s+}=1 . \qquad (10.1.16)$$

The end conditions

$$u''(1)=0=u'''(1)\ , \tag{10.1.17}$$

then yield

$$(r(x)u''(x))'=\begin{cases}-1 & 0\le x<s\\ 0 & s<x\le 1\end{cases}, \tag{10.1.18}$$

and

$$r(x)u''(x)=\begin{cases}s-x & 0\le x<s\\ 0 & s\le x\le 1\end{cases}. \tag{10.1.19}$$

Thus

$$u'(x)=\int_o^{x_o}\frac{(s-t)dt}{r(t)}\ , \tag{10.1.20}$$

where $x_o=\min(x,s)$, so that the displacement $u(x)$, i.e., the Green's function $G(x,s)$, is

$$G(x,s)=\int_o^{x_o}\frac{(x-t)(s-t)dt}{r(t)}\ . \tag{10.1.21}$$

Under general end-conditions of the form (10.1.9), (10.1.10), the Green's function has the following properties:

(1) $G(x,s)$ is, for fixed s, a continuous function of x and satisfies the end-conditions (10.1.9), (10.1.10).

(2) Except at s, the first four derivatives of $G(x,s)$ w.r.t. x are continuous in [0,1]. At $x=s$, the third derivative has a jump discontinuity given by

$$\left[\frac{\partial}{\partial x}\left(r(x)\frac{\partial^2 G(x,s)}{\partial x^2}\right)\right]_{x=s-}^{x=s+}=1\ . \tag{10.1.22}$$

(3) $B_x G(x,s)=0$ for $0\le x<s$ and $s<x\le 1$. (This may be rephrased as

(3^1) $B_x G(x,s)=0$ for $0\le x<s$ and $B_s G(x,s)=0$ for $s<x\le 1$.)

Theorem 10.1.3. Green's function is symmetric, i.e.,

$$G(s,t)=G(t,s)\ .$$

Proof: See Ex. 10.1.3.

Theorem 10.1.4. If $f(x)$ is piecewise continuous then

$$u(x)=\int_o^1 G(x,s)f(s)ds \tag{10.1.23}$$

is a solution of

$$B\,u=f(x)\,, \tag{10.1.24}$$

and satisfies the end conditions (10.1.9), (10.1.10). Conversely, if $u(x)$ satisfies (10.1.24) and the end-conditions (10.1.9), (10.1.10), then it can be represented by (10.1.23).

Proof: See Courant and Hilbert (1953), p. 362.

The construction procedure for the Green's function of the Sturm-Liouville equation can be generalized. It can be shown (Ex. 10.1.6) that

$$G(x,s)=\begin{cases}\phi(x)\theta(s)+\psi(x)\chi(s) & 0\le x\le s\\ \phi(s)\theta(x)+\psi(s)\chi(x) & s\le x\le 1\ ,\end{cases} \tag{10.1.25}$$

where $\phi(x)$, $\psi(x)$ are linearly independent solutions of $B\,u=0$ satisfying the conditions (10.1.9a), (10.1.10a) at $x=0$, while $\theta(x)$, $\chi(x)$ are linearly independent solutions satisfying the conditions (10.1.9.b), (10.1.10b) at $x=1$. Note that for (10.1.21), these functions are

$$\phi(x)=\int_o^x\frac{(x-t)dt}{r(t)}\ ,\quad \psi(x)=\int_o^x\frac{t(x-t)dt}{r(t)}\ ,\quad \theta(x)=x\ ,\quad \chi(x)=-1\ . \tag{10.1.26}$$

Theorem 10.1.4 allows us to replace the differential equation (10.1.3) and end conditions (10.1.9), (10.1.10) with the integral equation

$$u(x)=\lambda\int_o^1 a(s)G(x,s)u(s)ds\ . \tag{10.1.27}$$

In Section 10.2 we shall now show that $G(x,s)$ is an oscillatory kernel: the reader may omit that section and proceed straight to Section 10.3.

Example 10.1

1. $\phi_{0,1}(x)=1$ and $\phi_{0,2}(x)=cx+d$ will be orthogonal rigid body modes if

   $\int_o^1 a(x)\{cx+d\}dx=0$. Show that when $a(x)$ is symmetrical about $x=1/2$, then $\phi_{0,2}=c(x-1/2)$.

2. Show that two solutions $\phi_i(x)$, $\phi_j(x)$ of equation (10.1.3) satisfying the same end conditions of the form (10.1.9), (10.1.10) and corresponding to different eigenvalues λ_i, λ_j are orthogonal w.r.t. $a(x)$, i.e., we may choose $\{\phi_i(x)\}_o^\infty$ so that

   $$(\phi_i,\phi_j)\equiv\int_o^1 a(x)\phi_i(x)\phi_j(x)dx=\delta_{ij}\ .$$

   Show that $(\phi_i'',r\phi_j'') = \lambda_i\delta_{ij}$.

3. Show that the Green's function $G(x,s)$ for (10.1.3) under the end conditions (10.1.9), (10.1.10) is symmetric. (Adapt the argument of Theorem 8.3.2).

4. Show that the Green's function for a pinned-pinned beam is

$$G(x,s)=x(1-s)\int_x^1 \frac{t(1-t)dt}{r(t)}+xs\int_s^1 \frac{(1-t)^2dt}{r(t)}+(1-x)(1-s)\int_o^x \frac{t^2dt}{r(t)}$$

when $x \le s$. This shows that $G(x,s)>0$ when $0<x,s<1$. Show that $y(x)=G(x,s)$ satisfies $y'(0)>0$, $y'(1)<0$.

5. Write the Green's function of Ex. 10.1.3 in the form (10.1.25) and check that ϕ, θ, ψ and χ do satisfy the appropriate end conditions.

6. Establish equation (10.1.25). Use the fact that $u(x)=G(x,s)$ has the forms

$$u(x)=\begin{cases} c(s)\phi(x)+d(s)\psi(x) & 0\le x<s \\ e(s)\theta(x)+f(s)\chi(s) & s<x\le 1 \end{cases}.$$

Now use the facts that $u(x)$, $u'(x)$, $u''(x)$ are continuous at s, while the third derivative has the jump given by equation (10.1.22).

10.2 Oscillatory properties of Euler-Bernoulli kernels

We will prove that the Green's function of a positive Euler-Bernoulli system is oscillatory. First we prove some preliminary resuts.

Theorem 10.2.1. Under the end conditions (10.1.9), (10.1.10) for finite, positive h_1, h_2, k_1, k_2, the Green's function $u(x)=G(x,s)$ for $0\le s\le 1$ satisfies

$$M'(x)\equiv(r(x)u'')'=\begin{cases} -c & 0\le x<s \\ 1-c & s<x\le 1 \end{cases}, \tag{10.2.1}$$

where $0<c<1$.

Proof:

Equation (10.1.22) shows that equation (10.2.1) holds for *some* c; we prove that $0<c<1$.

Consider the values of $M(x)\equiv r(x)u''(x)$. The end conditions (10.1.10) preclude the case in which $M(0)<0$ and $M(1)<0$, for then $u'(0)<0$, $u'(1)>0$ so that $u''(x_o)>0$ i.e., $M(x_o)>0$ for some x_o in (0,1). But $M(x)$ is piecewise linear, so that $M'(0)=(M(x_o)-M(0))/x_o>0$, $M'(1)=(M(1)-M(x_o))/(1-x_o)<0$ which contradicts (10.1.22). Closer scrutiny actually shows that we may preclude $M(0)\le 0$ and $M(1)\le 0$.

Secondly, *if* $c<0$ or $c>1$, i.e., if $M'(x)$ has the same sign throughout [0,1], then the case $M(0)\ge 0$, $M(1)\ge 0$ is excluded. For then $M(x)\ge 0$ for all x, while the end conditions (10.1.10) yield $u'(0)\ge 0$, $u'(1)\le 0$, which is a contradiction. (Clearly $u(x)\equiv 0$ is excluded).

Suppose $c<0$, so that $M'(x)>0$, then $M(0)<0$ and $M(1)>0$ (since all other cases are excluded). Thus the end conditions (10.1.10) yield $u'(0)<0$, $u'(1)<0$, and since $M(x)$ is piecewise linear with only one zero, $u'(x)<0$ for all x in [0,1]. But $M'(x)>0$ and the end conditions (10.1.9) yield $u(0)<0$ and $u(1)>0$, which is a contradiction.

If $c>1$, so that $M'(x)<0$, then $M(0)>0$ and $M(1)<0$ and the end conditions (10.1.10) yield $u'(0)>0$ and $u'(1)>0$ and, as before, $u'(x)>0$. But now $u(0)>0$ and $u(1)<0$ which is again contradictory. ■

Corollary. $M(x)$ cannot have the same sign throughout [0,1]. For $M(x)<0$ has already been excluded, and $M(x)>0$ implies $M(0)>0$, $M(1)>0$ so that the end conditions (10.1.10) give $u'(0)>0$, $u'(1)<0$, which is contradictory.

The limiting cases when h_i, k_i are 0 or ∞ are the subject of Ex. 10.2.1.

Theorem 10.2.2. Under the conditions (10.1.9), (10.1.10), the Green's function of the beam satisfies

$$G(x,s)>0 \text{ for } x,s \text{ in } I$$

Proof:

Here I has the same meaning as in Chapter 8; it is the closed interval [0,1] if h_1, h_2 are finite, (0,1] if $h_1=\infty$, i.e., if $u(0)=0$, etc.

Theorem 10.2.1 and its corollary show that if $u(x)=G(x,s)$, then the functions $M'(x)$, $M(x)$, $u'(x)$ and $u(x)$ must have one of the three forms shown in Figure 10.2.1. ■

Theorem 10.2.3. Under the action of n forces $(F_i)_1^n$ acting at $(s_i)_1^n$ where $0\le s_1<s_2<\cdots<s_n\le 1$, the beam can reverse its sign at most $n-1$ times.

Proof:

We assume that the beam is a positive system, as described in Section 10.1. First assume that h_1, h_2, k_1, k_2 are finite and positive, $s_1>0$ and $s_n<1$. The deflection of the beam is

$$u(x)=\sum_{i=1}^{n} F_i F(x,s_i)\ , \tag{10.2.2}$$

and because of (10.1.22) it satisfies

$$M'(x)=\begin{cases} c_o, & 0\le x<s_1 \\ c_i, & s_i<x<s_{i+1}\ , \\ c_n, & s_n<x\le 1 \end{cases} \tag{10.2.3}$$

where

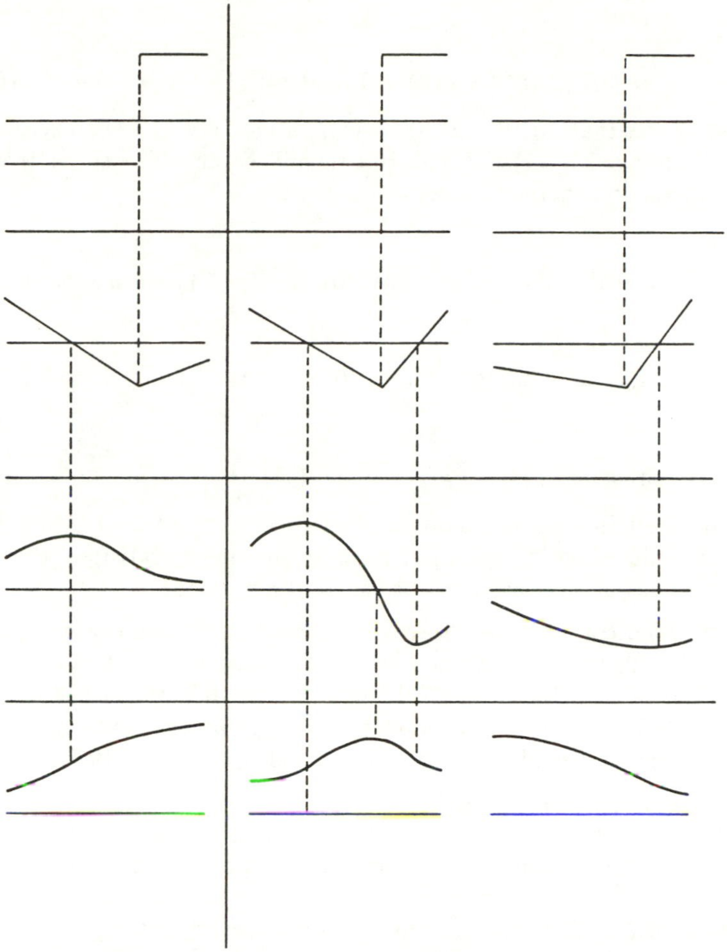

Figure 10.2.1 – The Formation of the Green's Function, Showing $M'(x)$, $M(x)$, $u'(x)$, $u(x)$ in [0,1].

$$c_i = c_o + \sum_{j=1}^{i} F_j \ , \quad i = 1,2,\ldots,n \ . \tag{10.2.4}$$

Thus $M'(x)$ has the property stated in the corollary to Theorem 8.7.3 so that $M(x)$ has at most $n+1$ zero places, at most one in each of $[0,s_1]$, $[s_1,s_2],\ldots,[s_n,1]$. Thus $M(x)$ and $u''(x)$, which are continuous in [0,1], have at most $n+1$ nodes in (0,1), so that by Theorem 8.7.3, $u'(x)$ has at most $n+2$ nodes, and $u(x)$ at most $n+3$ nodes. Now consider the sequences

$$M'(0), M(0), u'(0), u(0) \text{ and } M'(1), M(1), u'(1), u(1) \;,$$

i.e.,

$$-h_1 u(0), k_1 u'(0), u'(0), u(0) \text{ and } h_2 u(1), -k_2 u'(1), u'(1), u(1) \;.$$

Suppose first that $u(0)$, $u'(0)$, $u(1)$, $u'(1)$ are all non-zero, then always $u''(0)u'(0)>0$ and $u''(1)u'(1)<0$. But then Theorem 8.7.3 states that $u'(x)$ has at most n nodes, and therefore $u(x)$ at most $n+1$.

Now

$$M'(0)M(0) = -h_1 k_1 u'(0)u(0),\; M'(1)M(1) = -h_2 k_2 u'(1)u(1) \;,$$

so that

$$\text{either } M'(0)M(0)>0 \text{ or } u'(0)u(0)>0$$

and

$$\text{either } M'(1)M(1)>0 \text{ or } u'(1)u(1)>0 \;.$$

If one of the left hand inequalities is satisfied, then $M(x)$ has one less node than expected, while if one of the right hand inequalities is satisfied, then $u(x)$ has one less node. Thus in any case $u(x)$ has at most $n-1$ nodes.

A detailed consideration of special cases is left to the examples, but in the typical case $h_1=0$, k_1 finite and non-zero, we may argue as follows. $M'(0)=0$, so that $M'(x)\equiv 0$ in $[0,s_1]$, $M(x)=k_1 u'(0)$ in $[0,s_1]$, so that $M(x)$ has no node in $[0,s_1]$; $u''(0)u'(0)>0$ so that the remainder of the argument holds. Note: the analysis used in this proof provides an alternative proof of Theorem 10.2.2. ■

Theorem 10.2.3 holds the key to the proof that the Green's function is an oscillatory kernel. However, in order to prove the result we must continue the investigation of oscillatory systems of functions started in Section 8.8. First we introduce

Definition 10.2.1. The function $\phi(x)$ is said to reverse its sign k times in the interval I, and this is denoted by $S\phi = k$, if there are $k+1$ points $(x_i)_1^{k+1}$ in I such that $x_1 < x_2 < \cdots < x_{k+1}$ and

$$\phi(x_i)\phi(x_{i+1}) < 0 \quad \text{for } i=1,2,..,k$$

and there do not exist $k+2$ points with this property. Evidently, if $\phi(x)$ is continuous in $[0,1]$ and $S\phi = k$, then ϕ has k nodal places in $(0,1)$.

Before going further we state a basic composition formula. Suppose $\{\phi_i(x)\}_1^n$ are continuous on $[0,1]$, $M(x,s)$ is continuous on the square $0 \le x,s \le 1$ and

$$\psi_i(x) = \int_o^1 M(x,s)\phi_i(s)ds \;, \tag{10.2.5}$$

then (Ex. 10.2.2).

$$\Psi\begin{pmatrix} x_1, & x_2 & ,\dots, & x_n \\ 1, & 2 & ,\dots, & n \end{pmatrix} = \int\int_V \cdots \int M\begin{pmatrix} x_1, & x_2 & ,\dots, & x_n \\ s_1, & s_2 & ,\dots, & s_n \end{pmatrix}$$

$$\Phi\begin{pmatrix} s_1, & s_2 & ,\dots, & s_n \\ 1, & 2 & ,\dots, & n \end{pmatrix} ds_n ds_{n-1} \dots ds_1 \tag{10.2.6}$$

where V is the simplex defined by $s_1 s_2 < \cdots < s_n$; s_r in [0,1].

Theorem 10.2.4. Let $\phi(x)$ be continuous, $\phi \not\equiv 0$, and $S\phi \leq n-1$ in [0,1]. Let $M(x,s)$ be a continuous kernel with the property

$$M\begin{pmatrix} x_1, & x_2 & ,\dots, & x_n \\ s_2, & s_2 & ,\dots, & s_n \end{pmatrix} > 0 \quad 0 \leq \begin{matrix} x_1 < x_2 < \cdots < x_n \\ s_1 < s_2 < \cdots < s_n \end{matrix} \leq 1 ,$$

then

$$\psi(x) = \int_o^1 M(x,s)\phi(s)ds \tag{10.2.7}$$

does not vanish more than $n-1$ times in [0,1].

Proof:

By assumption there are points $(\xi_1)_o^n$ such that $0 = \xi_o < \xi_1 < \cdots < \xi_n = 1$ and $\phi(x)$ has one sign and is not identically zero on each interval (ξ_{i-1}, ξ_i) $(i = 1,2,\dots,n)$. Put

$$\psi_i(x) = \int_{\xi_{i-1}}^{\xi_i} M(x,s)\phi(s)ds \ , \tag{10.2.8}$$

then $\psi(x) = \sum_{i=1}^{n} \psi_i(x)$, and for all $(x_i)_1^n$ such that $0 \leq x_1 < x_2 < \cdots < x_n \leq 1$ we have (Ex. 10.2.3)

$$\Psi\begin{pmatrix} x_1, & x_2 & ,\dots, & x_n \\ 1, & 2 & ,\dots, & n \end{pmatrix} = \int_{\xi_{n-1}}^{\xi_n} \cdots \int_{\xi_o}^{\xi_1} M\begin{pmatrix} x_1, & x_2 & ,\dots, & x_n \\ s_1, & s_2 & ,\dots, & s_n \end{pmatrix}$$

$$\phi(s_1)\phi(s_2)\dots\phi(s_n)ds_1 ds_2 \dots ds_n. \tag{10.2.9}$$

The integrand is not identically zero and its non-zero values have one and the same sign, so that Ψ is strictly of one sign, and hence the $(\psi_i(x))_1^n$ form a Chebyshev system (Definition 8.8.1), and $\psi(x)$ does not vanish more than $n-1$ times in [0,1]. Note: the summation in equation (8.8.5) is from 0 to n. ■

An important kernel which satisfies the conditions of Theorem 10.2.4 is provided by

$$M_\varepsilon(x,y) = \frac{2}{\sqrt{\pi}\varepsilon} \exp\left\{ -\frac{(x-y)^2}{\varepsilon^2} \right\} \tag{10.2.10}$$

This kernel has a remarkable property. Let $\phi(x)$ be continuous in [0,1], and define

$$\psi(x,\varepsilon)=\int_o^1 M_\varepsilon(x,s)\phi(s)ds\ . \tag{10.2.11}$$

If we define $\phi(s)=0$ for $x>1$, then

$$\begin{aligned}\psi(x,\varepsilon) &= \frac{2}{\sqrt{\pi}\varepsilon}\int_o^\infty \exp\left\{-\frac{(x-s)^2}{\varepsilon^2}\right\}\phi(s)ds \\ &= \frac{2}{\sqrt{\pi}}\int_o^\infty \exp(-\xi^2)\phi(x+\varepsilon\xi)d\xi\ ,\end{aligned} \tag{10.2.12}$$

and

$$\lim_{\varepsilon\to o}\psi(x,\varepsilon)=\phi(x) \text{ when } 0\le x<1\ . \tag{10.2.13}$$

Using this kernel we may now prove

Theorem 10.2.5. Let $\{\phi_i(x)\}_1^n$ be linearly independent continuous functions in [0,1], and define

$$\phi(x)=\sum_{i=1}^n c_i\phi_i(x)\ .$$

The necessary and sufficient condition for $S\phi\le n-1$ in [0,1] for all c_i, not all zero, is that the determinant

$$\Phi\equiv\Phi\begin{pmatrix} x_1, & x_2 & ,\ldots, & x_n \\ 1, & 2 & ,\ldots, & n\end{pmatrix} \quad 0\le x_1<x_2<\cdots<x_n\le 1\ ,$$

which should have fixed sign, i.e., one and the same sign for those points for it is not zero. (Compare this with Theorem 8.8.2).

Proof: If for certain c_i, not all zero, $S\phi\le n-1$, and

$$\psi_i(x,\varepsilon)=\int_o^1 M_\varepsilon(x,s)\phi_i(s)ds\ , \tag{10.2.14}$$

then Theorem 10.2.4 shows that

$$\psi(x,\varepsilon)=\sum_{i=1}^n c_i\psi_i(x,\varepsilon) \tag{10.2.15}$$

vanishes in [0,1] not more than $n-1$ times. Conversely, equation (10.2.13) shows that if $\psi(x,\varepsilon)$ vanishes not more than $n-1$ times, then $S\phi\le n-1$. Thus $S\phi\le n-1$ for all c_i, not all zero, if and only if the functions $\{\psi_i(x,\varepsilon)\}_1^n$ form a Chebyshev system in [0,1], i.e., if and only if

$$\Psi_\varepsilon\equiv\Psi_\varepsilon\begin{pmatrix} x_1, & x_2 & ,\ldots, & x_n \\ 1, & 2 & ,\ldots, & n\end{pmatrix}$$

has strictly fixed sign when $0\le x_1<x_2<\cdots<x_n\le 1$. If Ψ_ε has strictly fixed sign then, since

$$\Phi = \lim_{\varepsilon \to o} \Psi_\varepsilon , \qquad (10.2.16)$$

Φ will have strictly fixed sign. On the other hand, since the $\{\phi_i(x)\}_1^n$ are linearly independent, Φ will not be identically zero. Thus equation (10.2.11) with $M = M_\varepsilon$ shows that if Φ has fixed sign, then Ψ_ε will have strictly fixed sign. ■

We have now established all the results needed to prove

Theorem 10.2.6. The Green's function of a positive Euler-Bernoulli system is oscillatory.

Proof:

There are three conditions to be fulfilled in the Definition 8.5.2. Theorem 10.2.2 yields condition (1), the argument via strain energy (Ex. 8.5.2) yields (3), and Theorem 10.2.5 then yields (2). ■

The following theorem states which of the determinants in (2) are zero and which non-zero; it is the analogue of Theorem 8.5.6 for the beam.

Theorem 10.2.7. The determinant

$$G\begin{pmatrix} x_1, & x_2 & ,\dots, & x_n \\ s_2 & s_2 & ,\dots, & s_n \end{pmatrix} > 0 \ , \quad 0 \le \begin{matrix} x_1 < x_2 < \cdots < x_n \\ s_1 < s_2 < \cdots < s_n \end{matrix} \le 1$$

if and only if $(x_i)_1^n$, $(s_i)_i^n$ are in I and $x_i < s_{i+2}, s_i < x_{i+2}$ for $i = 1,2,\dots,n-2$.

Proof:

First we show that the conditions are necessary for the minor to be positive. If one of x_1, x_n, s_1, s_n, e.g., x_1, is not in I, then $G(x_1,s_i) = 0$ for $i = 1,2,\dots,n$ and the determinant is zero. Suppose now that there is an index k such that $1 \le k \le n-2$ and $x_k \ge s_{k+2}$. Then $x_i \ge s_j$ when $i = k, k+1,\dots,n$ and $j = 1,2,\dots,k+2$. Consider the terms in the minor taken from columns $1,2,\dots,k+2$ and rows $k,k+1,\dots,n$ of the matrix $(G(x_i,s_j))$. Each will have the form

$$G(x_i,s_j) = \phi(s_j)\theta(x_i) + \psi(s_j)\chi(x_i) \ , \qquad (10.2.17)$$

so that the minor will have rank ≤ 2. If $n = 3$, then $k = 1$ and the minor is the complete matrix; which will thus have a zero determinant. If $n \ge 4$, then we evaluate the $n \times n$ determinant using Laplace's expansion with minors of order $k+2$ taken from the first $k+2$ columns; each such minor, having $k+2 \ge 3$ rows will be singular so that the determinant is zero.

Now suppose that $(x_i)_1^n$, $(s_i)_i^n$ are in I, and the $x_i < s_{i+2}, s_i < x_{i+2}$ for $i = 1,2,\dots,n-2$. We will prove that the determinant is positive, and will proceed by induction. When $n = 1$ the result holds on account of Theorem 10.2.2. Suppose that, if possible, it holds for $n-1$ but not for n, i.e., there exist $(x_i^o)_1^n$, $(s_i^o)_1^n$ in I and satisfying

$$x_i^o < s_{i+2}^o, s_i^o < x_{i+2}^o\ , \quad i=1,2,\ldots,n-2 \tag{10.2.18}$$

$$G\begin{pmatrix} x_1^o, & x_2^o & ,\ldots, & x_n^o \\ s_1^o, & s_2^o & ,\ldots, & s_n^o \end{pmatrix} = 0\ . \tag{10.2.19}$$

By the induction hypothesis

$$G\begin{pmatrix} x_1^o, & x_2^o & ,\ldots, & x_{n-1}^o \\ s_2^o, & s_2^o & ,\ldots, & s_{n-1}^o \end{pmatrix} > 0 \text{ and } G\begin{pmatrix} x_2^o, & x_3^o & ,\ldots, & x_n^o \\ s_2^o, & s_3^o & ,\ldots, & s_n^o \end{pmatrix} > 0\ . \tag{10.2.20}$$

Now choose arbitrary numbers $(x_i)_1^n$, $(s_i)_1^n$ such that

$$x_1^o \le x_1 < x_2 < \cdots < x_n \le x_n^o\ , \quad s_1^o \le s_1 < s_2 < \cdots < s_n \le s_n^o$$

and renumber the $(x_i^o)_1^n$ and $(x_i)_1^n$ in increasing order as $(x_i')_1^{2n}$, so that $x_i' \le x_2' \le \cdots \le x_{2n}'$; similarly renumber the $(s_i^o)_1^n$ and $(s_i)_1^n$ as $(s_i')_1^{2n}$. The $2n \times 2n$ matrix $(G(x_i', s_j'))$ is non-negative. Therefore applying Theorem 5.5.4 we conclude that it has rank $n-1$ so that

$$G\begin{pmatrix} x_1, & x_2 & ,\ldots, & x_n \\ s_1, & s_2 & ,\ldots, & s_n \end{pmatrix} = 0\ , \quad \begin{matrix} x_1^o \le x_1 < x_2 < \ldots < x_n \le x_n^o \\ s_1^o \le s_1 < s_2 < \ldots < s_n \le s_n^o \end{matrix} \tag{10.2.21}$$

If $n \ge 3$, then the inequalities (10.2.18) imply that the intervals (x_1^o, x_n^o) and (s_1^o, s_n^o) overlap, i.e., have a common internal point. They thus have n common interval points for which, on account of (10.2.21)

$$G\begin{pmatrix} x_1, & x_2 & ,\ldots, & x_n \\ x_1, & x_2 & ,\ldots, & x_n \end{pmatrix} = 0\ , \tag{10.2.22}$$

which contradicts the positive definiteness of (10.2.16). It remains to consider the case in which $n=2$ and (x_1^o, x_2^o), (s_1^o, s_2^o) have no common internal point. Without loss of generality we can take $x_2^o \le s_1^o$. Equation (10.2.21) gives

$$G\begin{pmatrix} x_1, & x_2 \\ s_1, & s_2 \end{pmatrix} = 0 \text{ for } \begin{matrix} x_1^o \le x_1 < x_2 \le x_2^o \\ s_1^o \le s_1 < s_2 < s_2^o \end{matrix}\ . \tag{10.2.23}$$

But since $x_2^o \le s_1^o$, we have $x_1 < s_1 < s_2$, $x_2 \le s_1 < s_2$ and

$$G\begin{pmatrix} x_1, & x_2 \\ s_1, & s_2 \end{pmatrix} = \begin{vmatrix} \phi(x_1)\theta(s_1)+\psi(x_1)\chi(s_1), & \phi(x_1)\theta(s_2)+\psi(x_1)\chi(s_2) \\ \phi(x_2)\theta(s_1)+\psi(x_2)\chi(s_1), & \phi(x_2)\theta(s_2)+\psi(x_2)\chi(s_2) \end{vmatrix}$$

$$= \begin{vmatrix} \phi(x_1) & \psi(x_1) \\ \phi(x_2) & \psi(x_2) \end{vmatrix} \cdot \begin{vmatrix} \theta(s_1) & \chi(s_1) \\ \theta(s_2) & \chi(_2) \end{vmatrix} = 0\ . \tag{10.2.24}$$

One or other of the factors in (10.2.24) must be zero. Suppose that for some s_1, s_2 the second factor is not zero, then the first must be identically zero, which is impossible since ϕ, ψ are linearly independent solutions of $B_x y = 0$ in $0 \le x < s$, i.e., in $x_1^o \le x \le x_2^o$. Similarly, if the first is not zero for some x_1, x_2 then the second must be identically zero, which is again impossible. Hence we have arrived at a contradiction. ■

Examples 10.2

1. Establish Theorem 10.2.1 when some of the h_i, k_i are 0 or ∞, but the system is still positive.
2. Verify equation (10.2.6) in the case $n=2$. Show that

$$\Psi\begin{pmatrix}x_1, & x_2\\ 1, & 2\end{pmatrix}=\int_o^1\int_o^1 M\begin{pmatrix}x_1, & x_2\\ s_1, & s_2\end{pmatrix}\phi_1(s_1)\phi_2(s_2)ds_2ds_1$$

$$=\frac{1}{2!}\int_o^1\int_o^1 M\begin{pmatrix}x_1, & s_2\\ s_1, & s_2\end{pmatrix}\Phi\begin{pmatrix}s_1, & s_2\\ 1, & 2\end{pmatrix}ds_2ds_1$$

$$=\int_o^1\int_o^{s_1} M\begin{pmatrix}x_1, & x_2\\ s_1, & x_2\end{pmatrix}\Phi\begin{pmatrix}s_1, & s_2\\ 1, & 2\end{pmatrix}ds_2ds_1 .$$

3. Establish equation (10.2.9); take $n=2$.

10.3 The eigenfunctions of the cantilever beam

For the cantilever beam the governing equations are

$$(r(x)u''(x))''=\lambda a(x)u(x)\ , \tag{10.3.1}$$

$$u(0)=0=u'(0)\quad ,\quad u''(1)=0=u'''(1)\ . \tag{10.3.2}$$

The theory of Section 10.2 shows that the Green's function for the beam is an oscillatory kernel on $I\equiv(0,1]$ so that the eigenvalues $(\lambda_i)_o^\infty=(\omega_i^2)_o^\infty$ are distinct and the eigenfunctions $\{\phi_i(x)\}_o^\infty$ have properties (1)-(4) stated in Theorem 8.8.3.

We need to strengthen this classical theorem. To do so we introduce $M(x)=r(x)u''(x)$. Equations (10.3.1)-(10.3.2) show that $M(x)$ satisfies

$$(b(x)M''(x))''=\lambda s(x)M(x)\ , \tag{10.3.3}$$

$$M''(0)=0=M'''(0)\quad ,\quad M(1)=0=M'(1)\ , \tag{10.3.4}$$

where $b(x)=1/a(x)$, $s(x)=1/r(x)$. Thus $M(x)$ is an eigenfunction of a second oscillatory kernel on $I^*\equiv[0,1)$, and the $M_i(x)$, and hence $\phi_i''(x)$ have the properties 1) - 4) of Theorem 8.8.3. We now extend Theorem 8.8.3 to the interval (0,1] and to the derivative $\phi_i'(x)$.

Theorem 10.3.1. If $\{\phi_i(x)\}_o^\infty$ are the eigenfunctions of the cantilever beam, then

1) $\phi_o(x)$, $\phi_o'(x)$ have no zeros in (0,1].

2) $\phi_i(x)$, $\phi_i'(x)$ have i nodes and no other zeros in (0,1].

3) If

$$\phi(x)=\sum_{i=k}^{m} c_i\phi_i(x) \ , \ 0\le k\le m \ , \ \sum_{i=k}^{m} c_i^2>0,$$

then $\phi(x)$ has not less than k nodes and not more than m zeros in (0,1].

Proof:

1) Suppose $\phi_o'(x_o)=0$ for some x_o in (0,1]. Since $\phi_o'(0)=0$, Rolle's theorem states that there is a ξ in $(0,x_o)$ such that $\phi_o''(\xi)=0$, which contradicts Theorem 8.8.3. for $\phi_o''(x)$. Thus $\phi_o'(x)\neq 0$ for all x in (0,1] and thus, since $\phi_o(0)=0$ we may deduce $\phi_o(1)\neq 0$. This extends part 1) of Theorem 8.8.3 to (0,1].

2) $\phi_i(x)$ $(i\ge 1)$ has i nodes $x_{i,j}$ $(j=2,3,...i+1)$ in $(0,\ell)$ and a zero at $x_{i,1}\equiv 0$. By Rolle's theorem $\phi_i'(x)$ has i zeros $\xi_{i,j}$ $(j=1,2,...,i)$ satisfying $x_{i,j}<\xi_{i,j}<x_{i,j+1}$. Since $\phi_i'(0)=0$, $\phi_i'(x)$ can have no other zeros in [0,1], since then $\phi_i''(x)$ would have more than i zeros in (0,1), contrary to 2) of Theorem 8.8.3 for $\phi_i''(x)$. We conclude in particular that $\phi_i(1)\neq 0$, $\phi_i'(1)\neq 0$.

Part 3) may be proved similarly. ■

Theorem 10.3.2. If $\phi_i(x)$ is an eigenfunction of a cantilever beam, then

$$\phi_i(1)\phi_i'(1)>0 \ .$$

Proof:

Theorem 10.3.1 shows that $\phi_i(1)\phi_i'(1)\neq 0$: we show that $\phi_i(1)$, $\phi_i'(1)$ have the same sign. Chose $\phi_i(x)$ so that $\phi_i(1)>0$. Differentiation of (10.1.27) gives

$$[\phi_i'(x)]_x^1=\lambda_i\int_x^1 \frac{ds}{r(s)}\int_x^{\ell}(t-s)a(t)\phi_i(t)dt \ . \tag{10.3.5}$$

Suppose x^* is the largest zero if $\phi_i(x)$; it will be a node if $i\ge 1$ and 0 if $i=0$. Since $\phi_i(1)>0$, we have $\phi_i'(x^*)>0$ and $\phi_i(x)>0$ for $x^*<x<1$. Thus equation (10.3.5) yields

$$[\phi_i'(x)]_{x^*}^1>0$$

so that $\phi_i'(1)>\phi_i'(x^*)>0$. ■

We are now in a position to start on the proof of the fundamental

Theorem 10.3.3. Suppose $a(x)$, $r(x)$ have derivatives of all orders. (This restriction could be relaxed, but the stated result is sufficient for our purposes), then the infinite matrix

$$\mathbf{P}=\begin{bmatrix} u_o & u_1 & u_2 & \cdots \\ \theta_o & \theta_1 & \theta_2 & \cdots \\ \lambda_o u_o & \lambda_1 u_1 & \lambda_2 u_2 & \cdots \\ \lambda_o \theta_o & \lambda_1 \theta_1 & \lambda_2 \theta_2 & \cdots \\ \lambda_o^2 u_o & \lambda_1^2 u_1 & \lambda_2^2 u_2 & \cdots \\ \cdot & \cdot & \cdot & \end{bmatrix}$$

is completely positive, i.e., all its minors are positive. Here $u_1 \equiv \phi_i(1)$, $\theta_i = \phi_i'(1)$, $\lambda_i = \omega_i^2$. The proof is long, and proceeds by way of a number of Lemmas, and may be omitted on a first reading.

Proof:

Because of the repetitive nature of the rows of **P**, the Corollary to Theorem 5.2.4 implies that a necessary and sufficient condition for **P** to be completely positive is that the minors

$$P\begin{pmatrix} 1, & 2 & ,\ldots, & p \\ i+1, & i+2 & ,\ldots, & i+p \end{pmatrix} \equiv \Phi(i,i+1,\ldots,i+p-1)\ ,$$

$$P\begin{pmatrix} 2, & 3 & ,\ldots, & p+1 \\ i+1, & i+2 & ,\ldots, & i+p \end{pmatrix} \equiv \Psi(i,i+1,\ldots,i+p-1)\ , \tag{10.3.6}$$

taken from rows $1,2,\ldots,p$, columns $i+1, i+2,\ldots i+p$, and from rows $2,3,\ldots p+1$, columns $i+1, i+2, \ldots i+p$, respectively, should be positive for all $i=0,1,2,\ldots$ and $p=1,2,\ldots$. The case $p=1$ has been proved; consider $p=2$. We prove

Lemma 1. If the eigenfunctions $\phi_i(x)$ are chosen so that $\phi_i(1)>0$, then there is an interval $(\xi,1)$ where $0<\xi<1$ (ξ depends on i), in which

$$\phi(x) \equiv \Phi\begin{pmatrix} x, & 1 \\ i, & i+1 \end{pmatrix} \equiv \begin{vmatrix} \phi_i(x) & \phi_{i+1}(x) \\ \phi_i(1) & \phi_{i+1}(1) \end{vmatrix} > 0\ . \tag{10.3.7}$$

Proof:

Part 3 of Theorem 2 shows that $\phi(x)$ has not less than i nodes in $(0,1)$ and not more than $i+1$ zeros in $(0,1]$. Since $\phi(1)=0$, $\phi(x)$ has just the i nodal zeros in $(0,1)$. Since $\phi_i(1)>0$, we have $(-)^i\phi_i(0+)>0$; similarly $(-)^{i+1}\phi_{i+1}(0+)>0$, and therefore $(-)^i\phi_i(0+)\phi_{i+1}(1) + (-)^{i+1}\phi_{i+1}(0+)\phi_i(1)>0$, so that $\phi(x)>0$ to the right of its last node, ξ. ■

Corollary 1. Since the $\phi_i'(x)$ have property 3) of Theorem 10.3.2 and $\phi_i'(1)>0$, there is an interval $(\eta,1)$ such that $0<\eta<1$ in which

$$\psi(x) \equiv \Psi\begin{pmatrix} x, & 1 \\ i, & i+1 \end{pmatrix} = \begin{vmatrix} \phi_i'(x) & \phi_{i+1}'(x) \\ \phi_i'(1) & \phi_{i+1}'(1) \end{vmatrix} > 0 . \tag{10.3.8}$$

Corollary 2. Since $\phi(1)=0=\phi''(1)=\phi'''(1)$, we have

$$\phi(1-x) = -x\phi'(1) + \frac{x^4}{4!}\phi^{iv}(1) + \ldots \tag{10.3.9}$$

for small x. Now

$$\phi^{iv}(1) = \frac{a(1)}{r(1)} \begin{vmatrix} \lambda_i\phi_i(1) & \lambda_{i+1}\phi_{i+1}(1) \\ \phi_i(1) & \phi_{i+1}(1) \end{vmatrix} < 0 \tag{10.3.10}$$

so that $\phi(1-x)>0$ for small x implies

$$-\phi'(1) = \begin{vmatrix} u_i & u_{i+1} \\ \theta_i & \theta_{i+1} \end{vmatrix} = \Phi(i,i+1) > 0 . \tag{10.3.11}$$

Because $\phi^{iv}(1)<0$, there cannot be equality in (10.3.11).

Corollary 3. Suppose $r(x)$ has a third derivative at $x=1$ then, since $\psi(1)=0=\psi'(1)=\psi''(1)$, we have

$$\psi(1-x) = -\frac{x^3}{3!}\psi'''(1) + \frac{x^4}{4!}\psi^{iv}(1) + \cdots \tag{10.3.12}$$

Again $\psi^{iv}<0$, so that $\psi(1-x)>0$ for small x implies

$$-\psi'''(1) = \frac{a(1)}{r(1)} \begin{vmatrix} \theta_i & \theta_{i+1} \\ \lambda_i u_i & \lambda_{i+1}u_{i+1} \end{vmatrix} = \Psi(i,i+1) > 0 . \tag{10.3.13}$$

We now proceed to the proof of the inequalities (10.3.6). The proof is inductive. Suppose the weak (≥ 0) inequalities hold for $p=2n$. Consider the case $p=2n+1$ and let

$$\phi(x) = \begin{vmatrix} \phi_i(x) & \phi_{i+1}(x) & \cdots & \phi_{i+2n}(x) \\ u_i & u_{i+1} & \cdots & u_{i+2n} \\ \theta_i & \theta_{i+1} & \cdots & \theta_{i+2n} \\ & & & \\ \lambda_i^{n-1}\theta_i & \lambda_{i+1}^{n-1}\theta_{i+1} & \cdots & \lambda_{i+2n}^{n-1}\theta_{i+2n} \end{vmatrix} . \tag{10.3.14}$$

Just as in Corollary 1, we may show that $\phi(1-x)\geq 0$ for small x. But $\phi^{(s)}(1)=0$ for $s=0,1,\ldots 4n-1$, so that for small x

$$\phi(1-x) = \frac{x^{4n}}{(4n)!}\phi^{(4n)}(1) - \frac{x^{4n+1}}{(4n+1)!}\phi^{(4n+1)}(1) + \cdots$$

from which we may deduce that $\phi^{(4n)}(1)\geq 0$. But, after an even number of row interchanges, $\phi^{(4n)}(1)$ may be written

$$\phi^{(4n)}(1)=(a(\ell)/r(\ell))^n\Phi(i,i+1,...i+2n)$$

so that $\Phi(i,i+1,...i+2n)\geq 0$. The inequality $\Psi(i,i+1,...1+2n)\geq 0$ may be proved in a similar way, as may the step from $2n+1$ to $2n+2$. This completes the inductive step for the weak (≥ 0) inequalities. We now show that if there is equality in some $\Phi\geq 0$, $\Psi\geq 0$ for a given p, then at least one Φ or Ψ for the next value of p, i.e., $p+1$ will be negative. Since we have already proved that all Φ, Ψ are positive when $p=2$ we shall concentrate on the step from 3 to 4, but it will be clear that the argument which we use is general.

We use a particular case of Sylvester's theorem on bordered determinants; this states that if $\mathbf{A}$ is an $n\times n$ matrix and b_{ik} is defined as the minor taken from rows $i_1,i_2,...i_p,i$ and columns $k_1,k_2,...k_p,k$, i.e.,

$$b_{ik}=A\begin{pmatrix} i_1,i_2,...,i_p,i \\ k_1,k_2,...,k_p,k \end{pmatrix} \tag{10.3.15}$$

then

$$B\begin{pmatrix} \ell_1 & \ell_2 \\ m_1 & m_2 \end{pmatrix}=A\begin{pmatrix} i_1,i_2,...,i_p \\ k_1,k_2,...,k_p \end{pmatrix}A\begin{pmatrix} i_1,i_2,...,i_p,\ell_1,\ell_2 \\ k_1,k_2,...,k_p,m_1,m_2 \end{pmatrix} \tag{10.3.16}$$

To illustrate this, take

$$\mathbf{A}=\begin{bmatrix} u_i & u_j & u_k \\ \theta_i & \theta_j & \theta_k \\ \lambda_i u_i & \lambda_j u_j & \theta_k u_k \end{bmatrix} \quad i<j<k \ . \tag{10.3.17}$$

If $p=1$, $i_1=2=k_1$, $\ell_1=1=m_1$, $\ell_2=3=m_2$, then

$$b_{11}=A\begin{pmatrix} 2 & 1 \\ 2 & 1 \end{pmatrix}=A\begin{pmatrix} 1 & 2 \\ 1, & 2 \end{pmatrix}=\Phi(i,j)$$

$$b_{13}=A\begin{pmatrix} 2 & 1 \\ 2 & 3 \end{pmatrix}=-A\begin{pmatrix} 1 & 2 \\ 2 & 3 \end{pmatrix}=-\Phi(j,k) \tag{10.3.18}$$

$$b_{31}=-\Psi(i,j) \ , \ b_{33}=\Psi(j,k)$$

Thus equation (10.3.16) becomes

$$B\begin{pmatrix} 1 & 3 \\ 1 & 3 \end{pmatrix}=\begin{vmatrix} b_{11} & b_{13} \\ b_{31} & b_{33} \end{vmatrix}=\theta_j A\begin{pmatrix} 2 & 1 & 3 \\ 2 & 1 & 3 \end{pmatrix}=\theta_j\Phi(i,j,k) \ , \tag{10.3.19}$$

or

$$\theta_j\Phi(i,j,k)=\Phi(i,j)\Psi(j,k)-\Phi(j,k)\Psi(i,j) \ . \tag{10.3.20}$$

We may show similarly that

$$u_j\Psi(i,j,k)=\lambda_k\Phi(j,k)\Psi(i,j)-\lambda_i\Phi(i,j)\Psi(j,k)\,. \tag{10.3.21}$$

We could write down the general relation linking the determinants of order p and $p+1$, but it is more instructive to display those for $p=3$:

$$\Psi(j,k)\Phi(i,j,k,m)=\Phi(i,j,k)\Psi(j,k,m)-\Phi(j,k,m)\Psi(i,j,k)\,, \tag{10.3.22}$$

$$\Phi(j,k)\Psi(i,j,k,m)=\lambda_m\Phi(j,k,m)\Psi(i,j,k)-\lambda_i\Phi(i,j,k)\Psi(j,k,m)\,. \tag{10.3.23}$$

We also use equation (5.2.17), which yields

$$\Phi(j,k)\Phi(i,j,m)=\Phi(i,j)\Phi(j,k,m)+\Phi(j,m)\Phi(i,j,k)\,. \tag{10.3.24}$$

We shall now show that the proven inequalities

$$\Phi(i,j,k,m)\geq 0\ \ ,\ \ \Psi(i,j,k,m)\geq 0\,, \tag{10.3.25}$$

can be satisfied for all i,j,k,m such that $i<j<k<m$ only if

$$\Phi(i,j,k)>0\ \ ,\ \ \Psi(i,j,k)>0\,, \tag{10.3.26}$$

for all i,j,k such that $i<j<k$. First we note that *if* there are integers i_o,j_o $(i_o<j_o)$ such that $\Phi(i_o,j_o,k)=0$ for *all* k, then repeated use of equation (10.3.24) shows that $\Phi(i,j,k)=0$ for *all* i,j,k. But then equations (10.3.22)-(10.3.23) show that $\Phi(i,j,k,m)=0=\Psi(i,j,k,m)$ for all i,j,k,m. But this means that all Φ (and Ψ) are zero for $p\geq 4$, and this implies that there is a linear combination

$$\chi(x)=\sum_{i=o}^{\infty}a_i\phi_i(x) \tag{10.3.27}$$

which, together with its derivations of all orders, is zero at $x=1$. This can happen only if $\chi(x)$ is identically zero, in contradiction to the orthogonality and consequent linear independence of the $\{\phi_i(x)\}_o^\infty$. We conclude that $\Phi(i_o,j_o,k)$ cannot be zero for all k.

For given i_o,j_o $(i_o<j_o)$, the indices k such that $k\neq i_o$, $k\neq j_o$ may be arranged in the sequence

$$0,1,\ldots\quad i_o-1,i_o+1,\ldots\quad j_o-1,j_o+1,\ldots\,.$$

Let k_R be the right hand neighbour of k in this sequence. Unless all $\Phi(i_o,j_ok)\neq 0$ (in which case $\Phi(i_o,j_o,k)>0$ for all $k>j_o$, as is to be proved) one or other of the sets S,T defined by

$$S=\{k;\Phi(i_o,j_o,k)=0\ ,\ \Phi(i_o,j_o,k_R)\neq 0\} \tag{10.3.28}$$

$$T=\{k;\Phi(i_o,j_o,k)\neq 0\ ,\ \Phi(i_o,j_o,k_R)=0\} \tag{10.3.29}$$

is non-empty. Suppose S is non-empty, then it has a least member k_o such that

$$\Phi(i_o,j_o,k_o)=0 \quad , \quad \Phi(i_o,j_o,k_R)\begin{cases} >0 & k_R<i_o \\ <0 & i_o<k_R<j_o \\ >0 & k_R>j_o \end{cases} , \tag{10.3.30}$$

where $k_R=(k_o)_R$. Equations (10.3.20), (10.3.21) show that $\Phi(i,j,k)$ and $\Psi(i,j,k)$ cannot be simultaneously zero when i,j,k are all different. Therefore

$$\Psi(i_o,j_o,k_o)\begin{cases} >0 & k_o<i_o \\ <0 & i_o<k_o<j_o \\ >0 & k_o>j_o \end{cases} , \tag{10.3.31}$$

so that on putting $m=k_R$ in equation (10.3.22) we find

$$\Psi(j_o,k_o)\Phi(i_o,j_o,k_o,k_R)=-\Phi(j_o,k_o,k_R)\Psi(i_o,j_o,k_o) . \tag{10.3.32}$$

But since $\Phi(i_o,j_o,k_o)=0$, equation (10.3.24) yields

$$\Phi(j_o,k_o)\Phi(i_o,j_o,k_R)=\Phi(i_o,j_o)\Phi(j_o,k_o,k_R) \tag{10.3.33}$$

which when combined with (10.3.32) gives

$$\Phi(i_o,j_o)\Psi(j_o,k_o)\Phi(i_o,j_o,k_o,k_R)=-\Phi(j_o,k_o)\Phi(i_o,j_o,k_R)\Psi(i_o,j_o,k_o) . \tag{10.3.34}$$

when $i_o<j_o<k_o<k_R$, then (10.3.30) and (10.3.31) show that $\Phi(i_o,j_o,k_R)$ and $\Psi(i_o,j_o,k_o)$ are positive, so that $\Phi(i_o,j_o,k_o,k_R)<0$, contradicting the proven inequality (10.3.25). In all other cases, $\Phi(i_o,j_o)>0$, and $\Phi(j_o,k_o)$, $\Psi(j_o,k_o)$ are either both positive (if $j_o<k_o$) or both negative (if $j_o>k_o$), and it may readily be verified that when i_o,j_o,k_o,k_R are rearranged in ascending order, i.e., $i_o,j_o,k_o,k_R \rightarrow i',j',k',m'$ $(i'<j'<k'<m')$, then again $\Phi(i',j',k',m')<0$, contradicting (10.3.25). We conclude that S is empty. If T is non-empty then, by applying an exactly similar argument to equation (10.3.23) we may show that $\Psi(i',j',k',m')<0$ for some $i'<j'<k'<m'$ in contradiction to (10.3.25). The whole of this argument may be modified for the general case. ■

10.4 The spectra of the beam

Suppose that the beam of equation (10.1.1), specified by length ℓ, cross-section $A(x)$ and second moment of area $I(x)$ is transformed into one with length ℓ^*, cross-section $A^*(x^*)$, second moment $I^*(x^*)$, where

$$x^*=\gamma x \ , \ I^*(x^*)=\alpha I(x) \ , \ A^*(x^*)=\beta A(x) \ , \ \ell^*=\gamma\ell , \tag{10.4.1}$$

then the spectra of the new beam under any combination of the end-conditions (10.1.5)-(10.1.8) will be the same as those of the original beam provided that

$$\gamma^4=\alpha/\beta . \tag{10.4.2}$$

With this relationship, equation (10.4.1) defines a two-parameter family of isospectral beams. Now consider a beam clamped at $x=0$ and acted on by a concentrated *static* load F and bending moment M at $x=\ell$. The deflection is given by

$$\frac{d^2}{dx^2}\left(I(x)\frac{d^2u}{dx^2}\right)=0\,. \qquad (10.4.3)$$

subject to

$$u(0)=0=u'(0)\,, \qquad (10.4.4)$$

$$\frac{d}{dx}\left(I(x)\frac{d^2u}{dx^2}\right)_{x=\ell}=-F\,,\quad I(x)\frac{d^2u}{dx^2}\bigg|_{x=\ell}=M\,, \qquad (10.4.5)$$

so that

$$I(x)\frac{d^2u}{dx^2}=F(\ell-x)+M\,, \qquad (10.4.6)$$

$$u(x)=F\int_o^x\frac{(x-s)(\ell-s)ds}{I(s)}+M\int_o^x\frac{(x-s)ds}{I(s)}\,, \qquad (10.4.7)$$

and the end displacement $u(\ell)$ and slope are given by

$$u(\ell)=G_2F+G_1M\,,\quad u'(\ell)=G_1F+G_oM\,. \qquad (10.4.8)$$

where the receptances are

$$G_i=\int_o^\ell\frac{(\ell-s)^i\,ds}{I(s)}\,,\quad i=0,1,2. \qquad (10.4.9)$$

For the transformed beam, the receptances will be

$$G_i^*=\frac{\gamma^{i+1}}{\alpha}G_1\,,\quad i=0,1,2,\,. \qquad (10.4.10)$$

We conclude that equation (10.4.2) and *any two* of the three equations

$$\ell^*=\ell\,,\quad G_i^*=G_i\,,\quad i=0,1,2, \qquad (10.4.11)$$

demand that

$$\alpha=1=\beta\,, \qquad (10.4.12)$$

so that the beams will be identical.

We shall now use the results of Section 10.3 to order the eigenvalues for a beam clamped at $x=0$ and subject to different end conditions at $x=1$. (We shall work with the dimensionless equation (10.1.3)). Consider the variational problem of finding the stationary values of the functional

$$J(u)=\frac{1}{2}\int_o^1 r(x)[u''(x)]^2dx-\frac{\lambda}{2}\int_o^1 a(x)u^2(x)dx-Fu(1)-Mu'(1), \qquad (10.4.13)$$

for $u(x)$ satisfying the clamped condition at $x=0$, i.e.,

$$u(0)=0=u'(0)\ . \tag{10.4.14}$$

The first variation of $J(u)$, computed in the usual way (Courant and Hilbert, Chapter IV) after an integration by parts is

$$\begin{aligned}\delta J &= \int_o^1 \{(ru'')''-\lambda au\}\delta u dx+[r(1)u''(1)-M]\delta u'(1)\\ &- [(r(x)u''(x))_1'+F]\delta u(1)\ ,\end{aligned} \tag{10.4.15}$$

so that the displacement which makes J stationary satisfies equation (10.1.3) and the end conditions

$$r(1)u''(1)=M\ ,\quad [ru''(x)]_1'=-F\ , \tag{10.4.16}$$

i.e., it is the displacement of the cantilever due to a concentrated force F and moment M applied at $x=1$. The eigenfunctions $\{\phi_i(x)\}_o^\infty$ of the cantilever are complete on (0,1). Therefore write

$$u(x)=\sum_{i=o}^{\infty}\alpha_i\phi_i(x)\ , \tag{10.4.17}$$

and use Ex. 10.1.1 to obtain

$$J(u)=\frac{1}{2}\sum_{i=o}^{\infty}\lambda_i\alpha_i^2-\frac{\lambda}{2}\sum_{i=o}^{\infty}\alpha_i^2-\sum_{i=o}^{\infty}\alpha_i\{Fu_i+M\theta_i\}\ , \tag{10.4.18}$$

where

$$\rho_i=\int_o^1 a(x)[\phi_i(x)]^2dx=1\ . \tag{10.4.19}$$

This is stationary if

$$(\lambda_i-\lambda)\alpha_i=Fu_i+M\theta_i\ , \tag{10.4.20}$$

i.e.,

$$u(x)=\sum_{i=o}^{\infty}\frac{\{Fu_i+M\theta_i\}}{(\lambda_i-\lambda)}\phi_i(x)\ . \tag{10.4.21}$$

When written in the form

$$u(x)=\alpha_{x1}F+\alpha_{x1'}M\ , \tag{10.4.22}$$

this equation yields the end receptances α_{x1}, $\alpha_{x1'}$, of the beam (Bishop and Johnson (1960), p. 336).

The eigenvalues of the clamped-pinned beam are the values of ω^2 for which the application of an end force F alone (i.e., $M=0$) produces no end displacement, i.e., $u(1)=0$. They are thus the roots of the equation

$$\sum_{i=o}^{\infty} \frac{u_i^2}{(\lambda_i - \lambda)} = 0\ . \tag{10.4.23}$$

We will denote the roots by $(\mu_i)_o^\infty$. Since $u_i \neq 0$, they satisfy

$$\lambda_i < \mu_i < \lambda_{i+1}\ . \tag{10.4.24}$$

Similarly, the eigenvalues $(\sigma_i)_o^\infty$ of the clamped-sliding beam are the values of ω^2 for which M alone ($F=0$) produces no end slope, i.e., $u'(1)=0$, and

$$\sum_{i=o}^{\infty} \frac{\theta_i^2}{(\lambda_i - \lambda)} = 0 \tag{10.4.25}$$

and, since $\theta_i \neq 0$, they satisfy

$$\lambda_i < \sigma_i < \lambda_{i+1}\ . \tag{10.4.26}$$

The anti-resonant eigenvalues $(\nu_i)_o^\infty$ at which F alone produces no end slope, or equivalently M alone produces no end displacement, are the roots of

$$\sum_{i=o}^{\infty} \frac{u_i \theta_i}{(\lambda_i - \lambda)} = 0\ , \tag{10.4.27}$$

and since $u_i\theta_i > 0$, they satisfy

$$\lambda_i < \nu_i < \lambda_{i+1}\ . \tag{10.4.28}$$

Now Corollary 1, i.e.,

$$u_i\theta_j - \theta_i u_j > 0\ ,\quad i < j\ , \tag{10.4.29}$$

and Theorem 7.4.2, which can clearly be extended to (convergent) infinite sums, yields

$$\sigma_i < \nu_i < \mu_i\ , \tag{10.4.30}$$

as in Section 7.4. The argument based on constraint that is used there shows that the clamped-clamped eigenvalues $(\lambda_i)_o^\infty$ satisfy

$$\sigma_i < \lambda_i < \sigma_{i+1}\ ,\quad \mu_i < \lambda_i < \mu_{i+1}\ . \tag{10.4.31}$$

Thus putting these inequalities together we deduce that

$$\lambda_o < \sigma_o < \nu_o < \mu_o < (\lambda_1, \gamma_o) < \sigma_1 < \nu_1 < \mu_1 < (\lambda_2, \gamma_1) < \cdots \tag{10.4.32}$$

In order to find the asymptotic form of the eigenvalues corresponding to different end conditions we use the W.K.B. approach, for example Carrier, Krook and Pearson (1966), p.291. The following analysis can be made rigorous.

First make the change in the independent variable

$$s=\int_x^1\left(\frac{a(t)}{r(t)}\right)^{1/4}dt \tag{10.4.33}$$

and write

$$b(s)=\left(\frac{a(x)}{r(x)}\right)^{1/4}, \quad c^2(s)=(r^3(x)a(x))^{1/4}, \tag{10.4.34}$$

then

$$\frac{d}{dx}=\frac{d}{ds}\cdot\frac{ds}{dx}=-\left(\frac{a(x)}{r(x)}\right)^{1/4}\frac{d}{ds}=-b(s)\frac{d}{ds}, \tag{10.4.35}$$

and

$$r(x)\frac{d^2}{dx^2}=r(x)b(s)\frac{d}{ds}\left(b(s)\frac{d}{ds}\right)=c^2(s)\frac{d}{ds}\left(b(s)\frac{d}{ds}\right). \tag{10.4.36}$$

Thus equation (10.3.1) becomes

$$b\frac{d}{ds}\left(b\frac{d}{ds}\left(c^2\frac{d}{ds}\left(b\frac{du}{ds}\right)\right)\right)=\lambda b^3c^2u \tag{10.4.37}$$

since $a^3(x)=b^3(s)c^2(s)$. Thus, putting $'\equiv d/ds$ we find

$$(b(c^2(bu')')')'=\lambda b^2c^2u, \quad 0\le s\le L, \tag{10.4.38}$$

where

$$L=\int_o^1\left(\frac{a(t)}{r(t)}\right)^{1/4}dt. \tag{10.4.39}$$

The transformation (10.4.33) has the effect of reversing the ends. Thus the end conditions (10.3.2) become

$$(bu')'(0)=0=(c^2(bu'))'(0), \tag{10.4.40}$$

$$u(L)=0=u'(L). \tag{10.4.41}$$

Now suppose that λ is large positive, put $\lambda=z^4$ and expand the left-hand side of equation (10.4.38) to give

$$p^2(s)u^{iv}(s)+2p(s)p'(s)u'''(s)+f_1(s)u''(s)+f_2(s)u'(s)=z^4p^2(s)u(s), \tag{10.4.42}$$

where

$$p(s)=b(s)c(s), \quad f_1(s)=(p^2(s))''+b(s)c^2(s)b''(s)+2b(s)c(s)b'(s)c'(s) \tag{10.4.43}$$

$$f_2(s)=(b(s)(c^2(s)b'(s))')' \tag{10.4.44}$$

For large z it will be the first two terms in (10.4.42) that will be dominant.

We now look for a solution of equation (10.4.42) having the form

$$u(s)=\exp(\int(z\phi_o(s)+\phi_1(s))ds)\,. \tag{10.4.45}$$

After substituting this into (10.4.42) and retaining only the terms involving z^4 and z^3 we find

$$\phi_o^4(s)=1\,,\;\; p^2(s)\{6\phi_o^2(s)\phi_o'(s)+4\phi_o^3(s)\phi_1(s)\}+2p(s)p'(s)\phi_o^3(s)=0\,, \tag{10.4.46}$$

so that $\phi_o(s)=\pm 1,\ \pm i$ and

$$\phi_1(s)=-p'(s)/[2p(s)]\,,\quad \exp\,[\phi_1(s)]=p^{-1/2}(s)\,. \tag{10.4.47}$$

Thus there are four solutions corresponding to the four values of $\phi_o(s)$, and we may write

$$u(s)=p^{-1/2}(s)\{A\cos zs+B\sin zs+C\cosh zs+B\sinh zs\}\,. \tag{10.4.48}$$

Apart from the factor $p^{-1/2}(s)$, this has exactly the same form as the deflection of a uniform beam, so that the eigenvalue equation will be the same as that for a uniform beam of equivalent length L. Thus for the cantilever, the four end conditions (10.4.40), (10.4.41) will yield the eigenvalue equation (Bishop and Johnson (1960), p. 382)

$$\cos zL\cosh zL+1=0 \tag{10.4.49}$$

so that

$$\cos zL=-sech zL\simeq-2\exp(-zL) \tag{10.4.50}$$

and

$$z_r L\simeq(2r+1)\pi/2\,. \tag{10.4.51}$$

or

$$\lambda_r\simeq(2r+1)^4\pi^4/(16L^4)\,. \tag{10.4.52}$$

In a similar way we find

$$\begin{aligned}
&\mu_r\simeq(4r+5)^4\pi^4/(256L^4)\\
&\nu_r\simeq(r+1)^4/L^4\\
&\sigma_r\simeq(4r+3)^4\pi^4/(256L^4)\\
&\gamma_r\simeq(2r+3)^4\pi^4/(16L^4)\,.
\end{aligned} \tag{10.4.53}$$

and we note that these do obey the interlacing condition (10.4.32). Note also that, taking account of the change of notation and numbering, the values of μ_r, ν_r, γ_r agree with those given by Barcilon (1982); his values of σ_r^2, ω_r^2 (our σ_{r-1}, λ_{r-1}) are incorrect.

10.5 Statement of the inverse problem

Inverse problems for the vibrating Euler-Bernoulli beam seem to have been studied first by Niordson (1967). He was not concerned with the reconstruction of a unique beam from sufficient data, in the sense to be described below. Rather, he was concerned with constructing a beam in a class having n arbitrary parameters so that it would have n specified eigenvalues, which would be perturbations of the eigenvalues of the uniform cantilever beam.

The proper study of the inverse problem for the vibrating Euler-Bernoulli beam began with the work of Barcilon. He realised that there are three questions to be answered. First, what spectral (and other) data are required to determine the properties (cross-sectional area $A(x)$, second moment of area $I(x)$) of the beam? In Barcilon (1974b, 1974c) he showed that three spectra, corresponding to three different end conditions, are required. Secondly, what are the necessary and sufficient conditions on the data to ensure that the beam properties will be realistic, i.e., $A(x)>0$, $I(x)>0$. Barcilon battled with this question in Barcilon (1982), but it was not fully answered until Gladwell (1986). Thirdly, how can the beam be reconstructed? Barcilon (1976) answered this question for the case in which the spectra were small perturbations on those for the uniform beam, but a proper reconstruction procedure was not available until McLaughlin (1984b).

As a result of the analysis described above we may state that there is only a two-parameter family of beams, of the type described in Section 10.3, which have three given spectra $\{\lambda_i,\mu_i,\sigma_i\}_o^\infty$ (or $(\lambda_i,\mu_i,\nu_i)_o^\infty$ or $(\lambda_i,\nu_i,\sigma_i)_o^\infty$). The particular member in the family may be found as in (10.4.11). The spectra $\{\lambda_i,\mu_i,\sigma_i\}_o^\infty$ will have to satisfy certain conditions, amongst which will be some asymptotic ones. The argument of Section 10.4 shows that to be given $\{\lambda_i,\mu_i,\sigma_i\}_o^\infty$ and some appropriate asymptotic conditions is equivalent to being given $\{\lambda_i,u_i,\theta_i\}_o^\infty$ and some other asymptotic conditions. We can, and shall, circumvent the asymptotic conditions altogether by supposing – and this is consistent with the way in which the problem will be posed in practice – that only $(\lambda_i,u_i,\theta_i)_o^n$ are given, while the remainder are chosen so that

$$(\lambda_i,u_i,\theta_i)_{n+1}^\infty=(\lambda_i^o,u_i^o,\theta_i^o)_{n+1}^\infty \tag{10.5.1}$$

where the ^{o} quantities relate to a known beam which, without loss of generality, may be taken to be a uniform beam.

In this case equation (10.4.24), for example, may be written

$$\sum_{i=o}^{n}\frac{u_i^2}{\lambda_i-\lambda}-\sum_{i=o}^{n}\frac{u_i^{o2}}{\lambda_i^o-\lambda}+\sum_{i=o}^{\infty}\frac{u_i^{o2}}{\lambda_i^o-\lambda}=0\,. \tag{10.5.2}$$

The infinite sum is the receptance of the uniform beam and may be expressed in a closed form, in fact (Bishop and Johnson (1960)) if $u_i^o=1$, then

$$\sum_{i=o}^{\infty}\frac{u_i^{o2}}{\lambda_i-\lambda}=\frac{\cos\phi\sinh\phi-\sin\phi\cosh\phi}{4\phi^3(\cos\phi\cosh\phi+1)}\ ,\ \phi=\lambda^{1/4}\ . \tag{10.5.3}$$

Thus the statement that equation (10.5.2) is satisfied by $\lambda=(\mu_i)_o^n$ yields $n+1$ linear simultaneous equations for $(u_i^2)_o^n$. The first set of necessary conditions is therefore

1) $\mu_j\neq\lambda_i$, $\mu_j\neq\lambda_i^o$ for $i,j=0,1,\ldots$. This ensures that the matrix of coefficients in the equations for $(u_i^2)_o^n$ is non-singular, and the right hand sides are well-defined. The second condition is

2) the solution $(u_i^2)_o^n$ must be positive.

Similarly, if $(\theta_i^2)_o^n$ are to be determined from euation (10.4.26) then we need

3) $\sigma_j\neq\lambda_i$, $\sigma_j\neq\lambda_i^o$ for all $i,j=0,1,\ldots$.

4) the solution $(\theta_i^2)_o^n$ must be positive.

Provided that these conditions are satisfied, i.e., $(u_i,\theta_i)_o^n$ may be found, then we shall show that the complete positivity of the matrix **P** of Theorem 10.3.3, which has been shown to be necessary, is also a sufficient condition for the construction of the unique beam.

10.6 The reconstruction procedure

The procedure is basically the same as the described in Section 9.6 for the vibrating rod, and is due to McLaughlin (1976, 1978, 1984a, 1984b). We suppose that we wish to reconstruct a cantilever beam, i.e., functions $r(x)$, $a(x)$, such that the equation

$$(r(x)u''(x))''=\lambda a(x)u(x)\ , \tag{10.6.1}$$

subject to the conditions

$$u(0)=0=u'(0) \tag{10.6.2}$$

$$u''(1)=0=u'''(1) \tag{10.6.3}$$

has specified eigenvalues $(\lambda_i)_o^\infty\equiv(\omega_i^2)_o^\infty$, and has eigenvalues $\{\phi_i(x)\}_o^\infty$, normalized w.r.t. $a(x)$, i.e., such that

$$\int_o^1 a(x)\phi_i(x)\phi_j(x)dx=\delta_{ij}\ ,\ i,j=0,1,2,\ldots\ , \tag{10.6.4}$$

which have specified values of $\{\phi_i(1),\phi_i'(1)\}_o^\infty$.

First make the change in the independent variable used earlier in Section 10.3, namely

$$s=\int_x^1\left(\frac{a(t)}{r(t)}\right)^{1/4}dt \tag{10.6.5}$$

and write

$$b^2(s)=\left(\frac{a(x)}{r(x)}\right)^{1/4}, \quad c^2(s)=(r^3(x)a(x))^{1/4}, \tag{10.6.6}$$

$$p(s)=b(s)c(s), \tag{10.6.7}$$

then equation (10.6.1) becomes

$$(b(c^2(bu')')')'-\lambda b^2c^2u=0, \quad 0\le s\le L, \tag{10.6.8}$$

where $'\equiv d/ds$ and

$$L=\int_o^1\left(\frac{a(t)}{r(t)}\right)^{1/4}dt, \tag{10.6.9}$$

while the end conditions (10.6.2), (10.6.3) become

$$(bu')'(0)=0=(c^2(bu')')'0, \tag{10.6.10}$$

$$u(L)=0=u'(L). \tag{10.6.11}$$

Without loss of generality, we assume that $b(0)=1=c(0)$.

Just as with the Sturm-Liouville reconstruction, we introduce an auxiliary problem. This is

$$(b_o(c_o^2(b_ov')')')'-\lambda b_o^2c_o^2v=0, \quad 0\le s\le L, \tag{10.6.12}$$

$$(b_ov')'(0)=0=(c_o^2(b_ov')')'(0), \tag{10.6.13}$$

$$v(L)=0=v'(L), \tag{10.6.14}$$

where $b_o(s)$, $c_o(s)$ are known (e.g., $b_o(s)\equiv 1\equiv c_o(s)$), $b_o(0)=1=c_o(1)$, and $p_o(s)=b_o(s)c_o(s)$. This auxiliary problem has a certain set of eigenvalues $(\lambda_i^o)_o^\infty$, and its eigenfunctions $v_i^o(s)\equiv\phi_i^o(s)$, normalized so that

$$\int_o^L p_o^2(s)\phi_i^o(s)\phi_j^o(s)ds=\delta_{ij}, \quad i,j=0,1,2,... \tag{10.6.15}$$

will have end values and derivatives $\{\phi_i^o(0),\phi_i^{o'}(0)\}_o^\infty$

For given values of λ, ξ, η we may define a unique function

$$v(s;\lambda,\xi,\eta)\equiv v(s) \tag{10.6.16}$$

which is the solution of equation (10.6.12) satisfying

$$v(0)=\xi,\ v'(0)=\eta,\ (b_ov')'(0)=0=(c_o^2(b_ov')')'(0). \tag{10.6.17}$$

Clearly

$$v(s;\lambda_i^o,\phi_i^o(0),\phi_i^{o'}(0))=\phi_i^o(s). \tag{10.6.18}$$

The eigenfunctions $\{\phi_i^o(s)\}_o^\infty$ are orthogonal with weight function $p_o^2(s)$, as shown by equation (10.6.15). The eigenfunctions $\{\phi_i(s)\}_o^\infty$ of equations (10.6.8)-(10.6.11) are to be orthogonal w.r.t $p^2(s)$, i.e.,

$$\int_o^L p^2(s)\phi_i(s)\phi_j(s)ds=\delta_{ij} \ , \ \ i,j=0,1,2,... \tag{10.6.19}$$

Therefore, following equation (9.6.8), we construct equation (10.6.8) so that the solution (10.6.16) of equation (10.6.12) is transformed into a solution of equation (10.6.8) satisfying

$$u(0)=\xi \ , \ \ u'(0)=\eta \ , \ \ (bu')'(0)=0=(c^2(bu')')'(0) \ , \tag{10.6.20}$$

by means of the equation

$$p(s)u(s)=p_o(s)v(s)+\int_o^s K(s,t)p_o^2(t)v(t)dt \ . \tag{10.6.21}$$

The eigenfunctions of equations (10.6.8)-(10.6.11) will be

$$\phi_i(s)=u(s;\lambda,\phi_i(s)\,|_{s=o} \ , \ \ \phi_i'(s)\,|_{s=o}) \tag{10.6.22}$$

and we note that

$$\phi_i(s)\,|_{s=o}=\phi_i(x)\,|_{x=1} \ , \tag{10.6.23}$$

$$\frac{d}{ds}\phi_i(s)\,|_{s=o}=-\frac{d}{dx}\phi_i(x)\,|_{x=1} \ . \tag{10.6.24}$$

If the eigenvalues λ_i and end values $\phi_i(0)$, $\phi_i'(0)$ (with variable s) are chosen so that

$$\lambda_i=\lambda_i^o \ , \ \ \phi_i(0)=\phi_i^o(0) \ , \ \ \phi_i'(0)=\phi_i^{o'}(0) \ , \ \ i=n+1,n+2,... \ , \tag{10.6.25}$$

then the system $\{\phi_i(s)\}_o^\infty$ will form a complete orthogonal set with weight $p^2(s)$ if and only if $K(s,t)$ satisfies equation (9.6.9), i.e.,

$$K(r,s)+\int_o^r p_o^2(t)K(r,t)F(t,s)dt+p_o(r)F(r,s)=0 \ , \ \ 0\le s\le r \ , \tag{10.6.26}$$

where

$$F(r,s)=\sum_{i=o}^{n}\{v_i(r)v_i(s)-v_i^o(r)v_i^o(s)\} \ , \tag{10.6.27}$$

and

$$v_i(s)=v(s;\lambda_i,\phi_i(0),\phi_i'(0)\} \ , \tag{10.6.28}$$

$$v_i^o(s)=v(s;\lambda_i^o,\phi_i^o(0),\phi_i^{o'}(0)\}\equiv\phi_i^o(s) \ . \tag{10.6.29}$$

We note that

$$u(s;0,1,0)=1 \ \ , \ \ v(s;0,1,0)=1 \ , \tag{10.6.30}$$

so that equation (10.6.21) gives

$$p(s)=p_o(s)+\int_o^s K(s,t)p_o^2(t)dt \ . \tag{10.6.31}$$

On the other hand, if

$$q(s)=\int_o^s \frac{dt}{b(t)}\ ,\quad q_o(s)=\int_o^s \frac{dt}{b_o(t)}\ , \tag{10.6.32}$$

then

$$u(s;0,0,1)=q(s)\quad v(s;0,0,1)=q_o(s)\ , \tag{10.6.33}$$

so that

$$p(s)q(s)=p_o(s)q_o(s)+\int_o^s K(s,t)p_o^2(t)q_o(t)dt\ . \tag{10.6.34}$$

The reconstruction procedure is therefore to solve equation (10.6.26) for $K(s,t)$ and then find $p(s)$, $q(s)$ from equations (10.6.31), (10.6.34); find $b(s)$, $c(s)$ from equations (10.6.7)-(10.6.32); and x, $a(x)$, $r(x)$ from (10.6.5)-(10.6.6). To justify this procedure we need to verify that when $p(s)$, $q(s)$ are given by equations (10.6.31), (10.6.34), then

1) $p(s)$, $q(s)$ are well-defined and positive, and $q(s)$ is an increasing function.
2) $U(s)$ satisfies the end conditions (10.6.10) at $s=0$.
3) $u(s)$ satisfies the differential equation (10.6.8).
4) $u_i(s)$ satisfies the end conditions (10.6.11) at $s=L$.

We shall consider these points in the order 2, 3, 4, 1.

Equation (10.6.31) yields $p(0)=p_o(0)=1$, while equation (10.6.21) yields $p(0)u(0)=u(0)=p_o(0)v(0)=v(0)=\xi$. On differentiating equation (10.6.31) we obtain

$$p'(0)=p_o'(0)+K(0,0) \tag{10.6.35}$$

while on differentiating equation (10.6.21) we find

$$\begin{aligned} p'(0)u(0)+p(0)u'(0) &= p_o'(0)v(0)+p_o(0)v'(0) \\ &+ K(0,0)p_o^2(0)v(0)\ , \end{aligned} \tag{10.6.36}$$

which yields

$$u'(0)=v'(0)=\eta\ . \tag{10.6.37}$$

By continuing this analysis we may establish the remainder of 2) and 3).

The solution of equation (10.6.26) has the form

$$K(r,s)=\sum_{i=o}^{n}\{F_i(r)v_i(s)-G_i(r)v_i^o(s)\}\ , \tag{10.6.38}$$

where, as in Section 9.3, $F_i(x)$, $G_i(r)$ satisfy

$$p_o(r)v_i(r)+F_i(r)+\sum_{j=o}^{n}\{b_{ij}F_j(r)-c_{ij}G_j(r)\}\ 0\ , \tag{10.6.39}$$

$$p_o(r)v_i^o(r)+G_i(r)+\sum_{j=o}^{n}\{c_{ji}F_j(r)-d_{ij}G_j(r)\}=0\ , \tag{10.6.40}$$

where

$$b_{ij}(r)=\int_o^r p_o^2(t)v_i(t)v_j(t)dt\ ,\quad c_{ij}(r)=\int_o^r p_o^2(t)v_i(t)v_j^o(t)dt\ ,$$
$$d_{ij}(r)=\int_o^r p_o^2(t)v_i^o(t)v_j^o(t)dt\ . \tag{10.6.41}$$

In considering point 4), we need to discuss two cases, $i\leq n$ and $i>n$. For the first we note that on substituting (10.6.38) into (10.6.21) and using (10.6.39) we may deduce

$$p(s)u_i(s)=-F_i(s)\quad i=0,1,2,...n\ . \tag{10.6.42}$$

But equation (10.6.39) with $r=L$ and the orthogonality condition $d_{ij}(L)=\delta_{ij}$ ($v_i^o(t)$ are normalized eigenfunctions of the auxiliary problem) yields

$$\sum_{j=o}^{n}c_{ji}F_j(L)=0 \tag{10.6.43}$$

since $v_i^o(L)=0$. Thus, *if* $\mathbf{C}=(c_{ji}(L))$ is non-singular, *and* $p(L)\neq 0$, then

$$F_j(L)=0=u_j(L)\quad j=0,1,2,...n\ . \tag{10.6.44}$$

On differentiating equation (10.6.32) we find, under the same proviso, that

$$F_j'(L)=0=u_j'(L)\quad j=0,1,2,...n\ . \tag{10.6.45}$$

We shall return to this proviso later. When $i>n$, then $v_i(L)\equiv v_i^o(L)\equiv\phi_i^o(L)=0$, so that equation (10.6.28) yields

$$p(L)u_i(L)=\int_o^L K(L,t)p_o^2(t)v_i^o(t)dt\ , \tag{10.6.46}$$

so that on substituting for $K(L,t)$ from equation (10.6.37) we find

$$p(L)u_i(L) = \sum_{j=o}^{n}\{F_j(L)\int_o^L p_o^2(t)v_j(t)v_i^o(t)dt$$
$$-\ G_j(L)\int_o^L p_o^2(t)v_j^o(t)v_i^o(t)dt\}\ . \tag{10.6.47}$$

But since $i>n$, $v_i^o(t)$ is orthogonal to all the $v_j^o(t)$ so that the second group of terms is zero. Therefore again, *if* $F_j(L)=0$, $j=0,1,2,...n$ *and* $p(L)\neq 0$, then $u_i(L)=0$. The satisfaction of $u_i'(L)=0$ may be verified similarly. (Ex. 10.6.4)

We now discuss point 1). The first step is the determination of the $F_i(s)$, $G_i(s)$ from equations (10.6.39), (10.6.40). The argument used in Section 9.4 shows that the matrix of coefficients in these equations is non-singular unless $r=L$. Thus there remains only the case $r=L$. Then the matrix of coefficients in (10.6.38)-(10.6.39) takes the form

$$\mathbf{A} = \begin{bmatrix} \mathbf{I}+\mathbf{B} \; , & -\mathbf{C} \\ \mathbf{C}^T & 0 \end{bmatrix} \qquad (10.6.48)$$

so that $|\mathbf{A}| = |\mathbf{C}|^2$. Thus $|\mathbf{C}| \neq 0$ is a necessary and sufficient condition for $F_i(r)$, $G_i(r)$, and hence $p(s)$ to be well-defined. We now enquire as to when and whether $p(s) > 0$. Suppose $p(s) = 0$ for some s in $[0,\ell]$, then (10.6.42) and the similar equation

$$p(s)u_i^o(s) \equiv p(s)u(s;\lambda_i^o,\phi_i^o(0),\phi_i^{o'}(0)) = -G_i(s) \quad i=0,1,...n \qquad (10.6.49)$$

show that $F_i(s) = 0 = G_i(s)$, $i=0,1,...n$, and hence, on account of (10.6.39), (10.6.40), $p_o(s)v_i(s) = 0 = p_o(s)v_i^o(s)$, $i=0,1,...n$. But $p_o(s)$, belonging to an actual beam, is always positive, and the only common zero of the $v_i^o(s)$, or the *only* zero of $v_0^o(s)$, is $s=L$. Thus the only possible zero of $p(s)$ is $s=L$. At $s=L$, equations (10.6.39), (10.6.40) reduce to

$$F_i(L) = 0 \; , \; \sum_{j=o}^{n} c_{ij}(L)G_j(L) = p_o(L)v_i(L) \; , \; i=0,1,...n \; , \qquad (10.6.50)$$

while equation (10.6.31) reduces to

$$p(L) = p_o(L) - \sum_{i=o}^{n} G_i(L) \int_o^L p_o^2(t)v_i^o(t)dt \; . \qquad (10.6.51)$$

After solving equation (10.6.51) for the $G_j(L)$ we may, with some manipulation, write equation (10.6.51) in the form

$$p(L) = p_o(L)\,|\mathbf{E}|\,/\,|\mathbf{C}| \; , \qquad (10.6.52)$$

where

$$e_{ij} = \int_o^L p_o^2(t)v_i^o(t)[v_j(t) - v_j(L)]dt \; , \quad i,j=0,1,...n \; . \qquad (10.6.53)$$

Thus $p(L) \neq 0$, provided that $|\mathbf{E}| \neq 0$ and $|\mathbf{C}| \neq 0$. If these conditons are satisfied, then the operations carried out to obtain $p(s)$, $q(s)$, and hence $b(s)$, $c(s)$, are such that $b(s)$, $c(s)$ will have derivatives of all orders. In particular, since $b(s)$, $c(s)$ are continuous, never zero, and $b(0) = 1 = c(0)$, we conclude that $b(s) > 0$, $c(s) > 0$ and hence $a(x) > 0$, $r(x) > 0$.

Examples 10.6

1. Verify the transformation from equations (10.6.1) - (10.6.3) to (10.6.8) - (10.6.11). Show also that the conditions (10.6.10) are equivalent to

$$u''(0) - q''(0)u'(0) = 0 = u'''(0) - q'''(0)u'(0) \; .$$

2. Show that if $v(s)$ satisfies

$$v''(0) - q_o''(0)v'(0) = 0 = v'''(0) - q_o'''(0)v'(0)$$

then $u(s)$ will satisfy the conditions in Ex. 10.6.1.

3. Use equations (10.6.12) - (10.6.14) to obtain an expression for $c_{ij}(L)$.
4. Establish the condition $u_i'(L)=0$ first for $i \leq n$, and then for $i > n$.

10.7 The positivity of matrix P is sufficient

In Section 10.3 we showed that the eigenvalues $(\lambda_i)_o^\infty$ and end values $\phi_i(\ell)=u_i$, $\phi_i'(\ell)=\theta_i$ of a cantilever beam make the infinite matrix **P** of Theorem 10.3.3 completely positive. In Section 10.6 we found some sufficient conditions for the resconstruction of an actual beam from such data. We now show that the positivity of the matrix **P** is not only necessary, but sufficient.

It was shown in Section 10.6 that the reconstruction will proceed provided that the matrices **C**, **E** are non-singular. Here

$$c_{ij}=\int_o^L p_o^2(s)v_i(s)v_j^o(s)ds \ , \tag{10.7.1}$$

$$e_{ij}=\int_o^L p_o^2(s)v_i^o(s)[v_j(s)-v_j(L)]ds \ , \tag{10.7.2}$$

and $i,j=0,1,\ldots n$. Suppose **C** is singular, then its rows will be linearly dependent, i.e., there are multipliers $(\alpha_i)_o^n$, not all zero, such that

$$\sum_{i=o}^n \alpha_i c_{ij}=0 \ , \ \ j=0,1,\ldots n \ . \tag{10.7.3}$$

Thus

$$\int_o^L p_o^2\{\sum_{i=o}^n \alpha_i v_i(s)\}v_j^o(s)ds=0 \ , \ \ j=0,1,\ldots n \ . \tag{10.7.4}$$

But, since the $\{v_j^o(s)\}_o^\infty$, being the eigenfunctions of the auxiliary problem, form a complete orthogonal set with weight $p_o^2(s)$, equation (10.7.4) means that the sum in the integrand is a linear combination of the remaining $\{v_j^o(s)\}_{n+1}^\infty$. Thus

$$f(s)\equiv \sum_{i=o}^n \alpha i v_i(s)+\sum_{i=n+1}^\infty \alpha_i v_i^o(s)\equiv 0 \tag{10.7.5}$$

In particular, therefore, $f(s)$ and all its derivatives (they all exist) must be zero at $s=0$, the free end. Consider the case $b_o(s)\equiv 1 \equiv c_o(s)$. Now

$$v_i(s)\,|_{s=o}=\phi_i(x)\,|_{x=1}=u_i \tag{10.7.6}$$

$$b(s)\frac{dv_i(s)}{ds}\,|_{s=o}=v_i'(s)=-\frac{d\phi_i(x)}{dx}\,|_{x=1}=-\theta_i \tag{10.7.7}$$

while equation (10.6.17) gives

$$v_i''(0)=0=v_i'''(0) \tag{10.7.8}$$

and equation (10.6.12) gives

$$v_i^{iv}=\lambda_i v_i(0)=\lambda_i u_i \ , \ \ v_i^{v}(0)=-\lambda_i \theta_i \tag{10.7.9}$$

Thus

$$v_i^{(m)}(0)=0=v_i^{o(m)}(0) \text{ for } m=2,3;6,7;... \tag{10.7.10}$$

so that $f^{(m)}(0)$ is identically zero for these values of m. The equations obtained by setting $f^{(m)}(0)$ to zero for the remaining values 0,1;4,5;8,9; etc. are therefore

$$\sum_{i=o}^{\infty}\lambda_i^j u_i \alpha_i=0 \ , \ \ \sum_{i=o}^{\infty}\lambda_i^j \theta_i \alpha_i=0 \ \ j=0,1,2,... \ , \tag{10.7.11}$$

and here we have used the fact that $u_i^o=u_i$, $\theta_i^o=\theta_i$ when $i=n+1$,.... But the matrix of coefficients for equations (10.7.11) is just the matrix **P** of Theorem 10.3.3 and every minor of **P** is positive so that **P** has infinite rank. Thus all the α_i are zero, contradicting the assumed singularity of **C**. Thus **C** is non-singular. When $b_o(s)$, $c_o(s)$ are not identically unity, the rows of the matrix are linear combinations of the rows of **P** so that the conclusion still holds.

A similar argument shows that if **E** is singular, then there are multipliers $(\beta_i)_o^\infty$, not all zero, such that

$$g(s)\equiv\sum_{i=o}^{n}\beta_i\{v_i(s)-v_i(0)\}+\sum_{i=n+1}^{\infty}\beta_i v_i^o(s)\equiv 0 \tag{10.7.12}$$

i.e.,

$$g'(s)\equiv\sum_{i=o}^{n}\beta_i v_i'(s)+\sum_{i=n+1}^{\infty}\beta_i v_i^{o\prime}(s)\equiv 0 \tag{10.7.13}$$

When $b_o(s)\equiv 1\equiv c_o(s)$, the matrix of coefficients for the equations obtained by setting $g_{(m)}(0)$ to zero for $m=1;4,5;8,9;$ etc. is just the matrix formed from rows 2,3,... of matrix **P**. We conclude that the β_i are identically zero and that **E** is non-singular. We conclude that the complete positivity of the matrix **P** is a necessary and sufficient condition for the reconstruction of a realistic beam. Note however that this conclusion is subject to the restriction (10.6.25). Barcilon (1986) has established the sufficiency without imposing this restriction by considering the Euler-Bernoulli beam to be limit, as $N\to\infty$, of the discrete system considered in Chapter 7. Note also that the complete positivity of the matrix **P** ensures only that $a(x)$, $r(x)$ will be *positive*; they may still vary wildly along the beam, and in this case the vibration of the beam will not be governed by the Euler-Bernoulli equation, which applies only for slender beams, i.e., ones for which $a(x)$, $r(x)$ do not differ much from the values for a uniform beam. This matter is discussed fully in Gladwell, England and Wang (1986), where examples of reconstruction are given.

10.8 Determination of feasible data

Consider the reconstruction of the beam in the case $n=0$. It is required to find the region in (u_0,θ_0,λ_0) space for which all the inequalities $\Phi>0$, $\Psi>0$ are satisfied. We remember that if these inequalities are satisfied for all sets of *consecutive* indices, then they will be satisfied for *all* sets of indices. Secondly, since $(u_i,\theta_i,\lambda_i)_1^\infty \equiv(u_i^o,\theta_i^o,\lambda_i^o)_1^\infty$ relate to an actual beam, they satisfy all the inequalities

$$\Phi(i,i+1,...i+n)>0 \quad \Psi(i,i+1,...i+n)>0 \quad i=1,2,...;\ n=0,1,... \tag{10.8.1}$$

Thus it is sufficient to require that $(u_o,\theta_o,\lambda_o)\equiv(u,\theta,\lambda)$ lie in the intersection of the regions Ω_m given by

$$\Omega_m\equiv(u,\theta,\lambda;\Phi(0,1,...m)>0\ ,\ \Psi(0,1,...m)>0)\ . \tag{10.8.2}$$

Thus Ω_o is the first quadrant, while Ω_1 is the intersection of the regions defined by

$$\Phi(0,1)\equiv u\theta_1-\theta u_1>0\ ,\ \Psi(0,1)\equiv\lambda_1 u_1\theta-\lambda u\theta_1>0\ . \tag{10.8.3}$$

If Ω_1 is to be non-empty, then $\lambda<\lambda_1$. In this case, if $\Phi(0,1)=0$, then $\Psi(0,1)>0$ while if $\Psi(0,1)=$, then $\Phi(0,1)>0$, so that the projection of Ω_1 on a λ-plane is the intersection of the shaded regions in Figure 10.8.1.

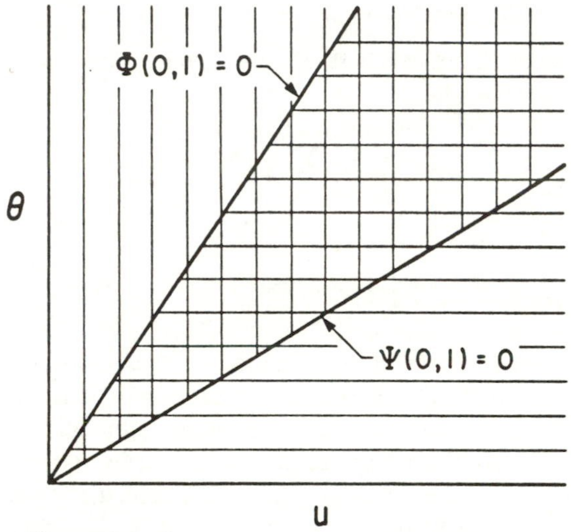

Figure 10.8.1 – The region Ω_1.

Equations (10.3.20), (10.3.21) with $i,j,k=0,1,2$ show that

1) when $\Phi(0,1)=0$, then $\Phi(0,1,2)<0$, $\Psi(0,1,2)>0$,

2) when $\Psi(0,1)=0$, then $\Phi(0,1,2)>0$, $\Psi(0,1,2)<0$,

and this means that Ω_2 lies inside Ω_1. The equations (10.3.22), (10.3.23) and their generalisation show that

$$\Omega_o\supset\Omega_1\supset\Omega_2\supset\ \cdots \tag{10.8.4}$$

When $\lambda=\lambda_0^o$, the zeroth eigenvalue of the auxiliary system, the intersection $\prod_{i=o}^{\infty}\Omega_i$ is not empty, as it contains $(u_0^o,\theta_0^o,\lambda_0^o)$. For other values of λ it may be empty.

For each value of m the determinants Φ, Ψ may be written in the form

$$\Phi(0,1,...m)=p_m u-q_m\theta\ ,\quad \Psi(0,1,...m)=r_m\theta-s_m u\ . \tag{10.8.5}$$

In particular $p_0=1$, $q_0=0$, $r_0=1$, $s_0=0$; $p_1=\theta_1$, $q_1=u_1$, $r_1=\lambda_1 u_1$, $s_1=\lambda\theta_1$. Put $\Phi(1,2,...m)=F_m$, $\Psi(1,2,...m)=G_m$, (all F_n, G_n are positive) then the general equations of the form (10.3.22), (10.3.23) yield

$$G_m(p_{m+1}u-q_{m+1}\theta)=G_{m+1}(p_m u-q_m\theta)-F_{m+1}(r_m\theta-s_m u)\ , \tag{10.8.6}$$

$$F_m(r_{m+1}\theta-s_{m+1}u)=\lambda_{m+1}F_{m+1}(r_m\theta-s_m u)-\lambda G_{n+1}(p_m u-q_m\theta). \tag{10.8.7}$$

Since these are identities in u, θ we deduce that

$$G_m p_{m+1}=G_{m+1}p_m+F_{m+1}s_m\ ,\quad G_m q_{m+1}=G_{m+1}q_m+F_{m+1}r_m\ , \tag{10.8.8}$$

$$F_m S_{m+1}=\lambda_{m+1}F_{m+1}s_m+\lambda G_{m+1}p_m\ ,\quad F_m r_{m+1}=\lambda_{m+1}F_{m+1}r_m+\lambda G_{m+1}q_m\ . \tag{10.8.9}$$

Equations (10.8.8a), (10.8.9a) enable us to express s_{m+1} in terms of p_m, p_{m+1}, i.e., s_m in terms of p_{m-1}, p_m. On substituting this expression for s_m into (10.8.8a) we find

$$p_{m=1}=a_m p_m+b_{m-1}p_{m-1}\ ,\quad q_{m+1}=a_m q_m+b_{m-1}q_{m-1}\ ,\quad m=1,2,...\ , \tag{10.8.10}$$

where

$$a_m=\frac{(F_{m-1}G_{m+1}+\lambda_m F_{m+1}G_{m-1})}{F_{m-1}G_m}\ ,\quad b_{m-1}=\frac{F_{m+1}(\lambda-\lambda_m)}{F_{m-1}}\ ,\ m=1,2,...\ , \tag{10.8.11}$$

and F_o, G_o are interpreted as 1. But equations (10.8.10) are just the recurrence relations for the numerator and denominator of a continued fraction. Thus

$$\frac{p_{m+1}}{q_{m+1}}=\frac{\theta_1}{u_1}\left\{1+\frac{b'}{a_1+}+\frac{b_1}{a_2+}\ \cdots\ \frac{b_{m-1}}{+a_m}\right\}\ ,\ m=1,2,...\ , \tag{10.8.12}$$

where $b'=b_o/\theta_1$. Since the b's are negative, the sequence $(p_m/q_m)_1^\infty$ is monotonic decreasing. It is bounded below, and therefore converges, to a limit $B>0$.

In an exactly similar way we deduce that

$$r_{m+1}=c_m r_m+d_{m-1}r_{m-1}\ ,\quad s_{m+1}=c_m s_m+d_{m-1}s_{m-1}\ ,\quad m=1,2,...\ , \tag{10.8.13}$$

where

$$c_m=\frac{(F_{m-1}G_{m+1}+\lambda_{m+1}F_{m+1}G_{m-1})}{F_m G_{m-1}}\ ,\quad d_{m-1}=\frac{G_{m+1}(\lambda-\lambda_m)}{G_{m-1}} \tag{10.8.14}$$

so that

$$\frac{r_{m+1}}{s_{m+1}}=\frac{\lambda_1 u_1}{\lambda\theta_1}\left\{1+\frac{d'}{c_1+}+\frac{d_1}{c_2+}+\ \cdots\ +\frac{d_{m-1}}{+c_m}\right\}\ ,\ m=1,2,...\ , \tag{10.8.15}$$

where $d'=d_0/u_1$. Again the sequence $(r_m/s_m)_1^\infty$ is monotonic decreasing, and bounded below and therefore converges to a limit $b>0$. Thus the feasible region of

θ/u is given by

$$\frac{1}{b} < \frac{\theta}{u} < B \ . \tag{10.8.16}$$

If instead of λ_0, u_0, θ_0, the second eigenquantities λ_1, u_1, θ_1 are to be changed, while the remaining λ_i, u_i, θ_i $(i \neq 1)$ are to be those of the auxilary system, then there are two upper bounds on θ_1/u_1 obtained from the sequencies

$$\Phi(0,1,...m)>0 \ , \quad m=0,1,2,... \ ; \ \Phi(1,2,...m)>0 \quad , \quad m=1,2,... \ , \tag{10.8.17}$$

and two lower bounds obtained from the sequences

$$\Psi(0,1,2,...m)>0 \ , \quad m=0,1,2,... \ ; \ \Phi(1,2,...m)>0 \quad , \quad m=1,2,... \tag{10.8.18}$$

Each of these bounds may be represented as infinite continued fractions, as before. A similar argument may be applied to the general case when $(\lambda_i, u_i, \theta_i)$ $(i>1)$ are changed, and the case when a number of sets $(\lambda_i, u_i, \theta_i)$ are changed.

BIBLIOGRAPHY

Ambarzumian, V. (1929) "Über eine Frage der Eigenwerttheorie", *Z. Physik*, *53*, 690-695, [193,194].

Anderson, L. (1970) "On the effective determination of the wave operator in the case of a difference equation corresponding to a Sturm-Liouville differential equation", *J. Math. Anal. Appl.*, *29*, 467-497, [215].

Backus, G.E. and Gilbert, J.F., (1967) "Numerical applications of a formalism for geophysical inverse problems", *Geophysical Journal of the Royal Astronomical Society*, *13*, 247-276.

Barcilon, V. (1974a) "Iterative solution of the inverse Sturm-Liouville equation", *J. Math. Phys.*, *15*, 429-436.

Barcilon, V. (1974b) "On the uniqueness of inverse eigenvalue problems", *Geophysical Journal of the Royal Astronomical Society*, *38*, 287-298. [245]

Barcilon, V. (1974c) "On the solution of inverse eigenvalue problems of high orders", *Geophysical Journal of the Royal Astronomical Society*, *39*, 143-154. [245]

Barcilon, V. (1976) "Inverse problem for a vibrating beam", *J. Appl. Math. Phys.*, *27*, 346-358. [245]

Barcilon, V. (1978) "Discrete analog of an iterative method for inverse eigenvalue problems for Jacobi matrices", *J. Math. Phys.*, *29*, 295-300. [66]

Barcilon, V. (1979) "On the multiplicity of solutions of the inverse problem for a vibrating beam", *SIAM J. Appl. Math.*, *37*, 605-613, [119,127].

Barcilon, V. (1982) "Inverse problems for the vibrating beam in the free-clamped configuration", *Phil. Trans. Roy. Soc. Lond. A*, *304*, 211-252. [119,127,244,245]

Barcilon, V. (1983) "Explicit solution of the inverse problem for a vibrating string", *J. Math. Anal. Appl.*, *93*, 222-234. [215]

Barcilon, V. (1986) "Sufficient conditions for the solution of the inverse problem for a vibrating beam", *J. Inverse Problems* (submitted).

Bellman, R. (1970) *Introduction to Matrix Analysis*, 2nd ed., New York: McGraw-Hill. [91]

Biegler-König, F.W. (1981) *Inverse Eigenwertprobleme*, Dissertation, Bielefeld. [137]

Biegler-König, F.W. (1980) "Construction of band matrices from spectral data", *Linear Algebra Appl.*, *40*, 79-87. [137]

Bishop, R.E.D., Gladwell, G.M.L. and Michaelson, S. (1965) *The Matrix Analysis of Vibration*, London: Cambridge. [1,16,19]

Bishop, R.E.D. and Johnson, D.C. (1960) *The Mechanics of Vibration*, London: Cambridge University Press.

Borg, G. (1946) "Eine Umkehrung der Sturm-Liouvilleschen Eigenwertaufgabe", *Acta Math.*, *78*, 1-96. [193,194]

Burridge, R. (1980) "The Gel'fand-Levitan, the Marchenko, and the Gopinath-Sondhi integral equations of inverse scattering theory, regarded in the context of inverse impulse-response problems", *Wave Motion*, *2*, 305-323. [196]

Carrier, G.F., Krook, M. and Pearson, C.E. (1966) *Functions of a Complex Variable*, New York: McGraw-Hill. [242]

Courant, R. and Hilbert, D. (1953) *Methods of Mathematical Physics*, Vol. I, New York: Interscience. [43,147,149,150,163,177,178,224,241]

Courant, R. and Hilbert, D. (1962) *Methods of Mathematical Physics*, Vol. II, New York: Interscience. [186,187]

de Boor, C. and Golub, G.H. (1978) "The numerically stable reconstruction of a Jacobi matrix from spectral data", *Linear Algebra Appl.*, *21*, 245-260. [64]

Fischer, E. (1905) "Über quadratische Formen mit reelen Koeffizienten", *Monatsch. f. Math. u. Phys.*, *16*, 234-249, [43].

Fix, G. (1967) "Asymptotic eigenvalues of Sturm-Liouville systems", *J. Math. Anal. Appl.*, *19*, 519-525. [183]

Forsythe, G.E. (1957) "Generation and use of orthogonal polynomials for data-fitting with a digital computer", *J. SIAM*, *5*, 74-88. [50]

Friedland, S. (1977) "Inverse eigenvalue problems", *Linear Algebra Appl.*, 17, 15-51. [137]

Friedland, S. (1979) "The reconstruction of a symmetric matrix from the spectral data", *J. Math. Anal. Appl.*, *71*, 412-422. [137]

Gantmakher, F.R. (1959) *The Theory of Matrices*, Vols. I & II, New York: Chelsea. [82]

Gantmakher, F.P. and Krein, M.G. (1950) *Oscillation Matrices and Kernels and Small Vibrations of Mechanical Systems*, Moscow-Leningrad: State Publishing House of Technical-Theoretical Literature. 1961 Translation, Washington D.C.,: U.S. Atomic Energy Commission. [45,59,69,78,168]

Garabedian, P.R. (1964) *Partial Differential Equations*, New York: Wiley. [216]

Gasymov, M.G. and Levitan, B.M. (1964) "On Sturm-Liouville differential operators with discrete spectra", (in Russian) *Mat. Sbornik, 63*, 445-448. *Amer. Math. Soc. Transl., Ser. 2, 68*, 21-23.

Gel'fand, I.M. and Levitan, B.M. (1951) "On the determination of a differential equation from its spectral function", (in Russian). *Izv. Akad. Nauk. SSSR., Ser. Mat. 15*, 309-360. *Amer. Math. Soc. Transl., Ser. 2, 1*, 253-304. [195,196,210]

Gladwell, G.M.L. (1962) "The approximation of uniform beams in transverse vibration by sets of masses elastically connected", *Proc. 4th U.S. National Congress of Applied Mechanics*, New York: Am. Soc. Mech. Eng., 169-176. [33]

Gladwell, G.M.L. (1984) "The inverse problem for the vibrating beam", *Proc. Roy. Soc. London A. 393*, 277-295. [119]

Gladwell, G.M.L. (1985a) "Qualitative properties of vibrating systems", *Proc. Roy. Soc. London A, 401*, 299-315.

Gladwell, G.M.L. (1985b) "The inverse mode problem for lumped-mass systems", *Q.J. Mech. Appl. Math.*, (to appear). [109]

Gladwell, G.M.L. (1986) "The inverse problem for the Euler-Bernoulli Beam", *Proc. Roy. Soc. London A*, (to appear). [245]

Gladwell, G.M.L., England, A.H. and Wang, D. (1986) "Examples of reconstruction of an Euler-Bernoulli beam from spectral data", *J. Sound Vibration*, (submitted). [253]

Gladwell, G.M.L. and Gbadeyan, J.A. (1985) "On the inverse problem of the vibrating string or rod", *Q.J. Mech. Appl. Math., 38*, 169-174. [75]

Golub, G.H. (1973) "Some uses of the Lanczos algorithm in numerical linear algebra", in J.H.H. Miller (Ed.), *Topics in Numerical Analysis*, New York: Academic Press. [63]

Golub, G.H. and Boley, D. (1977) "Inverse eigenvalue problems for band matrices", in G.A. Watson (Ed.), *Numerical Analysis*, Heidelberg, New York: Springer Verlag, 23-31. [66,135]

Golub, G.H. and Underwood, R.R. (1977) "Block Lanczos method for computing eigenvalues", in J.R. Rice (Ed.), *Mathematical Software III*, New York: Academic Press. [137]

Golub, G.H. and Van Loan, C.F. (1983) *Matrix Computations*, Baltimore: The Johns Hopkins University Press. [63]

Gopinath, B. and Sondhi, M.M. (1970) "Determination of the shape of the human vocal tract from acoustical measurements", *Bell System Technical Journal*, 1195-1214. [196]

Gopinath, B. and Sondhi, M.M. (1971) "Inversion of the telegraph equation and the synthesis of non-uniform lines", *I.E.E.E. Trans. Sonics and Utrasonics*, *59*, 383-392. [196]

Gould, S.H. (1966) *Variational Methods for Eigenvalue Problems*, Toronto: University of Toronto Press. [43]

Gradshteyn, I.S. and Ryzhik, I.M. (1965) *Tables of Integrals, Series and Products*, 4th ed., Moscow 1963. English translation, A. Jeffrey (Ed.), New York: Academic Press. [186,214]

Gray, L.J. and Wilson, D.G. (1976) "Construction of a Jacobi matrix from spectral data", *Linear Algebra Appl.*, *14*, 131-134. [63]

Hald, O.H. (1972) *On Discrete and Numerical Inverse Sturm-Liouville Problems*, Ph.D. Thesis, New York University, New York, N.Y. [215]

Hald, O.H. (1976) "Inverse eigenvalue problems for Jacobi matrices", *Linear Algebra Appl.*, *14*, 63-85. [63,64]

Hald, O.H. (1977) "Discrete inverse Sturm-Liouville problems", *Num. Math.*, *27*, 249-256. [215]

Hald, O.H. (1978a) "The inverse Sturm-Liouville problem with symmetric potentials", *Acta Math.*, *141*, 263-291. [194]

Hald, O.H. (1978b) "The inverse Sturm-Liouville problem and the Rayleigh-Ritz method", *Math. Comp.*, *32*, 687-705. [215]

Hald, O.H. (1984) "Discontinuous inverse eigenvalue problems", *Comm. Pure Applied Math.*, *37*, 539-577. [194,196]

Hochstadt, H. (1961) "Asymptotic estimates of the Sturm-Liouville spectrum", *Comm. Pure Appl. Math.*, *14*, 749-764. [183]

Hochstadt, H. (1967) "On some inverse problems in matrix theory", *Arch. Math.*, *18*, 201-207. [63]

Hochstadt, H. (1973) "The inverse Sturm-Liouville problem", *Comm. Pure. Appl. Math.*, *26*, 715-729. [194]

Hochstadt, H. (1974) "On the construction of a Jacobi matrix from spectral data", *Linear Algebra Appl.*, *8*, 435-446. [63]

Hochstadt, H. (1975a) "On inverse problems associated with Sturm-Liouville operators", *J. Diff. Equations*, *17*, 220-235, [194].

Hochstadt, H. (1975b) "Well posed inverse spectral problems", *Proc. Nat. Acad. Sci.*, *72*, 2496-2497. [194]

Hochstadt, H. (1976) "On the determination of the density of a vibrating string from spectral data", *J. Math. Anal. Appl.*, *55*, 673-685. [194]

Hochstadt, H. (1977) "On the wellposedness of the inverse Sturm-Liouville problem", *J. Diff. Equations*, *23*, 402-413. [194]

Hochstadt, H. (1979) "On the construction of a Jacobi matrix from mixed given data", *Linear Algebra Appl., 28*, 113-115. [69]

Hochstadt, H. and Kim, M. (1970) "On a singular inverse eigenvalue problem", *Arch. Rat. Mech. Anal., 37*, 243-254. [194]

Hochstadt, H. and Lieberman, B. (1978) "An inverse Sturm-Liouville problem with mixed given data", *SIAM J. Appl. Math., 34*, 676-680. [194]

Krein, M.G. (1933) "On the spectrum of a Jacobian matrix, in connection with the torsional oscillations of shafts", (in Russian), *Mat. Sbornik, 40*, 455-466. [59]

Krein, M.G. (1934) "On nodes of harmonic oscillations of mechanical systems of a certain special type", (in Russian), *Mat. Sbornik, 41*, 339-348. [59]

Krein, M.G. (1951a) "Determination of the density of a non-homogeneous symmetric cord from its frequency spectrum", (in Russian), *Dokl. Akad. Nauk. SSSR*, (N.S.) *76*, 345-348. [196]

Krein, M.G. (1951b) "On the inverse problem for a nonhomogeneous cord", (in Russian), *Dokl. Akad. Nauk SSSR*, (N.S.) *82*, 669-672. [196]

Krein, M.G. (1952) "Some new problems in the theory of Sturm systems", (in Russian), *Prikl. Matem. Mekh., 16*, 555-563. [59,196]

Lanczos, C. (1950) "An iteration method for the solution of the eigenvalue problem of linear differential and integral operators", *J. Res. Nat. Bur. Standards, Sect. B, 45*, 225-232. [63]

Landau, H.J. (1983) "The inverse problem for the vocal tract and the moment problem", *SIAM J. Math. Anal., 14*, 1019-1035. [196]

Levinson, N. (1949) "The inverse Sturm-Liouville problem", *Mat. Tidsskr. B.*, 25-30. [194]

Levitan, B.M. (1964a) *Generalized Translation Operators and Some of Their Applications*, Jersualem: Israel Program for Scientific Translations, Chapters IV, V. [179,194]

Levitan, B.M. (1964b) "On the determination of a Sturm-Liouville equation by spectra", (in Russian), *Izv. Akad. Nauk. SSSR Ser. Mat. 28*, 63-68. *Amer. Math. Soc. Transl. Ser. 2, 68*, 1-20. [196]

Lindberg, G.M. (1963) "The vibration of non-uniform beams", *Aero. Quart. 14*, 387-395. [33]

Marchenko, V.A. (1950) "On certain questions in the theory of differential operators of second order", (in Russian), *Dokl. Akad. Nauk. SSSR, 72*, 457-460. [196]

Marchenko, V.A. (1952) "Some problems in the theory of one-dimensioned second order differential operators I", (in Russian), *Trudy. Moskovskogo Mat. Ob., 1*, 327-420. [194]

Marchenko, V.A. (1953) "Some problems in the theory of one-dimensioned second order differential operators II", (in Russian), *Trudy. Moskovskogo Mat. Ob., 2*, 3-82. [194]

Mattis, M.P. and Hochstadt, H. (1981) "On the construction of band matrices from spectral data", *Linear Algebra Appl., 38*, 109-119. [137]

McLaughlin, J.R. (1976) "An inverse problem of order four", *SIAM J. Math. Anal., 7*, 646-661. [197,246]

McLaughlin, J.R. (1978) "An inverse problem of order four – an infinite case", *SIAM J. Math. Anal., 9*, 395-413. [197,210,246]

McLaughlin, J.R. (1984a) "Bounds for constructed solutions of second and fourth order inverse eigenvalue problems", in I.W. Knowles and T.R. Lewis (Eds.), *Differential Equations*, New York: Elsevier/North Holland, 437-443. [197,246]

McLaughlin, J.R. (1984b) "On constructing solutions to an inverse Euler-Bernoulli beam problem", in F. Santosa, et al., (Eds.), *Inverse Problems of Acoustic and Elastic Waves*, Philadelphia: SIAM, 341-347. [245,246]

McLaughlin, J.R. (1986) "Analytical methods for recovering coefficients in Sturm-Liouville equations", *SIAM Rev., 28*, 53-72. [194]

Newton, R.G. (1983) "The Marchenko and Gel'fand-Levitan methods in the inverse scattering problem in one and three dimensions", in J.G. Bednar, et al. (Eds.), *Conference on Inverse Scattering: Theory and Application*, Philadelphia: SIAM, 1-74. [193]

Niordson, F.I. (1967) "A method for solving inverse eigenvalue problems", in B. Broberg, J. Hults and F.I. Niordson (Eds.), *Recent Progress in Applied Mechanics: The Folke Odqvist Volume*, Stockholm: Almqvist and Wiksell, 373-382. [245]

Parker, R.L. (1977) "Understanding inverse theory", *Ann. Rev. Earth Plan. Sci., 5*, 35-64. [193]

Porter, B. (1970) "Synthesis of lumped-parameter vibrating systems by an inverse Holzer technique", *J. Mech. Eng. Sci., 12*, 17-19. [114].

Porter, B. (1971) "Synthesis of lumped-parameter vibrating systems using transfer matrices", *Int. J. Mech. Sci., 13*, 29-34. [114]

Sabatier, P.C. (1978) "Spectral and scattering inverse problems", *J. Math. Phys., 19*, 2410-2425. [190,193]

Sabatier, P.C. (1979a) "On some spectral problems and isospectral evolutions connected with the classical string problem. I. Constants of motion", *Lettere al Nuovo Cimento, 26*, 477-482. [194]

Sabatier, P.C. (1979b) "On some spectral problems and isospectral evolutions connected with the classical string problem. II. Evolution equations", *Lettere al Nuovo Cimento, 26*, 483-486. [194]

Sabatier, P.C. (1985) "Inverse problems – an introduction", *Inverse Problems*, 1, i-vi. [193]

Sondhi, M.M. (1984) "A survey of the vocal tract inverse problem: theory, computations and experiments", in F. Santosa, Y.-H. Pao, W.W. Symes and C. Holland (Eds.), *Inverse Problems of Acoustic and Elastic Waves*, Philadelphia: SIAM. [196]

Sondhi, M.M. and Gopinath, B. (1970) "Determination of vocal-tract shape from impulse response of the lips", *J. Acoust. Soc. Am., 49*, 1867-1873. [215]

Stieltjes, T.J. (1918) *Oevres Completes, Vol. 2*, Gröningen: Noordhoff. [59]

Sweet, R.A. (1969) "Properties of a semi-discrete approximation to the beam equation", *J. Inst. Math. Appl., 5*, 329-339. [119]

Sweet, R.A. (1971) "Oscillation properties of a semi-discrete approximation to the beam equation with a second-order term", *J. Inst. Math. Appls., 7*, 119-125. [119]

Temple, G. and Bickley, W.G. (1933) *Rayleigh's Principle and its Applications to Engineering,* London: Oxford University Press. [37]

Titchmarsh, E.C. (1962) *Eigenfunction Expansions. Part I*, Oxford: Oxford University Press.

Underwood, R.R. (1975) *An Iterative Block-Lanczos Method for the Solution of Large Sparse Symmetric Eigenproblems", Ph.D. Thesis, Stanford Unversity. [137]*

Vijay, D.K. (1972) *Some Inverse Problems in Mechanics*, M.A.Sc. Thesis, University of Waterloo. [109]

Washizu, K. (1982) *Variational Methods in Elasticity and Plasticity*, 3rd Edition, Oxford: Pergamon Press. [37]

Yoshida, K. (1960) *Lectures on Differential and Integral Equations*, New York: Interscience. [203]

Zienkiewicz, O.C. (1971) *The Finite Element Method in Engineering Science*, London: McGraw-Hill. [26]

Zikov, V.V. (1967) "On inverse Sturm-Liouville problems on a finite segment", (in Russian), *Izv. Akad. Nauk SSSR, Ser. Mat., 31*, *Math. USSR Izv., 1*, 923-934.